W0234949

ENTERING THE MULTIVERSE

The multiverse has portaled into the mainstream. *Entering the Multiverse* unpacks the surprising growth of the multiverse in media and popular culture today, and explores how the concept of alternate realities and parallel worlds has acted as a metaphor for centuries.

Edited by leading media and popular culture scholar Paul Booth, this collection explores the many different manifestations of the multiverse across different genres, media, fan-created works, and cultural theory. Each chapter delves into different aspects of the multiverse, including its use as a metaphor, as a scientific reality, and as a media-industry strategy. Addressing the multiplicity of multiversal meanings through multiple perspectives and always with an eye toward engagement with contemporary cultural issues, the chapters examine various distinctions and contradictions, in order to provide a strong basis for further thinking, writing, and research on the concept of the multiverse. Chapters in this collection tell the story of the multiverse in multiple realities: creative nonfiction, academic essay, screenplay, art, poetry, video, and audio essay.

A compelling read for students, researchers, and scholars of media and cultural studies, film and media culture, popular culture, comics studies, game studies, literary studies, and beyond.

Paul Booth is Professor at DePaul University, USA, where he teaches media and popular culture. He is the author or editor of fifteen books, including *Board Games as Media* (2021), *A Fan Studies Primer* (with Rebecca Williams, 2021), and *Adventures Across Space and Time: A Doctor Who Reader* (with Matt Hills, Joy Piedmont, and Tansy Rayner Roberts, 2023).

ENTERING THE MULTIVERSE

Perspectives on Alternate Universes and Parallel Worlds

Edited by Paul Booth

Routledge
Taylor & Francis Group

NEW YORK AND LONDON

Designed cover image: © vchal / Getty Images

First published 2025
by Routledge
605 Third Avenue, New York, NY 10158

and by Routledge
4 Park Square, Milton Park, Abingdon, Oxon, OX14 4RN

Routledge is an imprint of the Taylor & Francis Group, an informa business

ISBN: 978-1-032-77013-0 (hbk)
ISBN: 978-1-032-77011-6 (pbk)
ISBN: 978-1-003-48084-6 (ebk)

DOI: 10.4324/9781003480846

Typeset in Sabon
by SPi Technologies India Pvt Ltd (Straive)

Access the Support Material: www.routledge.com/9781032770116

To Katie, in all her multiversal splendor.

CONTENTS

FIGURES

TABLES

CONTRIBUTORS

Dustin Abnet is an Associate Professor of American Studies at California State University, Fullerton. He is a cultural and intellectual historian of capitalism who explores the intersections between work, leisure, and consumption and science and technology across American history. His first book, *The American Robot: A Cultural History*, was published by the University of Chicago Press in 2020 and chronicled the history of both real and imagined robots from the eighteenth century to the near present.

Maria K. Alberto (she/her) is an instructor in the Department of English at the University of Utah, where she just finished her PhD and wrote a dissertation about canon, *Dungeons & Dragons*, and drow. Her research more broadly draws on fan studies, game studies, and literary theory, to consider analog game-texts, transformative fanworks, and fandom community practices online. Her recent work has considered digital platforms, fandom communities, and queer readings of Tolkien's work.

Angélica Cabrera Torrecilla is a postdoctoral researcher at the School of Humanities of Osaka University and member of the National System of Researchers of Mexico specializing in the cultural studies of space and time, focusing mainly on the multiverse and techno-digital contexts in fiction and popular culture. She is the editor of *The Multiverse as Theory in Postmodern Speculative Fictional Narratives* (Routledge, 2024) and author of *What if a Multiverse? Literatura y ciencia en la obra de Grant Morrison* (2019), and more than twenty published articles and book chapters. She has research experience in Mexico, UK, Japan, Germany, and Spain. All her research can be found on her website www.angelicacabrera.com.

Amy Coles is a final year PhD student at the University of Buckingham. Her current thesis centers around the female double and its external influences across nineteenth and twentieth-century literature, examining the way in which the female double functions in critiquing the role of the woman in society and offering a path to self-discovery and redefinition of womanhood. She has recently published a chapter in the Critical Insights series on *Frankenstein* (Salem Press) which examines the texts' parodic adaptations and has a forthcoming article on the folkloric characters in the works of Shirley Jackson and Stephen King for *Shirley Jackson Studies*.

Blair Davis is Professor of Media and Popular Culture in the College of Communication at DePaul University in Chicago, Illinois. His books include *The Battle for the Bs: 1950s Hollywood and the Rebirth of Low-Budget Cinema* (2012, Rutgers University Press); *Movie Comics: Page to Screen/Screen to Page* (2017, Rutgers University Press); *Comic Book Movies* (2018, Rutgers University Press); *Comic Book Women: Characters, Creators and Culture in the Golden Age* (2022, University of Texas Press); *Christianity and Comics: Stories We Tell About Heaven and Hell* (2024, Rutgers University Press).

Sem Devillart is a cultural analyst focused on visual and online culture. For the last fifteen years she has collaborated with top organizations on innovative methods to turn cultural signals into high-value intelligence, media content, products, and business models. Sem was the chief architect at startup Sparks and Honey, a responsive, data-driven creative agency at Omnicom in New York City. Currently, she works with Harmony Labs, a media research studio working to understand the interaction of media narratives and audiences, and the inner mechanics of media influence. She teaches trend forecasting and cultural analysis at the Masters of Branding Program at the School of Visual Arts in NYC.

David Scott Diffrient is a Professor of Film and Media Studies in the Department of Communication Studies at Colorado State University. He is the co-editor of *Screwball Television: Critical Perspectives on Gilmore Girls* (2010) and *East Asian Film Remakes* (2023) and the author of *M*A*S*H* (2008), *Omnibus Films: Theorizing Transauthorial Cinema* (2014), *Comic Drunks, Crazy Cults, and Lovable Monsters: Bad Behavior on American Television* (2022), *Body Genre: Anatomy of the Horror Film* (2023), and (with coauthor Hye Seung Chung) *Movie Migrations: Transnational Genre Flows and South Korean Cinema* (2015) and *Movie Minorities: Transnational Rights Advocacy and South Korean Cinema* (2021).

Maria Faust received her Doctorate with Highest Distinction from the University of Leipzig, Germany researching "The Temporal Paradox of Time due to Digital Media in Germany and China," and is currently an Independent Scholar. Throughout her academic career, she spent time in Norway, China, the UK, and the US. In 2016 she won a Best Paper Award with the *International Association for the Development of the Information Society*. She has published with *NOMOS, East Asian Journal of Popular Culture, China Media Research, Kronoscope* and the *Journal of Multicultural Discourses*. In her Post Doc project she looks at "Spaces of Contemplation - The Epistemologies of Landscape in Times of Deep Mediatization."

Adalberto Fernandes, PhD, is a researcher at the Research Centre for Tourism, Sustainability and Well-Being (CinTurs), University of Algarve. Recent publications: "Has the Concept of Censorship Gone Astray?" (2024, *Journal of Iberian and Latin American Studies*); "Designing (the) Politics of Participation" (2023, *Journal of Science Communication*)

Lore FitzWhittemore is a PhD student in Communication Arts at the University of Wisconsin-Madison, and her research is interested in how storytelling possibilities are shaped by production cultures, policy and legality, and interactions between producers and audiences. Her contribution for this collection, "Expanding the Spider-Verse: Infinite Possibility and Fan Plausibility," was one component of the thesis she produced for her Master of Arts in Media and Public Engagement from the University of Colorado Boulder; a documentary also on the topic of Spidersonas was produced as part of this research project and can be found at expandingthespiderverse.github.io.

Teresa Forde is a Senior Lecturer in Film and Media at the University of Derby. Her research is within film, television, and curation. Her recent publications include chapters on *Star Trek: Discovery* and narrative representations in UK television drama as well as 'Co-curation as Feminist Practice' in the *Journal of Visual Art Practice*. She is currently writing a monograph on Women and Technology in SF Film and Television and planning a joint exhibition and symposium on women and surrealism for 2025.

Nev Fountain, born Steven John Fountain, is an English writer, best known for his comedy work with writing partner Tom Jamieson on the radio and television program *Dead Ringers*. He has written numerous audio dramas for *Doctor Who* through Big Finish, has worked on *Private Eye*, and has penned multiple novels. His newest novel is *The Fan Who Knew Too Much*.

Ross Garner is a Senior Lecturer in media and cultural studies at the School of Journalism, Media and Culture, Cardiff University. His research interests include the spatial and material cultures of media consumption, and he has published articles in edited collections by SUNY Press and journals in *Popular Communication, Tourist Studies*, and *Transformative Works and Cultures* exploring these issues. He is currently editing the collection *Pikachu's Transmedia Adventures*.

Tim Butler Garrett is a freelance academic, editor of the British UNIMA journal *Puppet Notebook*, and a visiting lecturer at Wimbledon College of Art. His research interests include the connections between puppets, the cinematic and contemporary Visual Theater; 'Vienna 1900'; and 1960s counter-culture. Recent writing for books and journals includes chapters on the UK's Suffragist movement; Jim Henson's *Labyrinth*; the Modernist avant-garde's engagement with female simulacra as objects of arousal; the Mitteleuropean sensibility of Kubrick's *Eyes Wide Shut*; and gender in the films of Arnold Schwarzenegger.

Lincoln Geraghty is Professor of Media Cultures in the School of Film, Media and Creative Technologies at the University of Portsmouth. He serves as editorial advisor for *The Journal of Popular Culture, Transformative Works and Culture, Journal of Fandom Studies* and *Journal of Popular Television* with interests in science fiction film and television, fandom, and collecting in popular culture. Major publications include *Living with Star Trek: American Culture and the Star Trek Universe* (IB Tauris, 2007), *American Science Fiction Film and Television* (Berg, 2009) and *Cult Collectors: Nostalgia, Fandom and Collecting Popular Culture* (Routledge, 2014).

Chris Gerrard is a video artist and film historian, focusing on how experimental aesthetics can facilitate knowledge exchange and political change. Their practical work focuses on collage, archival footage, and cinema history, and has been shown at galleries worldwide. Their written work explores the aesthetic development of cinema and television, as well as the impact of popular science-fiction. They are the course leader in Creative Media at Bath Spa University and co-chair of BSU's Imaginaries and Infrastructures research cluster.

Rebecca Gibson is an Assistant Professor at Virginia Commonwealth University in the Anthropology unit of the School of World Studies. Her published works include *Desire in the Age of Robots and AI: An Investigation in Science Fiction and Fact* (Palgrave Macmillan 2019) and *The Corseted Skeleton: A Bioarchaeology of Binding* (Palgrave Macmillan 2020), as well as two co-edited books on the supernatural (Lexington Books 2021 and 2022). She holds a PhD in Anthropology from American University, and when not writing or teaching can be found reading mystery novels amidst a pile of stuffed animals.

Marija Adela Gjorgjioska is a political scientist with a PhD in Social Representations and Communication. She lectures on Environmental and Political Communication at the Institute of Communication Studies in Skopje and currently serves as director of the Arete Institute for Sustainable Prosperity. Her research focuses on the social representations of AI in socialist and post-socialist countries. She is also a member of the Decolonial AI Collective, exploring the intersections of technology, power, and coloniality.

Matt Hills is an honorary Professor at the University of Bristol, and former Professor of Fandom Studies at the University of Huddersfield, UK. He is the author, editor or co-editor of ten books including *Fan Cultures* (Routledge 2002), *The Pleasures of Horror* (Continuum 2005), *Transatlantic Television Drama* (Oxford University Press 2019) and *Adventures Across Space and Time: A Doctor Who Reader* (Bloomsbury Academic 2023). He is currently working on a follow-up monograph to *Fan Cultures* for Routledge, entitled *Fan Studies*. Recent open-access publications include analyzing the late Professor Martin Barker's ground-breaking contributions to fan studies for the journal *Participations* (2023), and addressing comic-con's "event fans" in the collection *Media, Place and Tourism* (edited by Reijnders et al. 2024).

Rebecca Johns is director of the MFA and MA in Creative Writing and Publishing program at DePaul University. Her first novel, *Icebergs*, was a finalist for the Hemingway Foundation/PEN Award and a recipient of a Michener-Copernicus fellowship. Her second, *The Countess*, has been published in ten languages. Her work has appeared in *Ploughshares*, *StoryQuarterly*, the *Mississippi Review*, the *Harvard Review*, *Printer's Row Journal*, the *Chicago Tribune*, *Cosmopolitan*, *Mademoiselle*, *Ladies' Home Journal*, *Self*, and *Seventeen*, among others. She is a graduate of the Iowa Writers' Workshop and the Missouri School of Journalism.

Shayna Maskell is an Associate Professor in the School of Integrative Studies at George Mason University. Her book *Politics of Sound: Race, Class, and Gender in Washington DC Hardcore Punk 1978-1982* (2021) explores how and why cultural forms, such as music, produce and resist politics and power. Her areas of research include popular and youth culture, intersectionality, and social justice.

Rob McLaughlin is a Lecturer at Arden University and has a passion for all things comic book and horror. He has written about Tobe Hooper's *Poltergeist*, horror in Children's television, *Doctor Who* and *Deadpool*. He has presented papers on Hauntology, the demise of VHS, long-forgotten television and B-Movies.

Julia Neugarten is a PhD candidate in Arts and Culture Studies at Radboud University in Nijmegen, the Netherlands. She has an MA in Literary Studies from the University of Amsterdam. Her research focuses on fanfiction and fan communities using a mix of methods from literary studies, cultural analysis, and the digital humanities. In her creative non-fiction, Julia explores topics such as fandom, food cultures, and disability. Her essays have been published in *OFIC Magazine* and in *Ethical Eating: 32 Writers on Eating Ethically*, a collection of essays from New York University Press. She lives in Amsterdam with her partner.

Lauren R. O'Connor thinks absolutely everything is interesting and worthy of examination, and that includes K-pop bands and comics written for kids. Lauren is a Lecturer at Dominican University in River Forest, Illinois. Their research considers the history and social construction of adolescence, as well as popular media and visual culture more broadly. Her first book, *Robin and the Making of American Adolescence*, was published by Rutgers University Press in 2021. Lauren enjoys live music, iced coffee, and petting all the dogs in the neighborhood while out for a walk.

Ondine Park is an Assistant Professor of Sociology at the University of British Columbia, Okanagan. She is a cultural, environmental sociologist and critical, spatial theorist. Her publications include *Seasonal Sociology* (co-edited with TK Davidson; University of Toronto Press); *Ecologies of Hope: Placing Nostalgia, Desire and Hope* (co-edited with Davidson and R Shields; Wilfrid Laurier University Press).

Dusty Michael Perez is an Associate Professor of Humanities and Communication at Embry-Riddle.

Erica Ka-yan Poon is an independent scholar in East Asian cinemas, specializing in the relationship between Japan and Hong Kong. She received her PhD in Cinema and Media Studies from Hong Kong Baptist University, with her fieldwork study affiliated with the University of Tokyo, and her Master of Arts from the University of California, Los Angeles. Her writings have appeared in *Journal of Chinese Cinemas* and *Journal of Japanese and Korean Cinema*. Some of her research interests include transnational cinemas, anime, and media industries.

Nazario Robles-Bastida is an Assistant Professor (Teaching) of Sociology at the University of Calgary, where he teaches courses in media, ethnicity and racialization, and sociological theory. He is a Foucault super fan. Among his interests are popular culture, media subcultures, and critical theory.

Ian Sinnett is a PhD candidate in Cultural Studies at George Mason University. His primary research interests are in popular culture, popular music and hip hop studies, memory studies, political economy, and affect theory. His dissertation aims to explore the ways in which digital hip hop sampling's advent as the genre's primary means of composition in the 1980s connects to the various material conditions surrounding its rise to prominence. His work has appeared in the peer-reviewed journals *Riffs: Experimental Writing on Popular Music* and *Lateral*.

Louisa Ellen Stein is an Associate Professor of Film and Media Culture at Middlebury College. Louisa is the author of *Millennial Fandom: Television Audiences in the Transmedia Age* (University of Iowa Press, 2015) and co-editor of *A Tumblr Book: Platforms and Cultures* (University of Michigan Press, 2020), *Sherlock and Transmedia Fandom* (McFarland, 2012) and *Teen Television: Programming and Fandom* (McFarland, 2008). Louisa's work explores audience engagement in transmedia culture, with emphasis on cultural and digital contexts, gender, and generation. Louisa is mother of two fans, and in her spare time she edits fan video, video essays, and remix video.

Ana Tomičić, a social anthropologist with a PhD in Social Representations and Communication, currently serves as the Program Director and Researcher at ARETE Institute for Sustainable Prosperity, and as the Program Coordinator at the trade union Novi Sindikat. Her work primarily focuses on digitalization and AI. In addition to her professional roles, she is pursuing further studies in Asian Studies, specializing in Sinology.

Aglaia Maretta Venters has a PhD from Tulane University, studies History, French History, and French Literature, and is a cultural historian. She has taught as an instructor in the History Departments at Tulane University, Baton Rouge Community College, Dillard University, South Louisiana Community College, Xavier University, and Montclair State University. Her publications include works on Hegelian dialectic, monarchy in Renaissance France, Romanticism and David Bowie, Neil Gaiman's *American Gods* and Jesuit philosophy and proto-racial theory in the Americas, medieval French theatre and folktales and History of the Senses.

Ricardo Victoria-Uribe is an Industrial Designer with a PhD in Design with focus on Sustainability from Loughborough University. He has been an academic at the Autonomous University of the State of Mexico, School of Architecture & Design for 14 years. He is a member of the SFWA and has written a science fantasy book series, Tempest Blades, plus a dozen of SFF and horror stories and two academic books on design, and on sustainability. He collects toys, plays TTRPGs, and is a big fan of the TMNT. He co-authored the book *Green New Worlds: A Quick Guide to Sustainability Through Science Fiction and Fantasy*.

Matthew Voice is an Assistant Professor in Applied Linguistics at the University of Warwick, UK, where he teaches courses on linguistic research methods, sociolinguistics, and language and power. His research explores both critical and literary applications of cognitive linguistics. His work has been published in journals including *Language and Literature* and *Discourse Studies*, with topics such as military memoir, pop song lyrics, and forensic linguistic methods connected through cognitive approaches to language and style.

ACKNOWLEDGMENTS

In any universe, I am grateful to the following for their help and support during the editing of this volume: DePaul University and my colleagues in the College of Communication; all the contributors to this volume, for their thoughtful chapters (and patience with an anxious editor!); the peer reviewers, who provided thoughtful feedback; Chris Solis Green, for help reviewing the poetry selections; Baxter, Momo, and Toph, for the insane amount of pets solicited; and of course to Katie, for putting up with it all.

INTRODUCTION

Perspectives on Alternate Universes and Parallel Worlds

Paul Booth

The multiverse is, seemingly, everywhere all at once.

Once relegated to obscure science-fiction films or comics canon, the concept of alternate realities and parallel worlds has now portaled into the mainstream. The immense box-office and critical success of multiversal-focused films *Everything Everywhere All at Once* (2022) and *Spider-man: Into the Spiderverse* (2018; along with other Spider-films *No Way Home* (2021), *Across the Spiderverse* (2023), and *Beyond the Spiderverse* (TBA)), as well as many others, portends a shift in the common understanding of multiversal concepts. Maybe in a different universe, the multiverse was a common construct, but just twenty years ago, Marie-Laure Ryan (2006) noted that "the idea of parallel realities is not yet solidly established in our private encyclopedias and the text must give strong cues for us to suspend momentarily our intuitive belief in classical cosmology" (671). It took something as popular as the now-ubiquitous Marvel Cinematic Universe (MCU)—itself no longer confined to just the cinematic medium, having become super-vitalized in television, comics, video games, and streaming—to demonstrate how widespread the knowledge of the multiverse concept can be. The MCU's *Multiverse Saga* encompasses the series of films and shows from 2021 and beyond, and emphasizes the ever-expanding multiplicity of alternate realities that can be mined for content. That the MCU has become so comfortable discussing multiverse concepts in shows like *Loki* (2021–) and *What If...?* (2021–) highlights the fact that millions of viewers are also becoming fluent in concepts like parallel worlds, alternate realities, and mirror universes.

DOI: 10.4324/9781003480846-1

But it's not just films or the MCU, of course. Television series have also highlighted multiversal concepts. For more than a decade, the Greg Berlanti CW/DC *Arrowverse* (2012–2023) featured a multiverse across six live-action television shows and two animated series that crossing universes. From *Flash* to *Arrow* to *Supergirl* and back again before supper. These connections are more than just references (for example, the multiverse connections of the CW are more narratively significant than something like the infamous *St. Elsewhere* Westphall Universe theory[1]), but directly and explicitly hinge on the fact that the characters and situations *belong in different realities*. Although in literature the concept of the multiverse has been long-standing in speculative fiction, recent examples have also become incredibly popular: *This Is How You Lose the Time War* (2019) by Amal El-Mohtar and Max Gladstone and *Dark Matter* (2016) by Blake Crouch were both award-winning, bestselling books (both of which have been optioned for TV). But the concept goes back centuries: multiverse fiction like H.G. Well's *Men Like Gods* (1923) or C.S. Lewis's Narnia books (1950–1956) highlighted alternate worlds while more recent series like Roger Zelazny's *The Chronicles of Amber* (1970–1991) or Phillip Pullman's *His Dark Materials* (1995–2000, and TV series 2019–2022) deepen the concept for modern audiences.

The list can go on, and as many of the chapters in this collection illustrate, multiverse examples can be found in the strangest, most alien, or obscurest locations in contemporary culture. At the same time, the multiverse *concept* has resonance within and outside fiction. At a point in time when multiple claims on "truth" fight for vibrancy on social media; when multiple interpretations over the same event can contradict each other; when authorship can come into question with artificial intelligence; when history is specifically elided or questioned; when engagement with politics, media, fandom, and economy are inexorably wrapped up in questions of truth, reality, and meaning…the multiverse may be the best way to explain the fracturing of culture today.

If nothing else, it's clear that "The Darkest Timeline" has become so culturally significant as to become a meme (see Figure 0.1).

But what is the multiverse in cultural theory terms? And how does it resonate across different genres, media, and formats?

The authors in *Entering the Multiverse* aim to unpack the multiplicity of multiverse meanings through, naturally, multiple perspectives. The

---

1 Which posits that, as per the ending of the show, an immense number of television shows all belong to a single connected universe—which itself only exists in the mind of a child, Tommy Westphall. (https://www.unsupervisednerds.com/reads-full/2021/1/20/the-tommy-westphall-universe).

FIGURE 0.1   The Darkest Timeline, *Community*.

popularity and cultural significance of the multiverse is a story that can only be told through the edited collection: in which each chapter advances a new, different, unusual, or encompassing theory of the multiverse concept, while also retaining its unique philosophy, theory, or focus. Chapters in this collection tell the story of the multiverse across their own multiple realities: creative nonfiction, academic essay, screenplay, art, poetry, audio analysis, and video essay. As the chapters in this collection illustrate, definitions of the multiverse are, appropriately, multifarious. I am defining the multiverse in this Introduction as *a construct that allows exploration of alternate realities, multiple canons, and contradictory realities within the boundaries of one fictional narrative.* As commonly depicted, the multiverse is contained *within* media content rather than an external means to *organize* media texts. Traditional media theory has discussed concepts *linked* to the multiverse, including transmedia, paratextuality, and multimedia-shared universes, but the representation of the multiverse, as conceptualized here, is not the same thing.

Transmedia, as defined by Henry Jenkins (2006; see also Marsha Kinder 1991) is a style of narrative organization in which authors create one narrative across multiple media forms. Jenkins defines transmedia storytelling as the "art of world making" (26), where different aspects of a world are scattered across different media. Perhaps a film depicts one major aspect of a story while another aspect is told in a television show and a third is demonstrated in a video game. While the definition of *transmedia* has been stretched and adapted in the intervening twenty years since Jenkins popularized the term, the general sense of *one story told across multiple media* still holds true (see Pratten 2015; Harvey 2015; Freeman and Gambarato 2018). In that sense, while the process of transmedia storytelling *could* tell multiverse stories it, itself, is not *necessarily* multiversal. That is, transmediation is a framework in which to tell stories, not a type of story itself.

Similarly, paratextuality, as defined by Jonathan Gray (2010, from Gerard Genette (1997)), is a relationship between texts by which a piece of a text comments on, surrounds, and/or reflects aspects of the main text. As stated by Švelch (2020), paratextuality examines the "seemingly ancillary parts of a literary text" and questions "the centrality of [the] text" (para. 1). In game studies, too, paratextuality has become a significant part of how scholars and players understand games, though things like walkthrus, game guides, Twitch streams, and other ancillary products (see Consalvo 2017). As I have previously shown, games can also be paratextually linked to other media forms (see Booth 2015). Therefore, while a useful concept for understanding a text, the paratext is not necessarily multiversal, although as an organizing structure, it could contain multiversal content.

Finally, shared universe fiction tells stories across media texts that all take place within the same media universe: for example, *Doctor Who* (1963–, itself a show that illustrates multiverse concepts) shares a universe with a number of spin-off shows, including *Torchwood* (2006–2011), *Sarah Jane Adventures* (2007–2011), *Class* (2016), and *K9 and Company* (1981). Movies have sequels and prequels, television shows have spin-offs, and comics may take a minor character and produce a whole range based on them—all within a shared universe. To continue the *Doctor Who* example, the audio production company Big Finish has created entire ranges based on characters who only appeared in one or two episodes of the original television series (see Booth and Jones 2020). In many ways, the shared universe concept is the opposite of the multiverse, as all these media texts by definition *share* the *same* fictional universe rather than existing with separate ones.

The many different chapters in this book unpack these distinctions, often complicating, sometimes contradicting, but always with an eye toward engagement with contemporary cultural issues. But before we get into the chapters, let's dive a bit deeper into the history and development of the multiverse concept.

Welcome to the Multiverse

The word *multiverse* has, appropriately enough, multiple meanings. And like the concept in contemporary comic literature and film, in which it is possible to travel from one universe to another seemingly with ease, the history of the concept bounces between science, fiction, science-fiction, and back again, with each stop shaping and honing the term in ways that echo with similarity while retaining unique elements. It's as if the concept has been lying there, waiting to be uncovered, and writers, scientists, and philosophers have all hit upon it in zeitgeistish simultaneity.

The term "multiverse" itself originated from philosopher William James in his 1895 lecture (and subsequent 1896 book) *Is Life Worth Living*, in which he wrote:

> We of the nineteenth century, with our evolutionary theories and our mechanical philosophies, already know nature too impartially and too well to worship unreservedly any god of whose character she can be an adequate expression. Truly all we know of good and beauty proceeds from nature, but none the less so all we know of evil. Visible nature is all plasticity and indifference, a moral multiverse, as one might call it, and not a moral universe. To such a harlot we owe no allegiance.
>
> *(26)*

For James, modernity has allowed too much knowledge of nature, and we have forgotten that there is much that we *don't* know: science is a corruption of our natural awe. In his conception of the term, *multiverse* refers to the overwhelming human experience of nature, the "capriciousness and incomprehensibility posed by Mother Nature" (Cepelewicz 2017). For James, a multiverse exists within us all, as our own experiences shape our interpretation of the mysteries of the world around us.

Within narrative theory, the multiverse concept exists in various forms to rationalize, as Marie-Laure Ryan (2006) demonstrates, "texts that report contradictory versions of events" (668). Because, she argues, the human mind really wants to connect the dots, to avoid contradictions, and give coherence (re: preference) to *one* reality, then readers have to make sense of the multiverse in a variety of ways. Ryan gives us six interpretations of how the multiverse functions in a narrative *outside of actually being multiple universes*:

1 Mentalism: the multiverse is "the products of dreams, hallucinations, the imagination or they are the symptoms of mental conditions" (669).
2 Virtualization: the multiverse is "nonactualized possibilities" (669).
3 Allegory and Metaphor: the multiverse is "an idea rather than [an] objectively happening course of events" (669).
4 Meta-textualism: the multiverse is "different developments that the author is contemplating" (670).
5 Magic: the multiverse is "a veiled admission of the irrational, or fantastic, nature of this world" (670).
6 Do It Yourself: the multiverse is "material for [audiences] creating their own stories" (671).

For Ryan, writing 20 years ago, it's only when all those "interpretations are eliminated" that readers "rationalize the text though the idea of plural realities" (671)—and the scientific explanation that is the "multiverse" becomes relevant. Today, I believe audiences are more able to conceptualize the idea of the multiverse as an *actual* reality, a real possibility, rather than a justification when all else fails.

Yet, literature had explored the concept of the multiverse even before it had a name. In her overview of the multiverse concept for *Nautilus*, a science magazine that connects the sciences to philosophy, culture, and art, Jordana Cepelewicz (2017) cites "the many and branching stories of William Blake's *The Four Zoas*, written in the late 1700s and early 1800s" as a multiverse concept. And, the parallel universe concept itself can be traced back to the Puranas from ancient Hindu mythology, which discusses an infinite number of universes, or "The Adventures of Bulukiya" from the Persian *One Thousand and One Nights*.

Melding philosophy and literature, Argentinian writer Jorge Luis Borges uses the concept, if not the name, of the multiverse in his famous 1941 short story "The Garden of Forking Paths," which tells of a novel in which a character makes, not one choice, but rather an infinite number of choices that all diverge into a different chain of events/causes. Each chain leads to a new possible timeline, each a new possible world. In the literary world his work has become nearly synonymous with magical realism, philosophical fantasy, and labyrinthian narratives.

But let's switch metaphysical universes for a moment and look at the science of the multiverse.[2] Borges's work is said to be an inspiration for physicist Hugh Everett III's "many-worlds" interpretation of quantum physics (1957; see Cepelewicz 2017), which also builds upon Erwin Schrödinger's 1935 work of feline cruelty: the famous cat thought-experiment. To understand the experiment, and the multi-worlds theory, though, we need to get a basic grasp of quantum superposition. Bear with me for a moment.

In traditional physics, what's called Classical physics, we can know both the position and the momentum of a particle; how fast it moves and where it is at any particular moment. If we don't know that, it's simply a matter of measurement to find out. In Quantum physics, though, which details the movement of subatomic particles specifically, a particle can be in a *superposition*—in multiple states at the same time—*until* it is measured. Sounds

---

2 To be fair, not all physicists agree (a) that multiple universes exist; (b) or if they do, on what they would look like. What follows is a very cursory look at the way quantum physics has been applied/understood as a concept to help demonstrate multiple universe theory in media texts. For more on this controversy in narrative specifically, see Ryan (2006).

fictional, but this sort of multiple-state existence has been measured in experiments using light beams since 1801. Light functions as both a wave (it can be reflected, refracted, and diffracted, just like any wave could be) and a particle (photons that have energy but no mass). In the famous Light Slit experiment, Thomas Young aimed a beam of light at a barrier with two vertical slits cut into it. The light passed through and was recorded on a photographic plate. Young found that if one slit was covered, the pattern was obvious: a single line of light; light, in this case, acting as if made of particles. However, if both slits were open, multiple, overlapping light waves appeared—not, as might be expected, two lines of light. As Sara Metwalli (2023) has summarized, "When the light waves go through the slits, that wave breaks and forms patterns, which is what we see on the back wall." Further testing has revealed that if individual particles like electrons are sent through the slits one at a time, an interference pattern still emerges, which suggests that each particle somehow goes through both slits simultaneously, the superposition state. And, to make things even more bizarre, if one observes which slit the particle *actually* goes through, that superposition collapses, the interference pattern disappears, and instead two distinct bands forms, as if the particle now behaves like a classical object and only goes through one slit or the other.

The point of quantum superposition is that until we look at the wall, until it is measured, the light acts *both* a particle *and* a wave. It's only in the observation that it manifests one or the other.

So, back to that dead cat. Or, rather, not(?)-dead.

In this experiment, Schrödinger hypothesized a box containing a cat, a flask of poison, and a radioactive source of energy. If a single atom decays, the flask is shattered, releasing the poison and killing the cat. Because we don't know when that decay will happen, and because in quantum super-position *both states can be in effect at the same time*, the cat is both alive and dead simultaneously. It's only when the box is opened and we observe the poor creature that we know for sure. (A philosophical conundrum: have we killed the cat by opening the box?)

To be clear, no cats were actually harmed in this experiment; it's just a thought experiment.

Now, back to Everett's many-worlds theory, which holds that all pos-sibilities within a quantum superposition universe are real. Two different states—two different *universes*—come into being. In one, the cat is alive and, presumably, angry at being locked in a box with some poison. In the other, the cat is dead. Two different states have led to two de-coherent universes, each made *real* by opening the box. But by opening the box, the observer becomes "entangled" with the cat, so that the observation of an alive cat and the observer are part of the same universe; they cannot meet the evil twin that killed the cat.

Such is born the multiple universe theory: with every decision, with every chance encounter, different possibilities exist and spin off into their own different universes, each separate from each other and, ultimately, incompatible. Cosmologist George F. R. Ellis (2011) calls this the "level 2" multiverse theory (although he doubts it actually exists). What he claims is more probable is a "level 1" multiverse, which simply states that, if the edge of the known universe is 42 billion light-years (the "cosmic visual horizon"), then what lies *beyond* that may be "infinitely many domains much like the one we see" (38). The multiverse may not represent new universes popping up all the time, but simply bubbles of unexplored universes at the edge of known reality.

That being said, the "level 2" multiverse theory is the one that has captured the minds of science-fiction authors and philosophers. Karen Hellekson (2001) ties this type of narrative to alternate histories, which "revolve around the basic premise that some event in the past did not occur as we know it did, and thus the present has been changed" (2). The change can be from scientific means (e.g., time travel) or just exist as a fictional state. Richmond (2001) calls this "eschatological time-travel," in which fictional events allow "objective contradictions and try to hold everything else constant" through the creation of new worlds (312). Indeed, from a philosophical perspective, almost all time travel stories must inherently rely on many-worlds theories as a way of preserving the past and also demonstrating "human uncertainty about future action" (316). If every action we make results in a new universe branching off, and if time travel stories are ultimately about the effect of the choices we make in our lives, then they serve as the perfect literary litmus test for the multiverse concept.

Perhaps no medium has embraced the concept of the multiverse better than comics. For Karin Kukkonen (2010), "Superhero comics ... take a multiworld model of reality—the multiverse—largely as an ontological given" (40). In the Introduction to her edited collection *The Superhero Multiverse*, Lorna Piatti-Farnell (2020) defines the term multiverse as "a constructed space that draws strength from parallel and alternate narratives, as well as different intertextual, paratextual, and intermedial connections between different incarnations of popular icons" (3). Indeed, texts like the Marvel Cinematic Universe have popularized the term "multiverse" in films like *Doctor Strange in the Multiverse of Madness* (2022). But the concept goes back decades in the comics industry and stems from the most ironic of rationales: coherence. Given the number of people who have worked on various comic titles over the years, inconsistences crept into each superhero volume or collection. To try to "unify" all the various inconsistencies, DC and Marvel each created "multiverse" concepts "to clear up the problems with narrative continuity"—and to sell more volumes (Kukkonen 2010; see also Wolfman 2000). Kukkonen uses *Crisis on Infinite Earths* (1985–1986) as an exemplar of how this unification occurred. And just as the diegetic fracturing of the comics in a multiverse led to a coherence in the narrative, so too did it "cue … readers to establish a mental model of multiple parallel realities, enabling them to navigate the multiverse and cope with its violations of classical cosmology" (Kukkonen 2010, 42). The multiverse led to new reading strategies and opened up new ways of understanding textuality.

In her Introduction, Piatti-Farnell uses the metaphor of the multiverse to discuss "the multi-faceted nature of the process of adapting and re-adapting these iconic comic book figures" (2) rather than looking at the concept of the multiverse in and of itself. That is, her book looks more at the adaptation and transmediation of comic figures, not the way those figures play in and out of alternate narratives within the texts themselves. This is perhaps the true power of the multiverse in comics: the multiverse becomes a metaphor for something greater than itself.

In the same book, Whitney Harden and Julia Kiernan (2020) describe the branding advantages of using a multiverse concept in comics (and television/film) as a way of maintaining a text's distinctiveness within a heavily transmediated framework. They use the example of *Legion* (2017–2019), a superhero television series based on a Marvel franchise but *not* part of the Marvel Cinematic Universe. For Harden and Kiernan, the multiverse concept allows *Legion* "to retain its stand-alone status and to indulge in complex, and sometimes confusing, acts of narrative exploration" (35). The Marvel Cinematic Universe is known for particular types of stories and spectacles—a common criticism of the films/television series is the *sameness* they all eventually portray. Oh, look, another flying CGI-battle in a multi-colored sky. From a brand standpoint, this similitude is a strength of the MCU: audiences know what to expect. But through its use of the multiverse, new types of stories like *Legion*, can be told that don't fit into that branding. In other words, the multiverse becomes more than a metaphor; it is an opportunity to drive new types of stories that wouldn't be brand-specific and can engage different audiences.

Beyond branding and narrative, the concept of the multiverse has found itself in different disciplines. Christian Pipal, Hyunjin Song, and Hajo G. Boomgaarden (2023) use the term to describe social science research and the need to process "all reasonable options" of data modeling—in other words, rather than concentrating on just one arbitrary dataset, they recommend a researcher "runs all analyses using each instance of the multiverse" of datasets (258). The use of the term isn't about multiple realities but about the multiple possibilities of research, each one opening up a new set of datapoints.

Another discipline that uses the term multiverse at times is fan studies, which not only examines fans of the types of media texts that tend to *have* multiverses (cult media and science fiction) but also presents fan work as a type of multiversal addition, offering new possibilities/worlds that don't occur within the original text but that *could* happen in a multiverse of possibilities. Even the word "multiverse" echoes with the rhetoric of fandom: it comes together like two characters' names, smooshing "multi" (as in many) and "verse" (as in universe)—a shipname where opposites come together, that is, *multiple + one* (uni-)—an ideal fan pairing! For Anne Kustritz (2014):

> transmedia narratives may unfold in different sequences and across a different timeframe for each audience member; yet fan narratives not only publish events out of sequence, but also contain numerous alternate interpretations and versions of the same events. However, these complications need not negate the role of seriality; rather they call for a reevaluation and expansion of the serial form and function.
>
> *(para. 6)*

Fans often assert their own reading of the text as a primary one, seeing "fan works as legitimate parts of the overall transmedia story network, wherein they may either exist in parallel, or actually change and usurp the audience's understanding of the professional text" (para. 12). Fan work is thus a parallel expansion of the fictional world, relevant and significant in the eyes of the fandom.

Each Chapter, a Multiverse of Possibility

The chapters in this book take many forms. They approach the multiverse in a variety of ways. Each is its own little universe of content which may match what others articulate or may differ significantly in form, content, or opinion. That is the way it should be; the multiverse cannot, by definition, be singular. My hope is that educators can use the different genres represented in this book as unique rabbit-holes with which to explore new aspects of the multiverse.

We start the book with the section "Skipping Stones over the Waters of the Multiverse," which features ten short chapters that each explore one facet of the multiverse in detail. Think of these as bite-sized theories that, when taken together, explore the many sides of the multiverse, a tesseract of ideas. We start with Tim Butler Garrett's "The Irresistible Rise of the

Multiverse in the fin-de-Millennium Media Zeitgeist," where the multiverse allows us to explore the concept of "the darkest timeline"—and helps us see the possibilities in ours. Aglaia Venters' "'The Unsupportable Pride of Mankind': Seventeenth-Century Utopias, Humility, and the Multiverse," which walks readers through the history of the multiverse in seventeenth-century utopian literature and feeds into our present-day understanding of the concept. Angélica Cabrera Torrecilla's "Depictions of the Multiverse as a Labyrinth in the Stories of Jorge Luis Borges" follows a similar structure, only this time looking at the influence of Borges on both literature and quantum science in the twentieth century. Following this, Chris Gerrard updates the timeline to the twenty-first century by examining "A Creative Deconstruction of the Cultural Strands in the Multiple Timelines of *The Flash* (2023)"—and in an extensive hypertextual database, explores the multiple types of multiverses that appear in DC media. Turning to cultural influences, Sem Devillart's "The Multiverse as a Stylistic Structure" examines the connection between culture and technological developments and views the multiverse as a leitmotif of today's transmedial logics. Marija Adela Gjorgjioska and Ana Tomičić's "Capitalism and the Multiverse" asks the important question, is it possible to imagine a universe without capitalist structures? More reimagining happens in Dusty Michael Perez's "A Non-Binary Multiverse Poem," which explores the way the multiverse allows new views of gender. Adalberto Fernandes's "What Does It Mean to Be Eternally Alone in the Multiverse?" explores through philosophy how the multiverse can serve as a metaphor for elements that both separate and connect us. Maria Faust's collection of photographs and poems, "Lux in Tenebris," continues the section with a more positive view of alternate realities and the possibilities they engender. We conclude the section with the academic essay "Tales of Two Multiverses," by Matt Hills, who critiques the binaristic model of multiversal thinking.

Section II follows, with its exploration of "Multiversal Constructs Across Media." We open with three short essays that use multimedia to explore the multiverse concept—Ian Sinnett's "It's a Trip. It's Got a Funky Beat" deploys the multiverse as a way of understanding hip-hop music, and on our companion website, he dives into the Beastie Boys's album *Paul's Boutique* as a demonstration. Similarly, Louisa Ellen Stein's chapter on "The Shared Multiversal Power of K-pop Lore" focuses on the album concept, music videos, photoshoots, story films, comic books, web series, and novels that surround K-pop bands, and in a video essay on our companion website, she outlines the multiverse of connections they reveal. Following this, fiction author Nev Fountain continues the section with a deep, humorous dive into the (fictional) multiverses of *Doctor Who* companion Ian Chesterton.

Lauren O'Connor follows Fountain's piece with a deep dive into one particular K-pop band, BTS, and the Bangtan Universe that surrounds their work. She views this as a way of theorizing adolescence both within and beyond the South Korean context. Continuing in this vein, Shayna Maskell's "'The Infinite Knowledge and Power of the Multiverse'" uses Black feminist thought and intersectionality to discuss the way the multiverse in *Everything Everywhere All At Once* reflects not only the constitutive nature of intersectionality, but also the complex emotional toll these interlocking systems of oppression have. In Robert McLaughlin's chapter "Marvel's Multiverse of Madness," the multiverse becomes a way to understand the industrial connections and limitations of media franchising. Further unpacking Marvel's relationship to the multiverse in a new medium, Teresa Forde uses a feminist collage to depict *WandaVision*'s detailed understanding of the multiverse, while Lore FitzWhittemore explores the Spiderverse as an intense collaboration between fan and media producer. Moving to a more open view of the multiverse concept, Ricardo Victoria-Uribe and Nazario Robles-Bastida unpack the *Teenage Mutant Ninja Turtle*-verse in "Turtleverse! The Teenage Mutant Ninja Turtles Multiverse in Media as Tool to Explore Diverse Genres"; in this case the Turtleverse helps readers develop a greater understanding of genre.

In Section III, "Multiversal Constructs Across Literature and Games," the chapters uncover a rich vein of multiversal concepts in both written and playable works. Blair Davis's "The Comic Book Multiverse" opens the section with a discussion of the history of the multiverse in comics literature, arguing that it serves two functions: to create coherence and to create industry dominance. Ondine Park's "Multiversality in Children's Picture Books" unpacks the way multiverses appear in children's works, especially as a way of offering new possibilities. Following this, Rebecca Gibson's "The Atlast, the Resurrection, the Rising, and the Time War" shows how the concept of *love* appears in—and crosses—multiple multiverses. Further detailing love across the multiverse but in a blisteringly tragic manner, Rebecca Johns's screenplay "Forget-Me-Not" explores what it means to love, and lose, in the multiverse. Finishing our literary analyses, Amy Cole's "Finding Yourself: Multiversal Identity Crisis in Ted Chiang's 'Anxiety Is the Dizziness of Freedom'" highlights the dangers of the multiverse on a psychological level, and the damage it can have on an individual's psyche.

As a synthesis of literature, media, and individual play, games can have unique attributes in multiversal thought. Ross Garner's "On Wormholes, Link Cables, and the Limitations of Transmedia" asks the question, what happens when different national, production, and consumption contexts are emphasized in the analysis of a multiversal media mix like *Pokémon?* Maria K. Alberto takes a similar question to task in "Playing Across Planes with the D&D Multiverse," which argues that characterizing the role-playing game *Dungeons & Dragons* in terms of multiverse functions creates the impression of planned, diegetic cohesion across disparate genres, fictional settings, and mytho-historical pantheons—but can also be deployed for industry-specific profit. Finally, David Scott Diffrient's "Cards Against Monstrosity" takes the nearly-infinite number of combinations of cards in Living Card Games like *Arkham Horror: The Card Game* as multiverses in and of themselves, and their intersection with the multiversal concepts in Lovecraft's universes too.

Finally, Section IV examines "Multiversal Constructs Across the Self and Others," as the chapters in this section look at how audiences and fans construct multiversal ideas. Dustin Abnet's "Teaching the American Multiverse" opens the section by unpacking a class he taught on multiversal theories and American culture. Following this, Erica Ka-yan Poon's "Reaffirming Japanese Identity through the Multiverse" looks at nation-building through multiversal games, especially following natural disasters. Matthew Voice's "How to Travel the Multiverse" brings a more theoretical exploration of Text World Theory—a way in which worlds are accessed by readers and viewers of multiverse narratives. Julia Neugarten's creative non-fiction piece "The Multiverse of Fanfiction" follows, and puts Voice's theories to the text in her exploration of how her own fanfiction writing reflect different (multiversal) versions of the self. Finally, Lincoln Geraghty's "'Long Live the Empire!'" concludes the section with a deep dive into *Star Trek's* Mirror Universe, one of the most famous multiverses in science fiction, and how it ultimately arrives through fan work.

Conclusion

The multiverse cannot be singular: this is why this book is important now, at this moment in media history. Whether or not we will ever be able to hop between realities, travel between universes, of see the branches that are created by every decision we make, the multiverse *concept* is real. Young viewers are growing up in a world where *multiverse* is as common a term as *sequel* or *franchise* or *merchandise*. And the concept has so many possibilities to offer, from new mental models of knowledge to greater metaphors of meaning to unique understands of authorship.

Welcome to the multiverse. Please watch your step.

References

Booth, Paul. 2015. *Game Play: Paratextuality in Contemporary Board Games.* New York: Bloomsbury.

Booth, Paul, and Craig Owen Jones. 2020. *Watching Doctor Who: Fan Reception and Evaluation.* London: Bloomsbury.

Cepelewicz, Jordana. 2017. "The Multiverse as Muse." *Nautilus.* Feb 02. https://nautil.us/the-multiverse-as-muse-2-236410/

Consalvo, Mia. 2017. "When Paratexts Become Texts: De-centering the Game-as-Text." *Critical Studies in Media Communication*, 34, no. 2: 177–183.

Ellis, George F. R. 2011. "Does the Multiverse Really Exist?" *Scientific American* 305, no. 2: 38–43.

Freeman, Matt, and Renira Gambarato, eds. 2018. *The Routledge Companion to Transmedia Studies.* New York: Routledge.

Genette, Gerard. 1997. *Paratexts: Thresholds of Interpretation*s, trans. Jane E. Lewin. Cambridge: Cambridge University Press.

Gray, Jonathan. 2010. *Show Sold Separately: Promos, Spoilers, and Other Media Paratexts.* New York: New York University Press.

Harden, Whitney, and Julia Kiernan. 2020. "The Multiverse Paradigm and the Reinvention of *Legion*," in *The Superhero Multiverse: Readapting Comic Book Icons in Twenty-First-Century Film and Popular Media*, ed. Lorna Piatti-Farnell, 33–46. Lanham: Lexington.

Harvey, Colin. 2015. *Fantastic Transmedia: Narrative, Play and Memory Across Science Fiction and Fantasy Storyworlds.* Houndsmills, UK: Palgrave MacMillin.

James, William. 1896. *Is Life Worth Living?* Philadelphia: St. Burns Weston. Via: https://archive.org/details/islifeworthlivin00jameuoft/page/n7/mode/2up

Jenkins, Henry. 2006. *Convergence Culture: Where Old and New Media Collide.* New York: New York University Press.

Kinder, Marsha. 1991. *Playing with Power in Movies, Television, and Video Games.* Berkeley: University of California Press.

Kukkonen, Karin. 2010. "Navigating Infinite Earths: Readers, Mental Models, and the Multiverse of Superhero Comics." *Storyworlds: A Journal of Narrative Studies* 2: 39–58.

Kustritz, Anne. 2014. "Seriality and Transmediality in the Fan Multiverse: Flexible and Multiple Narrative Structures in Fanfiction, Art, and Vids." *TV/Series.* http://journals.openedition.org/tvseries/331.

Metwalli, Sara A. 2023. "What is Superposition?" *Builtin.* Jan 18. https://builtin.com/software-engineering-perspectives/superposition

Piatti-Farnell, Lorna. 2020. "Introduction," in *The Superhero Multiverse: Readapting Comic Book Icons in Twenty-First-Century Film and Popular Media*, ed. Piatti-Farnell, Lorna, 1–13. Lanham: Lexington.

Pipal, Christian, Hyunjin Song, and Hajo G. Boomgaarden. 2023. "If You Have Choices, Why Not Choose (and Share) All of Them? A Multiverse Approach to Understanding News Engagement on Social Media." *Digital Journalism* 11, no. 2: 255–275.

Pratten, Robert. 2015. *Getting Started in Transmedia Storytelling: A Practical Guide for Beginners* 2nd Edition. USA: CreateSpace Independent Publishing Platform.

Richmond, Alasdair. 2001. "Time-Travel Fictions and Philosophy." *American Philosophical Quarterly* 28, no. 4: 305–318.

Ryan, Marie-Laure. 2006. "From Parallel Universes to Possible Worlds: Ontological Pluralism in Physics, Narratology, and Narrative." *Poetics Today* 27, no. 4: 633–674.

Švelch, Jan. 2020. "Paratextuality in Game Studies: A Theoretical Review and Citation Analysis." *Game Studies* 20, no. 2: https://gamestudies.org/2002/articles/jan_svelch

Wolfman, Marv. 2000. "Introduction." *Crisis on Infinite Earths*. New York: DC Comics.

PART I

Skipping Stones over the Waters of the Multiverse

1

THE IRRESISTIBLE RISE OF THE MULTIVERSE IN THE FIN-DE-MILLENNIUM MEDIA ZEITGEIST

Tim Butler Garrett

The idea of an alternate history, an alternate timeline, a branching universe, is nothing new. If you are a child of the 80s, your mind might go to *Back to the Future: Part II*; in the 90s it might have been *Sliding Doors* or *Groundhog Day*; with the Marvel Cinematic Universe in full swing, the concept of a multiverse is now not understood solely by people who read Philip K. Dick. Whether the multiverse exists only as a cultural idea, or whether it is (or will be, at some point in the future) a scientifically verifiable reality seems to me to be one of the less important questions around it: that is to say, we could perhaps adapt one of Voltaire's favorite aphorisms—"If God did not exist, it would be necessary to invent him."—and suggest that the idea of the multiverse has become in some way era-defining at this point of our cultural history. Whether or not the multiverse exists *out there*, we need to be curious about why it has become such a powerful and pervasive idea of our age.

We may not fully agree with Wilde's dictum "Life imitates art far more than art imitates life" (1891), while also recognizing that it is often an era's works of art that are the vanguard of its *zeitgeist*—the *spirit of the age*. As other pieces in this collection show, across film, television, video games, novels, and comics the concept of alternate universes and parallel worlds has blossomed in our time; my title borrows Deakin's (2019) coinage *fin-de-millennium* to describe this era, with its nod to the fin-de-siècle—that previous fulcrum moment so full of both cynicism and optimism. It seems

DOI: 10.4324/9781003480846-3

to me that one of the vital thoughts behind the idea of the multiverse is the very ambivalent question "Could it have been different?" In the opening scenes of *Doctor Strange in the Multiverse of Madness* (2022), an old colleague of the protagonist asks (of the *blip*, which, in MCU continuity, turned half the inhabitants of the universe into dust, before being reversed by the Avengers) "Did it have to happen that way? Was there any other path?" Just as Robert Frost's iconic poem, *The Road Not Taken* considers the question of what might have been, and how it compares with what actually is, so this question lies at the heart of the fascination with the idea of the multiverse. The multiverse then takes the concept further, by suggesting that, theoretically, all these things that *could be*, actually *are*, in some other timeline/universe/dimension. And, in a very late-capitalist way, there is the creeping suggestion that perhaps, as timeline consumers, we could have a choice about which timeline we might wish to inhabit. *Back to the Future*'s Marty McFly is the prototypical timeline consumer: in *Back to the Future Part II* (1989), by a quirk of time travel, he is able to compare and contrast two versions of 2015: one future where his family's situation is precarious but ultimately redeemed and one future where his nemesis, Biff Tannen, has been elected to political office, and his whole community is now a dystopia.

In one of those Wildean moments of life imitating art, 2016 saw the United States (US) electing a simulacrum of Biff Tannen, the United Kingdom (UK) voting for the freedom to commit an act of self-harm, and the deaths of so many artists of note (including David Bowie, Prince, George Michael, and Leonard Cohen) that it felt like the universe had tipped over from indifferent to cruel. *Community* (2009–15), Dan Harmon's trope-loving sitcom, popularized the notion of *the darkest timeline* in a 2011 episode, drawing on a *Star Trek: The Original Series* episode "Mirror, Mirror," which featured a mirror universe with evil versions of Kirk and Spock—with the goatee of the evil Spock later becoming a trope used to identify "evil" versions of characters. In 2016, internet users found ever greater reasons to claim that we were living in *the darkest timeline*. Alongside a multitude of online comments and memes to this effect, there also emerged at this time a fascination with the so-called "Mandela Effect" (Broome, 2020), which posited that there are swathes of the population who remember a different history—be that big or small: believing that *The Berenstain Bears* used to be called "The Berenstein Bears," or that Nelson Mandela died (in prison) years earlier than he actually did (as a free man and having been leader of South Africa).

A 1977 talk given by Philip K. Dick entitled "If You Find This World Bad, You Should See Some of the Others" anticipated both the idea of *the darkest timeline* and the Mandela Effect: Dick claimed that he himself had experience of other, alternate, co-existent realities:

> The world of *Flow My Tears* is an actual (or rather once actual) alternate world, and I remember it in detail. I do not know who else does. Maybe no one else does. perhaps all of you were always — have always been — here. But I was not.

Dick believed that this timeline was not the darkest one (he remembered darker), nor was it the best one: he had been given glimpses of a "wonderful garden world," an "Arcady of the Greco-Roman pagan world" which could be "available as our future, in which all lost things shall be restored" Questions like these, around "actual" reality versus "perceived" reality go back at least as far as the realist/anti-realist debates of philosophers such as Hume (1711–1776) and Locke (1632–1704), though Dick's utopian vision seems drawn more from the same psychedelic milieu as Timothy Leary's concept of the "reality tunnel" (Leary, 1988).

This question of loss is central to the idea of the multiverse. We are increasingly aware of this nexus of possibilities: *Could it have been different? Could it have been better/worse?* And then the realization which flows from these thoughts: *Our reality might never have happened.* The 2019 film *Yesterday* picked up on this zeitgeisty idea, positing a world where all but a few people instantaneously forget the existence of The Beatles. The most pressing idea in the minds of the few who remember them is to reintroduce their music to the world: that the existence of the music, the knowledge of the music, is paramount. Beyond telling an inventive story, one thing *Yesterday* does is express something of the rise of Fan Studies, depicting this modern breed of cultural consumer who, in large part because of the internet, has been able to connect to others with similar niche interests, and discover, disseminate and curate knowledge of obscure bands, films, comics, television shows and other works of art. The internet, with its wealth of information and media, all existing in a sort of Platonic timeless now, has made us acutely aware, more than ever, of what was (even though it has sometimes been almost forgotten) but also of what could have been. This "could have been" is, arguably, experienced as a loss. More film fans than ever are now able to be aware that we could have had a world where Orson Welles' lost cut of *The Magnificent Ambersons* (1942) was released to great acclaim; likewise, Tom Schiller's unreleased masterpiece *Nothing Lasts Forever* (1984); or Stanley Kubrick's unfilmed *Napoleon*; it is easy to imagine an alternate reality where Brian Wilson completes his magnum opus *SMiLE* in 1967, and it equals or eclipses The Beatles' *Sgt. Pepper's Lonely Hearts Club Band*. These *could-have-beens* are tantalizingly close, and the notion that they might actually exist, somewhere "out there," is both frustrating and strangely comforting.

The Internet, which holds so many fragments from our actual and could-have-been past, has given us a surfeit of choice and a dislocation from previous fixed and shared narratives. For most of the history of pop music, there were national and international charts, there was a progression for new artists, and a dropping off of old ones. Now, new artists have to compete for streams with legacy artists, many of whom are dead and have not made new music in decades. While there are still new music trends and a dominant narrative of what is popular, it holds nowhere near the same power it once had. Music culture, like film and television culture, has become atomized, cut adrift from many of the previous fixed points of the shared culture. Television still has "'water cooler moments'"—notably, live events such as the World Cup, the Super Bowl, or the Eurovision Song Contest—but not to the same extent as before, and rarely for scripted television; the 105 million viewers for the final episode of *M*A*S*H*, 80 million for *Cheers*, or 50 million for *Friends* will never be equaled. As a society, as a world, we are moving toward overlapping but separate cultural points of reference.

Politically, we are already living in alternate realities. Since at least the point where we split off into the darkest timeline, we have been living in a world that recognizes "alternative facts." A proportion of the U.S. electorate believes the Sandy Hook tragedy didn't happen, the 2020 election was stolen, and the moon landings were faked. If these, and a host of other beliefs, might seem like recently adopted paranoid conspiracy theories to some, to those who hold them they seem to feel like lived experience, lived history of being lied to and humiliated by an arrogant ruling class. Angry members of both the left and the right now follow their own version of the (Marxist-derived) ideology that could be described as "critical constructivism" (Kincheloe, 2005)—privileging "lived experience" over objective or consensus reality: a sort of modern-day form of the Lockean/Humean realist/anti-realist debate. No wonder that gaslighting—a form of psychological abuse where a person is made to question their memories or perception of reality—has become such a hot topic. Inhabitants of each reality find it almost impossible to imagine living with the other's perception and indeed are affronted at the suggestion that their worldview might be less valid, or, more bluntly, not based on what is real.

Artistically and culturally, alternate versions of artworks we think we know can be damaging to our sense of fixedness and reality. Writing about the different cuts of *Heaven's Gate* (1980) available on different streaming services (and with no explanation of why it was a different cut or even that it was a different cut), David Hepworth (2023) suggests that "There was a time that when things were finished, they were finished...Streaming has fixed that. In the future everything will be provisional" (111). The furor around George Lucas's later tinkering with his *Star Wars* films—beyond updating special effects and into story elements that affect meaning and character, such as the "Han shot first/Greedo shot first" controversy—has only entrenched many fans in the belief in the superiority of ownership of physical media: superiority because, in the universe they grew up in, Han shot first and it is permanent.

If to many this timeline feels like the *darkest timeline*, the idea of retreating to a cultural/historical/media safe space of a universe we understand seems very inviting. And the idea that we could have a do-over—that there is a universe where we get to have another go, fix the timeline, or just exist in a bubble universe that is untouched by all the aspects of contemporary life that disturb us—has become an alluring myth of our age. In *Yesterday*, the Beatles never existed, which also means that John Lennon was never shot, and the film is able to have its cake and eat it by also bringing their music back via the few who remember. Yet, for the most part, the problem Robert Frost identified still remains: the choice we make opens up myriad possibilities; however, it also shuts down perhaps an equal number. What does this mean? As Doc Brown avers in *Back to the Future Part III*, "It means that your future hasn't been written yet. No one's has. Your future is whatever you make it. So make it a good one." We forever live on the precipice of a multiverse: our seemingly inconsequential decisions setting us on one particular path; and for the universe we create together, that has made all the difference.

References

Broome, Fiona. 2020. *The Mandela Effect—Theories and Explanations*. Independently published. https://fionabroome.com

Deakin, Pete. 2019. *White Masculinity in Crisis in Hollywood's Fin de Millennium Cinema*. Lanham, Maryland: Rowman and Littlefield.

Dick, Philip K. 1977. "If You Find This World Bad, You Should See Some of the Others." *The Shifting Realities of Philip K. Dick: Selected Literary and Philosophical Writings* (1996). New York: Vintage Books.

Frost, Robert. 1915. "The Road Not Taken." *The Road Not Taken: A Selection of Robert Frost's Poems* (1991). New York: H. Holt and Co.

Hepworth, David. 2023. 'Streams of Consciousness' *Radio Times*. 7–13 October.

Kincheloe, Joe L. 2005. *Critical Constructivism*. New York: Peter Lang.

Leary, Timothy. 1988. *Neuropolitique*. Los Angeles: New Falcon Publications.

Voltaire (François-Marie Arouet). 1769. *Epistle to the Author of "The Three Impostors"* in *OEuvres complètes de Voltaire*, ed. Louis Moland [Paris: Garnier, 1877–1885], tome 10, pp. 402–5.

Wilde, Oscar. 1891. 'The Decay of Lying—An Observation' in *Intentions* (first published 1913). London: Methuen.

2

"THE UNSUPPORTABLE PRIDE OF MANKIND"

Seventeenth-Century Utopias, Humility, and the Multiverse

Aglaia Maretta Venters

Though some might think of Multiverse literature as a feature of recent culture, the origins of the genre stretch back for centuries. Some scholars posit that the mythology of the ancient world, with its stories of epic voyages, presaged the early modern utopian literature that would become popular with Thomas More's 1516 *Utopia* as well as today's science-fiction genre (Evans 2017, 13). The shape-shifting nature of all facets of the universe became a critical similarity between early modern utopian literature and the multiverse media of today. David Seed (2011), citing William James, defines *multiverse* as: "the variety of Nature, here suggesting the pluralization and malleability of reality itself" (26). Indeed, as discoveries of the Scientific Revolution, the Age of Exploration, and New World colonization presented the early modern European elite with new views of the world, life, and religion, readers were confronted by newfound fluidity between the realms of reality, imagination, and spirituality. This article explores the works of Cyrano de Bergerac and Margaret Cavendish as examples that highlight the relationship between scientific advancement, world exploration, and fluidity of human rational and imaginative capabilities. Because of their emphasis on fluidity, Cyrano de Bergerac's works: *Cyrano de Bergerac's Comical History, Containing the States and Empires of the Moon* (1657) and *New Works of Cyrano de Bergerac, Including the Comical History of the States and Empires of the Sun* (1662) and Margaret Cavendish's *The Blazing World* (1666) are examples of utopian literature that cross into the territory of multiverse literature. Social commentary

DOI: 10.4324/9781003480846-4

communicated through the works of Cyrano and Cavendish criticized the European elite (the majority of their audiences), setting precedents for later multiverse literature to include themes that encouraged humility among its readers.

Savinien de Cyrano de Bergerac wrote the two books of *Other Worlds: The Comical History of the States and Empires of the Moon and Sun* in 1657 and 1662. Adam Roberts (2016) comments that the two trials Cyrano faces in the story (the first on the Moon, then on the Sun) are eidetic of seventeenth-century Europeans' uncomfortable tension between fantastic visions of science and the weakening of religion's authority (64). While Robert's interpretation does ring true, tension between science and religion are not the only conflict Cyrano faces. In another twist of inversion of assumptions, Cyrano argues that humility is the ideal, most blessed position for humans to take.

> To this may be added the unsupportable Pride of Mankind, who perswade themselves that Nature hath only been made for them; as it were likely that the Sun, a vast Body Four hundred and thirty four times bigger than the Earth, had only been kindled to ripen their Medlars and plumpen their Cabbage.
>
> *(Cyrano)*

Cyrano asks readers to consider that God's creation of each immortal soul meant that God also could create infinite worlds and realties, his clearest prediction of later multiverse literature. Early modern European literature saw the rise of a genre that dealt with the use of voyages to strange and unreal worlds as social commentary, which included the scientific discoveries of an era in which human understanding of the world would be challenged by experiments based on scientific method (Evans 2017, 15). Like the work of scientists like Copernicus, Galileo, and Newton, Cyrano's voyages to the Moon and the Sun undercut man's concept of himself as the center of the Universe and the highest of God's creations. Hence, while expressions of tension between religion and science were prominent themes in these utopian works, it appears that Cyrano felt that humans needed humbling regardless of the existence of the divine and/or any other kind of authority.

Cyrano's embrace of Cartesian mathematics led to anticlerical sentiment (Evans 2017, 17). The works begin with the main character, also named Cyrano, hearing his friends posit mystical ideas about the composition of the Moon. Some offered that the Moon was a window of Heaven, and others that the Moon was the Sun looking in to check on the world at night. Cyrano instead argued that the Moon was another planet, and its inhabitants saw the Earth as their Moon and found as comical to believe there are people on the Earth (Kindle edition). Hence, from the beginning, Cyrano makes clear that his intention is to guide his readers through an alternate world. Cyrano spends the remainder of the story discovering that his explanation was the correct one; other celestial bodies are inhabited by beings who find humans to be ridiculous in their assumptions. Seed (2011) notes that Cyrano's characters' space voyage allows them to view the world around humans, to see their societies on Earth as outsiders looking in, enabling them to find the "irony" of human assumptions (6). Cyrano's utopias also challenged readers to find in Copernicus's scientific discoveries potential for space voyages and flying machines while communicating his criticisms of the dogma of the Gallican Church.

Cyrano also finds himself at odds with the clergy and traditional Christianity throughout his voyages. Using the Sun's rays to evaporate dew that he trapped in glass containers around his body, Cyrano floats to a strange land, which he soon finds out is Canada (Cyrano). Recalling the trope of noble savage popular in early modern Europe, the nude people Cyrano meets there reflect the influence of the Age of Exploration on the author's imagination. The influence of the Scientific Revolution continues to permeate the story as Cyrano debates in favor of the heliocentric universe and the movement of the planets with the governor of New France (Cyrano). When Cyrano finally reaches the Moon, he meets the four-legged Moon men. The Moon men explain to Cyrano a somewhat inverted philosophy of the place of humans in Nature:

> Besides, consider a little how they have the Head raised toward Heaven; it is because God would punish them with scarcity of all things, that he hath so placed them; for the supplicant Posture shews that they complain to Heaven of him that Created them, and that they beg Permission to make their best of our Leavings. But we, on the contrary, have the Head bending downwards, to behold the Blessings whereof we are the Masters, and as if there were nothing in Heaven that our happy condition needed Envy.
>
> *(Cyrano)*

The passage offers readers a literal inversion of everything they assume about their place in the natural world and the cosmos and shows them that such overturning of their biases could have its own logic in an alternate reality. Perhaps the aspect that most clearly foretells multiverse culture is the fact that inverted worlds do have a logic of their own. Also, Cyrano (the author) attacks the pride of humans by showing that four-legged beings (normally considered beneath humans) may feel superior to humans. He also ridicules human ideas of their elevated place in God's esteem by portraying them as ungrateful and as having received divine punishment. Perception of scarcity is the punishment for human's ingratitude, whereas the constantly humbled are the most blessed of all.

Still, critiques of accepted authority structures were a feature of many early modern utopias, including Cavendish's *The Blazing World*. Roberts (2016) highlights Cavendish's work as utopian and feminist literature, philosophical commentary, and scientific treatise, especially as it was largely dismissed and ignored until the twentieth century (63). Seed counted *The Blazing World* as not only an example of the earliest alternate-world literature but also a feminist utopia within the history of science-fiction (Seed 2011, 4). Like Cyrano's narrator, Cavendish's protagonist merged with the author to engage in social and philosophical commentary. Meanwhile, Cavendish embraced scientific exploration of the natural world, and in the process questioned social assumptions that men were naturally predetermined to dominate women. Cavendish's work also reflected the trends of the Age of Exploration and early modern colonization of other continents. Her work begins with a merchant who traveled to another land fell in love with a lady and kidnapped her. The lady was the only survivor of a shipwreck in the North Pole and ends up sailing to a strange other land where the Emperor mistakes her for a goddess. The Emperor then marries her and makes her the absolute ruler of the world (Cavendish, 10). Cavendish's story predicts later multiverse literature because a world ruled by a woman deliberately overturns early modern social order and assumptions about male dominance over women as a natural and rational absolute truth.

As Empress, the lady learns from Bird-men, Worm-men, and Fly-men. In passages in which the Empress seeks whatever wisdom others can give her, Cavendish exalts the genius she found in the natural world. Cavendish continuously upends early modern assumptions about human mastery over nature, thereby stressing that the world she created is an alternate to the one in which her readers live. Like Cyrano, Cavendish's heroine discovers that a particular logic, different from the world from which she came, supports the alternate reality surrounding her. The fact that the heroine is willing to learn and understand the logic of the world in which she now finds herself is a clear sign that Cavendish too predicted multiverse culture. When speaking to Galenical Physicians about herbs and drugs the Empress finds that:

> their operations and vertues were generally caused by their proper, inherent, corporeal, figurative motions, which being infinitely various in Infinite Nature, did produce infinite several effects. And it is observed, said they, that Herbs and Drugs are wise in their operations, as Men in

their words and actions, nay wiser; and their effects are more certain than Men in their opinions; for though they cannot discourse like Men, yet they have Sense and Reason, as well as Men; for the discursive faculty is but a particular effect of Sense and Reason in some particular Creatures, to wit, Men, and not an principle of Nature, and argues often more folly than wisdom.

(Cavendish, 31–2)

Cavendish undermined the traditional view that reason, which elevated humans above animals, created greater wisdom than in herbs and drugs. In doing so, she disputed religious conventions that men could claim dominance over women (who were assumed to be less capable of rational thought).

Cavendish also undermined the idea that humans should dominate nature, one of the justifications for the conquest of the New World. Another aspect of the Age of Exploration comes in when, after hearing that her native land was under attack, the Empress returns as conqueror:

when her Country-men perceived at a distance, their hearts began to tremble; but coming something nearer, she left her Torches, and appeared only in her Garments of Light, like an Angel, or some Deity, and all kneeled down before her, and worshipped her with all submission and reverence.

(Cavendish, 81)

For humans to be peaceful, they must be willing to humble themselves.

Yet, Cavendish ultimately chooses to end with humility, by advising her readers to understand that the Blazing World is her creation, and that they had the same capacity for creating their own worlds:

[I]f any should like the World I have made, and be willing to be my Subjects, they imagine themselves such, and they are such, I mean in their Minds, Fancies, or Imaginations; but if they cannot endure to be Subjects, they may create Worlds of their own, and Govern themselves as they please.

(Cavendish, 97)

This passage from the epilogue has the most prominent premonitions of multiverse culture because Cavendish concedes that as many worlds exist as human imaginations can conceive of them. Like Cyrano, while exploring tensions between religion and science and communicating social commentary, Cavendish had reasons for encouraging humility and chastising human pride.

Questions remain regarding why early modern Europeans were interested in the humbling experience of reading works like those of Cyrano and Cavendish. It is possible that pondering humility helped some early modern Europeans adapt to widespread discomfort with how far science could take humanity, questioning the divine, and tensions regarding the moral implications of transatlantic slavery and violent conflict with the natives. Perhaps social commentary requires humility from its writers and readers. Further complicating matters, the tension between authority, religion, and science, coupled with the availability of travel narratives from the transatlantic exploration and the utopian dreams of the colonists, gave writers of the seventeenth century creative inspiration. One of the ways in which early modern utopian literature by authors like Cyrano and Cavendish shaped western society was to create multiverse culture as a humbling experience for its authors and readers.

References

Bergerac, Cyrano. 2016. *The Other World, or The Comical History of the States and Empires of the Moon*. Translated by Archibald Lovell. Dicot House: Kindle.

Cavendish, Margaret. 2020. *The Description of a New World, Called the Blazing World by the Thrice Noble, Illustrious, and Excellent Princesse The Duchess of Newcastle Margaret Cavendish*. US: Beehive Books.

Evans, Arthur. 2017. "The Beginnings: Early Forms of Science Fiction." In *Science Fiction: A Literary History*, edited by Roger Luckhurst, 11–44. London: The British Library.

Roberts, Adam. 2016. *The History of Science-Fiction*. 2nd ed. London: Palgrave MacMillan.

Seed, David. 2011. *Science-Fiction: A Very Short Introduction*. Oxford: Oxford University Press.

3

DEPICTIONS OF THE MULTIVERSE AS A LABYRINTH IN THE STORIES OF JORGE LUIS BORGES

Angélica Cabrera Torrecilla

During my research career on the subject of the multiverse (see Cabrera Torrecilla 2021, 2020, 2019a, 2019b, 2018, 2024a), I have found it fascinating that seemingly opposing nature of the sciences and the humanities has been overcome. Given contemporary interest in the multiverse, my work has aimed to recover these two intertwined strands that construct the idea of parallel universes in our culture and, above all, in our fictions. Thus, I have been able to lay the foundations of what I call the "Multiverse Theory for Speculative Fictions," useful for the analysis of all fictional narratives whose theme is the multiverse (see Cabrera 2024b). Moreover, I have shown how a singularly scientific and academic concept has acquired an easily identifiable cultural dimension that demonstrates the need for transdisciplinarity as a way of thinking and organizing knowledge itself. As Basarab Nicolescu established in his study *Transdisciplinarity: Theory and Practice*: "in transdisciplinarity, each discipline is integrated into a new field of study, with new tools and approaches, without suppressing the specialized training of each discipline (in Cabrera Torrecilla 2017, 11 my translation).

A transdisciplinary methodology, therefore, better represents the complex relationship between disciplines and allows us to create a new conceptual framework (unlike inter- or multidisciplinary) to approach a common object of research. In this sense, a transdisciplinary approach to the multiverse is not only capable of relying on isolated scientific or literary discourses, but also on how both can be intertwined to provide new knowledge. Therefore, I have defined the multiverse as a *cultural icon* that bridges

DOI: 10.4324/9781003480846-5

different disciplinary approaches, converging harmoniously in literary and cultural manifestations (Cabrera Torrecilla 2019a, 18). To exemplify this, in this chapter, I examine three stories by the Argentine writer Jorge Luis Borges: "The Garden of Forking Paths" (1941), "The Aleph" (1949), and "Tlön, Uqbar, Orbis Tertius" (1940). Studying them, I discovered that the rich polysemic sense of the topos of the labyrinth used by Borges helps to describe other possible ways of understanding the multiverse beyond the cosmological realm, embracing, on the contrary, its more transdisciplinary sense. Today, parallel worlds are part of the *lingua franca* of fiction, but if there is a writer to whom we owe the introduction of this idea in literature, it is undoubtedly Jorge Luis Borges in his story "The Garden of Forking Paths." This story is the first to formulate the idea of the multiverse in a form closer to the multiverse theory of quantum physics, first expounded by Hugh Everett III in 1957. In fact, when Everett's thesis was published in 1973, its first page contained an epigraph referring to Borges' "The Garden," a clear tribute from science to fiction (Cabrera Torrecilla 2018, 200).

Borges manages to reference the quantum proposal through the use of one of the recurrent symbols within his literary design: the labyrinth. In "The Garden" the labyrinth is evident not only as a symbolic space, but also as a figure of language and archetypal image, the representation of a chaos that contains its own explanation. The labyrinth works from four basic dimensions: length, height, depth, and time—the same dimensions that make up our own universe. But specifically in the labyrinth created for "The Garden," the dimension that predominates is *temporal*. "The Garden" does not construct its multiverse as *spatial*, which would imply a limit and, therefore, a finite existence. On the contrary, it is the dimension of time that gives the labyrinth of "The Garden" the characteristic of infiniteness. Moreover, by being bifurcated in time, the labyrinth of "The Garden" allows all possibilities to exist:

in all fictions, when a man is faced with several alternatives, he chooses one and eliminates the others; in that of the almost inextricable Ts'ui Pen, he chooses—simultaneously—all of them. He thus creates different futures, different times, which also proliferate and bifurcate.

(Borges 156, my translation)

The multiverse that Borges creates in "The Garden" is reminiscent of Everett's quantum multiverse, according to which Schrodinger's cat is not alive *or* dead, but alive *and* dead, because each of these realities belongs to different universes.

"The Garden" proposes a labyrinth that allows us to recognize the existence of an expanding reality. We become aware that our knowledge about the world is unlimited. In this sense, the shape of the labyrinth created in "The Garden" is fractal: it is impossible to choose a single path as each is intercepted by a multiplicity of others, which extend and repeat themselves infinitely. Let us imagine that we have infinite time to execute all possibilities and make all decisions. If that were to happen, I could *theoretically* be all of you because my decisions would be identical to yours. In other words, our paths would not only converge, but they would be practically the same "and an identical path but lived in different temporalities and spaces is nothing more than a parallel universe" (Cabrera Torrecilla 2018, 195, my translation).

The labyrinth also appears in "The Aleph," but opposite the one created in "The Garden." While in "The Garden" the labyrinth is complex due to the multiple bifurcations, in "The Aleph" the structure is extremely simple, reduced to the figure of the point. Borges refers to the aleph as: "one of the points in space containing all the points [...whose diameter] would be two or three centimeters, but the cosmic space was there, without decrease in size" (Borges, 2004, 625, my translation).

In "The Aleph" there are two parallel realities moving together through time and space, one is that of the "real" universe of the characters, and the other is the eternal reality of the aleph. But when the two realities converge, a "bubble" is created from that single point to be placed outside the "real" spatio-temporality.

The multiverse created in "The Aleph," an inflationary multiverse that emerges from a single point, differs from the quantum multiverse of "The Garden." It is "a minuscule figure which, being the simplest of all, serves to constitute any other structure, for in the point all forms are contained and detached" (Cabrera Torrecilla 2021, 126, my translation).

Whereas for physics it is many worlds that constitute the multiverse, for logics a single world contains in itself the multiverse. This is what happens with the story "Tlön," in which the labyrinth's bifurcations are imaginary, mere possibilities, although very real ones since "every possible universes exist and do not differ in kind from the real universe" (Cabrera Torrecilla 2019b, 41, my translation). The idea of "possible worlds" has its basis in the 17th and 18th century postulates of philosopher Gottfried Leibniz, but the philosopher David Kellogg Lewis (1986) formulated possible worlds theory during the 1960s arguing that the term *real* in *real world* is only a subjective distinction; any subject can declare their world to be the real one (70). On these postulates (on the real universe as a subjective distinction) the reality of "Tlön" is built, whose story, like a set of nesting dolls (matryoshka), has several layers of reality.

"Tlön" first depicts a fictional world (#1), the one we read and where all the characters are placed (among them a fictionalization of Borges himself).

This first world contains identical elements to the reader's real world, namely the one outside the diegesis (#-1), so, in a sense, world #1 somehow invades the reader's world, #-1 world. But world #1 is invaded by a second fictional world created inside the story, called Uqbar (#2 world), which in turn is invaded by a third fictional world, Tlön (#3 world). All these sub-universes are interrelated and the fictionality of all of them invades all the others.

The multiverse proposed by this story consists of all imaginary or possible universes existing in the same place, which is the same as saying that anyone is capable of making a possible world real. And this dialectical game is the challenge of the multiverse in "Tlön": the universes of Tlön and Uqbar emerge as possibilities that end up legitimizing themselves, while the primordial world from which they emerge (#1) begins to seem a mere possibility, that is, the imaginary imposes itself over the real, as Borges says: "Almost immediately, reality gave way in more than one sense. The truth is that it was anxious to yield" (Borges 106–7, my translation).

In these three stories, "The Garden of Forking Paths," The Aleph," and "Tlön, Uqbar, Orbis Tertius," Borges reflects the ideas of multiverse theory, not only about discoveries or scientific proposals, but also about philosophical assumptions brought together from the poetics of fiction. With these stories, Borges reminds us that only in fiction is the human being capable of experiencing the absolute paradox and impossibility to which thinking about other universes leads. Therefore, given the current and growing popularity of parallel universes, "it is necessary to recover the idea of multiverse from its most humanistic proposition" (Cabrera Torrecilla 2019a, 18). These stories are three stagings that reflect on the real and an enormous and still unknown reality that may, without a doubt, harbor many more universes than the one we know.

References

Borges, Jorge Luis. 2004. *Obras completas I (1923–1949)*, C.V. Frías (ed.), Buenos Aires: Emecé.

Cabrera Torrecilla, Angélica and Sáez de Adana, Francisco (eds.). 2024a. *The Multiverse as Theory in Postmodern Speculative Fiction and Narratives*, New York: Routledge, 2024.

Cabrera Torrecilla, Angélica. 2024b. "Towards Multiverse as Theory for Speculative Fictions, a Proposal," *The Multiverse as Theory in Postmodern Speculative Fiction and Narratives*, A. Cabrera Torrecilla and F. Sáez de Adana (eds.), New York: Routledge, 2024.

Cabrera Torrecilla, Angelica. 2017. "Las direcciones del pensamiento: La transdisciplina como alternativa de estudio," *Elementos: Revista de Ciencia y Cultura* 24, no. 108: 9–13.

Cabrera Torrecilla, Angelica. 2018. "Del gusto por el extravío: representaciones del multiverso en 'El jardín de los senderos que se bifurcan,'" *Variaciones Borges* 45: 187–206.

Cabrera Torrecilla, Angelica, 2019a. *What if a... Multiversity? Literatura y Ciencia en la obra de Grant Morrison*. Marmotilla.

Cabrera Torrecilla, Angelica. 2019b. "Mapas abstractos: Los posibles multiversos lógicos en 'Tlön, Uqbar, Orbis Tertius' de Jorge Luis Borges", *Thémata: Revista de Filosofía* 60: 117–132. DOI: 10.12795/themata.2019.i60.7.

Cabrera Torrecilla, Angelica. 2020. "Topología onírica. El sueño como mundo paralelo," *Diferencia(s). Revista de Teoría Social Contemporánea* 11: 59–68.

Cabrera Torrecilla, Angelica. 2021. "Visiones del multiverso: Microcosmos y totalidad en 'El Aleph' de Jorge Luis Borges," *Alpha: Revista de Artes, Letras y Filosofía*, 53: 121–139.

Lewis, D.K. 1986. *On the Plurality of Worlds*. Oxford: Blackwell.

4

A CREATIVE DECONSTRUCTION OF THE CULTURAL STRANDS IN THE MULTIPLE TIMELINES OF *THE FLASH* (2023)

Introduction to the Online Resource "Multiversal Spaghetti"

Chris Gerrard

Please visit www.routledge.com/9781032770116 and click on Support Material to access the digital supplement of this chapter.

In 2023's *The Flash*, the lead character Barry Allen travels back in time to save his murdered mother. After this intervention in the past, Barry expects time to simply continue on a new course, but instead discovers that numerous changes have occurred in his world, both before and after his change to the timeline: Eric Stoltz played Marty McFly, Bruce Wayne has a different face, Superman never reached Earth. Upon visiting Bruce Wayne/Batman, he receives the explanation that instead of curving time off into another direction, he has instead stepped from one path at a crossroads onto another. This is metaphorically displayed through the use of spaghetti: two strands being placed upon each other at 90 degrees, with Barry's moment of intervention being the conjunction of the two. Instead of changing time, he has crossed from one strand of spaghetti to another; from one universe to another. This is described as being a simplistic explanation of the nature of reality, with the true version being more like a cooked bowl of spaghetti made up of dozens, hundreds, infinite strands intermixed, meeting at some points, being wildly disparate at others. This is the film's idea of the multiverse, a web of interconnected timelines that have some similarities but diverge in many ways.

This spaghetti metaphor feels somewhat different from the usual depiction of the multiverse in fiction. Other sources generally display multiple

DOI: 10.4324/9781003480846-6

realities as layered upon the same spatiotemporal grid, or as a catalog of Earth-1, Earth-2, etc. *The Flash*'s world is messy, disorganized, chaotic; with multiple strands waving in and out of connection with one another. This metaphor seemed ripe for a piece of creative academia. Many of the themes, aesthetics, and production issues of the film are similarly intermixed and chaotic and called for a research work that followed these elements through the same structure as the film's multiverse.

Therefore, the digital resource, "Multiversal Spaghetti" that I produced for this book is a web of interlinked ruminations on various elements of the film. The overarching structure that holds these is a blog, where each thought is formatted as a post. However, instead of each being a self-contained essay on an element of the film, they are interlinked. Each offers several links at the beginning and end allowing for the reader to interact with the structure in their own unique way, following whichever line of thought, whichever strand of multiversal spaghetti, that they personally choose. A commentary on the film as a metaphor for 9/11 can lead them to thoughts on historical trauma, the role of the time period the film is set in or to the implications of telling a story that revives the aesthetics of the War on Terror in 2023. The initial post that a reader arrives at is chosen at random, so no journey even begins in the same way. Each post is both self-contained (exploring a specific element) and interlinks with those around it. These posts take multiple formats: traditional articles, general ruminations, short provocations to further thought. At the time of writing, the resource contains twenty-five separate thoughts, but the plan for the future is to expand. I plan on contributing further points into the web of interconnections, and additionally, there is a submit function, where other academics, film theorists, and members of the public can write further posts to be added into the matrix. Currently, the majority of the posts are in a written format, but in future, I and others may add videos, audio, artworks, etc. The idea is to create a virtual pinboard of connections, with the hyperlinks providing the proverbial red string to link the important trains of thought to each other. These thoughts and connections explore much of the tissue of the film's internal workings and external context, while existing as an experiment in structure, attempting to parallel the format of criticism and that which is being critiqued.

Beyond the format, the question remains of why *The Flash* is worth discussing in such depth. Critically it was panned, and the financial returns were minimal. It was simply a film that most people were not interested in seeing or critically engaging with. My own screening was the only time I have ever been the only member of the audience. It also has not built an online or cult following since its release, being widely forgotten, even as a meme (compare this to *Morbius* (2022), a film with similar critical and financial failings, but one that has at least acquired some status as a joke online). However, it is this very lack of relevance that is one of the film's most intriguing features. It is essentially the culmination of Warner Bros attempt to replicate the success of the Marvel Cinematic Universe, the final gasp before they were forced to give up and start again (while Aquaman: The Lost Kingdom (2023) post-dates it, that was released after WB had already announced their planned reboot). Even the idea of the cinematic universe as a whole is beginning to look past its prime with the success of more individual and original properties at the expense of the previously omnipresent superhero narratives. The focus on the 9/11 metaphor comes from a bygone era, calling back to *Man of Steel* which even upon its release in 2012 felt somewhat behind the times. Many aspects of *The Flash* are therefore phantoms of cinematic, cultural, or societal movements that were over by the time of its release, leading to a work that is, to draw from Mark Fisher (2012), "haunted" by signs and signifiers that lack their original relevance or context. The web of multiversal spaghetti thereby allows for the drawing out of these disparate elements, showing their connections, their past, and their potential lack of future, and how this all converges on a single chaotic cultural artifact.

The aim of this resource is therefore twofold. First, in its structure, it presents a possible route for more creative Humanities research methods, allowing for the combination of creative writing and media criticism. Second, in its content, its purpose is to unpick the multiple threads of cultural significance that intersect within *The Flash* and use the film as a convergence point, a single moment to capture the development of multiple strands of sociocultural spaghetti. Hopefully, with my own additions and the submit function, it will continue to grow and cover more powerful intersections within the film, and I would suggest all readers to consider adding their own additional thoughts into the web.

References

Aquaman: The Lost Kingdom. 2023. *Directed by James Wan*. USA: Warner Bros, DC Entertainment, Domain Entertainment, Atomic Monster, The Safran Company, Québec Production Services Tax Credit, Australian Film, Icelandic Film.

Fisher, Mark. 2012. "What Is Hauntology?" *Film Quarterly* 66, no. 1: 16–24. https://doi.org/10.1525/fq.2012.66.1.16.

Morbius. 2022. *Directed by Daniel Espinosa*. USA: Columbia Pictures, Marvel Entertainment, Avi Arad Productions, Matt Tolmach Productions.

The Flash. 2023. *Directed by Andy Muschietti*. USA: DC Comics, DC Entertainment, New Zealand Film Commission, Québec Production Services Tax Credit, The South Australian Film Corporation, Warner Bros.

5

THE MULTIVERSE AS A STYLISTIC STRUCTURE

Sem Devillart

Seemingly overnight, the idea of multiple, discontinuous worlds coexisting has gone from niche to culturally mainstream. We see this blatantly in popular entertainment, where it's become common to see characters chasing their fates across proliferating worlds, each with different rules, different laws of physics, and different iterations of themselves, their friends, and enemies. Popular entertainment titles—*Loki* (2021–2023), *Spider-man Across the Spider-verse* (2023), *Everything Everywhere All At Once* (2022), *Severance* (2022), and *His Dark Materials* (2019–2022)—depict a narrative shift from one world to many.

The multiverse's dramatic and sudden infiltration prompts curiosity as to where else it can be seen in contemporary culture. Where else in culture might the concept of a multiverse be lurking? And further, what does culture's current infatuation with the multiverse say about our relationship with our world and the reality we live in?

To investigate, we can translate the abstract notion of a multiverse into something empirical. This is where a practice like socio-cultural logic analysis comes in. By distilling the multiverse into a clear stylistic structure, we can see it in places both subtle and pervasive. Seeing the multiverse as a stylistic structure gives our study breadth (the ability to look across disciplines from aesthetics to politics) and depth (we can observe it in phenomena from the micro to the macro and perspectives both short and long term).

DOI: 10.4324/9781003480846-7

Multiverse as Leitmotif

A socio-cultural method of studying cultural phenomena identifies stylistic structures and patterns, and stems from trend studies, cultural analysis, and media philosophy. Detecting commonalities, like a multiversal structure, between different cultural artifacts, a socio-cultural analysis uses repeated structures to theorize about how we make sense of the world and ourselves.

While cultural manifestations present externally through economic trade flow or popular fiction storylines, they also develop—consciously or unconsciously—from internal thoughts, motivations, curiosities, and need states of people. A pattern that repeatedly permeates culture becomes a motif, a dominant or influential idea. There can be an abundance of motifs during any given time period. And yet, when a pattern becomes so potent and powerfully explanatory, as we observe with the multiversal structure, it represents not just one signal but a definitive or leading idea of a time. These social "leitmotifs" epitomize shifting systems of contemporary values and interests. They reveal new ways of relating to our world and reality.

If we postulate the multiverse to be a leitmotif, how might it shed light into how we—on a collective level—are processing the human experience during this unique period of time? Cross-cultural, multidisciplinary analysis of a multiversal structure suggests we are experiencing a deep cultural shift from a networking logic to a logic of multimodality, across multiple fields and disciplines, in and among multiple types of structures. This socio-cultural, leitmotif analysis reveals the multiverse as the new organizing principle of our time. Herein, realities and modalities are now plural. And with them, possibilities—and our grip on them—are ever-expanding as we play out multiple systems and scenarios that require multiple literacies to navigate.

A Shift from One World to Many Worlds

Consistent with the multiverse narratives we see in film and TV, a multiversal structure can be defined as a shift from one world to many worlds. *Everything Everywhere All At Once* (2022) depicts parallel realities and encounters with alternative selves—the characters appear human in one world and rocks in another. In the Marvel world, divergent timelines in multiple parallel dimensions explore the implications of altering the fabric of reality. Even shows like *Severance* (2022) derive their action from a kind of multiverse in which work and life are split into discontinuous worlds by design and altered biology.

Importantly, each world appears self-contained, distinctly different, and defined by its own law and logic. Worlds are not a part of the same interconnected network.

A multiverse model entails many worlds that appear disconnected. They run in parallel and, though these worlds can occasionally be penetrated, the paths or links between them are neither obvious nor easy to navigate. Traversing them requires special abilities, insider knowledge, or finding hidden portals.

Information technology outputs, like software and the internet, can be ever-expanding, but they follow a network model of interconnection—and thus multiple environments are in sync with the idea of "one world." We are now witnessing a shift from that one-world model of the World Wide Web toward a many-world model of Virtual Reality (VR) and Augmented Reality (AR). Industry leaders like Microsoft, Google, Meta, and Apple are implanting terms such as virtual reality, augmented reality, extended reality, mixed reality, and a "metaverse" into common vernacular. From the network-based interactions of social media, we see the multi-dimensional virtual and physical worlds of spatial computing. From the one-world monetary model of hard currencies, we see the emergence of cryptocurrencies that exist in multiple financial modes expanding in parallel. All these formats signify a shift toward a system of multiple worlds.

Further, in psychology and identity analysis, we see a shift from one world to many in how humanity classifies and partitions itself. Labels used to describe cognitive ability are shifting from a one-way-of-thinking model, from which neurological disorders or disabilities were derived, toward a model of neurodiversity that sees cognitive differences as valuable and distinct ways of thinking in their own right (Baumer and Frueh 2021). Likewise, in place of the one world of gendered pronouns, neopronouns such as "ae" and "ze" represent a multiplicity of self-demarcations governed by new rules of self-expression (CNN 2023). Thus instead of moving toward classifications of self and space that unite humanity in a single world, humanity's new organizing principle is one of distinct labels, logic, and environmental divides. As people continue to challenge and reckon with colonialist history, the frame of humanity as a "melting pot" has shaped into a multiverse (Hochschild 2020).

When we look into the publishing sector, we observe a growing array of books exploring an "interspecies" theme. Instead of one type of intelligence, texts like *An Immense World*, *Entangled Life*, *The Language of Plants*, and *Alles Fühlt* expound on the many forms of intelligence that only plants or animals have access to. Here, other species have consciousness and communication systems that are obtuse and invisible in our world. There are hints that a level of sentience exists beyond what we humans can grasp. And despite a human desire to understand or derive meaning for these other worlds—*perhaps a case for greater humility, fuel for scientific investigation or new philosophical doctrines*—the divide between our world and theirs remains elusive.

Similarly, when we look toward geopolitics, we see the shift from one dominating world to many parallel world powers. Pre-9/11, perceptions of the United States's political sway, military reach, economic force, and cultural appeal were unparalleled and extended across the globe (Daalder and Lindsay 2023). America's dominance during this time went hand in hand with the spread of globalization and the Internet—all phenomena containing the theoretical premise of interconnection and the formation of a single network. Even the fall of the Berlin Wall signified this melding of divisions—east and west. Yet shortly after the turn of the millenia, we see a shift from the unipolar model of America as a superpower toward a new multipolar world order where multiple important global powers now exist (Ashford and Cooper 2023). In tandem, the "splinternet" and a fragmentation of economies began to emerge (Merrill and Komaitis 2020; Wolf 2022).

In step, today's global trade routes and migration paths no longer resemble a single-world model (Griffiths and Trebilcock 2023). The previous sense of a "liquid modernity" reinforced by perceptions of the world as a "melting pot" has been reduced back down to solids (Bauman 2000). With emergent national policy and "protectionism," socio-economic exchanges no longer flow freely (World Bank 2023). As a result, we see a disconnected distribution of people and goods that resembles many, territorial worlds.

The multiversal effect goes on. Even the way in which we perceive and process veracity has gone from one truth to a multiplicity of potential truths. As seen in the surge of misinformation, alternative facts, fake news, and deep fakes, there is no collective grip on a singular reality (Arslan 2018). Related, the surfacing of the psychological concept "dialectic"— which allows for the idea that two seemingly opposing things can be true at the same time—has surfaced as a tool for processing complex issues and ideologies (Veraksa et al. 2020; de Waal 2024).

While we observe these phenomena across disciplines, sectors, and categories, they each contain something akin to a shared cultural DNA. When aggregated, they relate as if they were part of the same family. And in their pervasiveness there is power.

An Exploration Instrument of Time

Just like multiversal structures, the practice of identifying and categorizing repetitive cultural phenomena and synthesizing derivative leitmotifs has an expansive quality. This method of analysis opens up our understanding of the most powerful ideas shaping and shifting beliefs, interests, and behaviors. And it illuminates how those ideas slot into, stem from and give way to a chronology of other era-defining stylistic structures.

This form of analysis is also expansive in that its logic connects to different disciplines and theoretical backgrounds. By aggregating cross-sections of cultural phenomena, different disciplines gain access to new sources of enrichment and intellectual fodder. And although the world seems to be moving from a liquid modernity to one of multiversal divides, this may also create opportunities for connecting researchers and theoretical frameworks that belong to radically different disciplines, styles, and worldviews.

References

Arslan, Russell C. 2018. *Alternative Facts*. CreateSpace Independent Publishing Platform.

Ashford, E., and E. Cooper. 2023. "Yes, the World Is Multipolar." *Foreign Policy*, October 5. Accessed March 18, 2024. https://foreignpolicy.com/2023/10/05/usa-china-multipolar-bipolar-unipolar/.

Bauman, Zygmunt. 2000. Liquid Modernity. *Polity*.

Baumer, Nicole, and Julia Frueh. 2021. "What Is Neurodiversity?" *Harvard Health Publishing*, November 23. Accessed March 18, 2024. https://www.health.harvard.edu/blog/what-is-neurodiversity-202111232645.

CNN. 2023. "What Are Neopronouns? A Guide to These New Terms for Gender Identity." August 12. Accessed March 18, 2024. https://edition.cnn.com/us/neopronouns-explained-xe-xyr-wellness-cec/index.html.

Daalder, I. H., and J. M. Lindsay. 2023. "The Globalization of Politics: American Foreign Policy for a New Century." *Brookings Institution*, January 1. Accessed March 18, 2024. https://www.brookings.edu/articles/the-globalization-of-politics-american-foreign-policy-for-a-new-century/.

de Waal, Alex. 2024. "World War Three or World Counter-Revolution? Slavoj Žižek's Dialectics on Ukraine and Gaza." *Reinventing Peace, World Peace Foundation*, Fletcher School at Tufts University, February 7. Accessed March 18, 2024. https://sites.tufts.edu/reinventingpeace/2024/02/07/world-war-three-or-world-counter-revolution-slavoj-zizeks-dialectics-on-ukraine-and-gaza/.

Griffiths, C., and J. Trebilcock. 2023. "Continued and Intensified Hostility: The Problematisation of Immigration in the UK Government's 2021 New Plan for Immigration." *Critical Social Policy* 43, no. 3: 401–422. https://doi.org/10.1177/02610183221109133.

Hochschild, Adam. 2020. "The Fight to Decolonize the Museum." *The Atlantic*, January/February. Accessed March 18, 2024. https://www.theatlantic.com/magazine/archive/2020/01/when-museums-have-ugly-pasts/603133.

Merrill, Nick, and Konstantinos Komaitis. 2020. "The Consequences of a Fragmenting, Less Global Internet." *Brookings*, December 17. Accessed March 18, 2024. https://www.brookings.edu/articles/the-consequences-of-a-fragmenting-less-global-internet/.

Veraksa, Nikolay, et al. 2020. "Dialectical Thinking: A Proposed Foundation for a Post-modern Psychology." *Frontiers in Psychology* 13. https://doi.org/10.3389/fpsyg.2022.710815.

Wolf, Zachery B. 2022. "This Is Globalization in Reverse." *CNN*, May 24. Accessed March 18, 2024. https://edition.cnn.com/2022/05/24/politics/globalization-trade-tariffs-what-matters/index.html.

World Bank. 2023. "Protectionism Is Failing to Achieve Its Goals and Threatens the Future of Critical Industries." Accessed March 18, 2024. https://www.worldbank.org/en/news/feature/2023/08/29/protectionism-is-failing-to-achieve-its-goals-and-threatens-the-future-of-critical-industries.

6

CAPITALISM AND THE MULTIVERSE

An Interdisciplinary Exploration of
Socio-Economic Influence on
Cosmological Concepts

Marija Adela Gjorgjioska and Ana Tomičić

This chapter unravels the complex interplay between the multiverse and the socio-economic ideologies that shape our understanding of it. Integrating cosmology, socio-economic theory, and speculative fiction, we explore how capitalist ideologies color our theories, imaginaries, and perceptions of the multiverse (Fisher 2009, 7). Through the lens of Liu Cixin's (2014) *The Three-Body Problem*[1], we examine how fiction reproduces capitalist epistemologies onto the concept of the multiverse, challenging the assumption of its inherent diversity (Harvey 1989).

In physics and cosmology, the multiverse refers to a hypothetical set of multiple universes, each with its own set of physical laws, constants, and properties, and arises from theories such as string theory and quantum mechanics, which suggest the possibility of parallel universes existing alongside our own (Greene 2011). In communication studies, the multiverse refers to the diverse range of media platforms, narratives, and communication channels through which information, ideas, and culture are

---

1 *The Three-Body Problem* was originally written within the Chinese context, reflecting aspects of Chinese society during the period covered in the novel/show. It's important to note that during this time, Chinese society was undergoing a transition towards socialism. The Second Centenary Goal of the CPC (Communist Party of China) aims to "build a modern socialist country that is prosperous, strong, democratic, culturally advanced, and harmonious" by 2049. This societal backdrop adds depth to the narrative, as it explores themes of science, society, and ideology against the backdrop of China's evolving socio-political landscape.

DOI: 10.4324/9781003480846-8

exchanged within society. In this context, the multiverse emphasizes the multiplicity and complexity of communication networks and the interconnectedness of different media landscapes (Jenkins 2006). Despite both definitions revolving around the idea of multiple entities existing concurrently, they operate within distinct domains and epistemologies.

The Three-Body Problem embodies aspects of both definitions. As a fictional work, it is presented across various platforms (a novel, TV series, and productions by both Tencent and Netflix). Although its science fiction theme, exploring distant solar systems, resonates with the cosmological notion of the multiverse, the novel intersects with communication studies by engaging diverse interpretations surrounding the narrative. Thus, it serves as an intersection of both definitions, showcasing the complex interplay between artistic expression and scientific speculation (Dunne & Raby 2013; Jenkins 2006).

The Three-Body Problem: A Literary Metaphor

The Three-Body Problem intertwines Chinese history with advanced physics and alien contact. The story begins during China's Cultural Revolution and follows physicist Ye Wenjie, who becomes involved with a secret government project searching for extraterrestrial intelligence. She contacts an alien civilization, the Trisolarans, who live in a chaotic solar system and are seeking a new home. The novel explores the complexities of communication and the potential for both collaboration and conflict in the face of the unknown, setting the stage for a thought-provoking saga about the future of humanity and its place among the stars. *The Three-Body Problem* explores multiverse-like alternate existences within a unified narrative. Intertwining fiction, multiverse narrative techniques, and capitalist themes, both implicitly and explicitly, we argue that this narrative can be viewed as a canvas where the ideologies of capitalism—competition, expansion, and cyclical growth and decay—are projected onto the fabric of the multiverse, suggesting that our understanding of possible worlds is influenced by capitalist paradigms (Jameson 2005, 150). Moreover, it invites us to reflect on whether the narrative reproduces or critiques capitalist ideology.

In the novel, Ye Wenjie's character complexity stems from her childhood trauma alongside societal turmoil and familial betrayal, during the Cultural Revolution. Her worldview leads her to question humanity's kindness and rationality, and ultimately its instinct for self-preservation. Her personal disillusionment crystallizes into a nihilistic view of humanity, marked by moral decay, irrationality, and a critical reassessment of its survival instincts. Liu employs Ye's perspective to illustrate what happens when we overlook the way broader socio-political systems shape individuals' morality, rationality, and survival.

Indeed, her rationalization of her experiences do not include an account for the complex interplay of historical, social, and political factors that define human behavior and societal dynamics. Driven by her profound disappointment in humanity, Ye establishes the Earth-Trisolaris Organization (ETO) with the aim of facilitating contact and collaboration with the Trisolarans. The ETO embodies her belief that Earth's salvation may lie in the intervention of a more advanced civilization. This reflects her disillusionment with humanity's capacity to resolve its own crises. Her actions are driven by a hope that an external, more advanced civilization might enforce a reset on Earth's trajectory, which she sees as spiraling toward self-destruction.

Led by her nihilism and pessimism Ye confirms that for her "it's easier to imagine the end of the world than the end of capitalism" (Jameson, 2003). This narrative reveals that the challenges faced by humanity, particularly under the socio-political structure of capitalism, are perceived as endemic rather than circumstantial. Ye's disillusionment, rooted in the specific epoch spanning the latter half of the twentieth century and the dawn of the twenty-first, under capitalism's sway, could also be perceived as a critique targeting the repercussions of this economic system rather than humanity *per se*. However, Ye's perspective conflates capitalism with human nature, without considering the social construction of this socio-political system, suggesting the extinction of humanity due to the failures of capitalism. Ye's sense of defeatism echoes the sentiment that there is no alternative (TINA)[2] to the current socio-political system.

Trisolaris is considered to be superior to Earth, based on their technological advancements. There is no regard for the socio-political system of Trisolaris, or its relevance to its civilizations' (supposed) superiority in relation to Earth. The narrative primarily explores the Trisolarans's efforts to survive and their interactions with Earth, rather than delving deeply into their social makeup or internal societal organization. However, the Trisolarans do refer to a figure known as the "Princeps," which can be likened to a king or ruling leader in their civilization. This title indicates a hierarchical structure within Trisolaran society, suggesting elements of centralized leadership. The awareness of the socio-political system's significance is notably absent in the consideration of both Earth and Trisolaris, underscoring a critical gap in acknowledging the foundational role these systems play in

2 TINA: "There Is No Alternative" – a phrase popularized by British Prime Minister Margaret Thatcher in the 1980s to signify the belief that the free market economy is the only viable system for economic development, leaving no alternative to capitalism.

shaping civilization's trajectory and ethical frameworks. This stance inadvertently endorses supremacism/racism and imperialism, suggesting that a supposedly superior civilization, like Trisolaris, has the right to conquer and impose its systems on Earth, mirroring the mechanisms of colonialism and imperialism as "the highest stage of capitalism" (Lenin 1917, 72).

Quantum Insights and Observations: Reimagining Earth's Socio-Political Pathways

Ye's disillusionment, set against the backdrop of her belief in the superiority of Trisolaris, could be viewed as a critique of humanity's socio-political stagnation under capitalism. This critique, grounded in the story's cosmic scale, paves the way for a deeper exploration of alternatives, invoking quantum theory to underscore the potential for radical change and diversity in the fabric of our socio-political realities (Barad 2007, 132–5).

In the context of quantum theory, superposition suggests that, until observed, particles exist in all possible states simultaneously (Rosenblum & Kuttner 2006, 87–92). Applied to a macroscopic object like Earth, being "observed" or "measured" by another planet from another solar system could mean that Earth's state becomes defined or "fixed" in relation to that observation. In quantum mechanics, observation collapses the superposition into a single state. By introducing Trisolaris as an external observer, akin to quantum theory's concept of superposition, Liu could imply that Earth's fixation at a specific socio-political point—capitalism—symbolizes a pivotal decision in human history's trajectory, challenging readers to contemplate alternative socio-political realities beyond the status quo. The portrayal of Trisolaris's challenges juxtaposed with Earth's suggests a perspective that Earth's issues may be less insurmountable than perceived. However, the narrative emphasis on scientific solutions hints at an oversight of socio-political and ethical considerations fundamental to addressing civilizations' crises. By presenting characters and scenarios from divergent realities, the narrative invites readers to envision broader socio-political possibilities, challenging the assumption that the current system is immutable and prompting a reimagining of alternative societal structures beyond capitalism. One implication for the spectator could be that witnessing scenarios from different realities within *The Three-Body Problem* might lead to a reevaluation of their understanding of reality. The narrative's contrast between Trisolaris's astrophysical dilemma and Earth's challenges, highlights the self-inflicted nature of Earth's issues, prompting reflection on their potential resolutions.

The story of Trisolaris, illustrated in "The Three-Body Problem," with its civilization's cyclic rises and falls in response to the unpredictable orbits of its three suns, draws a parallel to the fluctuating dynamics characteristic of capitalist markets—highlighted by cycles of booms and busts, and periods of prosperity followed by downturns. This allegory reflects the cyclical nature of both the fictional Trisolaris and real-world capitalist systems, emphasizing cycles of growth and contraction, alongside the inevitability of innovation becoming obsolete. Jameson's critique on the cultural logic under late capitalism (Jameson 1991) suggests that our perception of diversity among universes, or economic systems, might be more about variations of recurring themes, governed by similar underlying principles such as competition, survival, and adaptation.

Trisolaris's environmental and societal upheavals serve as a metaphor for the competitive essence of capitalism, where civilizations and economies alike are forced to navigate chaos and adapt to survive. This struggle reflects the competitive drive in capitalist societies to innovate in the face of economic challenges and market volatility. However, the narrative extends this metaphor to critique the conventional portrayal of the multiverse in popular culture. Rather than presenting a realm of boundless diversity, it posits a cosmos where civilizations like Trisolaris face inevitable collapse. This choice serves as a commentary on the limitations of the multiverse concept as envisioned through capitalist ideologies, suggesting that differences between realities are superficial, masking an underlying uniformity. Furthermore, Trisolaris's repetitive cycles of crisis and recovery underscore the critique of perceived uniqueness within the multiverse, arguing against the notion that each universe is fundamentally distinct. Instead, it suggests our conceptualization of the multiverse—and our view of capitalist economies—may be influenced by a tendency to replicate familiar patterns. This insight, echoed by Unger (2014), challenges the depth of our imagination within theoretical models of the multiverse, potentially constrained by capitalist thought to a framework defined by repetition and competition. Finally, the narrative suggests that the recurring cycles of collapse and rebirth in Trisolaris, where survival strategies ultimately lead to failure, reflect a profound commentary on our inability to transcend our existing socio-political and economic paradigms when conceiving of multiverses. This narrative underscores the notion that despite our efforts to envision alternate realities as distinct and varied, they may actually be bound by the same conceptual and philosophical frameworks that limit our reality, challenging us to question the extent of our imaginative capabilities and the possibility that true innovation in conceptualizing alternate realities requires a radical departure from current frameworks.

The Capitalist Multiverse

Capitalism, driven by the imperative of capital accumulation, shapes social relations, class dynamics, and economic development. It is characterized by market competition, the pursuit of individual profit, and the commodification of resources. In the process, it promotes a homogenization of cultural and product diversity under the guise of variety, where differences are more superficial than substantial. David Harvey illuminates how capitalism molds societal and individual perspectives, creating a landscape where economic growth and competition drive deeper cultural and product homogenization. Our imaginations are intricately linked to the structures we create within capitalist frameworks, underscoring the extent to which capitalist ideologies permeate our visions of the future. The imaginative colonization by capitalism, deeply ingrained in our collective consciousness, implies that it not only dictates the dynamics of our economic and social systems but also extends its influence into our speculative theories of the multiverse. Conversely, the multiverse might represent a space where we erect a new vision of the future, distinct from capitalism, in our imagination before we bring it into reality.

Exploring Diversity through Capitalism

Within capitalism, the illusion of diversity is prevalent. Markets seem to offer an endless variety of products and services. However, diversity often reveals itself to be superficial—variations on a theme rather than genuine uniqueness. Many products are differentiated by branding rather than by substantial differences in functionality or quality, reflecting a deeper homogeneity underlying the apparent diversity. The concept of superficial diversity in capitalism offers a useful framework for reevaluating our expectations of diversity in the multiverse. It prompts us to question whether our assumptions about the uniqueness of each universe (or media platform) are influenced by the same mechanisms that create the illusion of diversity in markets. This reflection challenges us to consider the possibility that the principles underlying our imaginaries might be more uniform than previously imagined (Frank 1997, 45).

Exploring the parallels between capitalist dynamics and the fictional multiverse within literature, we aim to delve into the notion that our narrative pathways hinge on a pivotal choice: to either navigate through the enduring presence of capitalism, which overlooks the potential for alternatives such as socialism, or to face a nihilistic future marked by humanity's self-destruction, challenging the conceptual boundaries of diversity itself.

Concluding Reflections

This exploration, bridging the realms of cosmology, socio-economic theory, and speculative fiction, reveals the profound impact of our socio-economic systems on both our scientific inquiries and cultural narratives. The reflection on the parallels between capitalist dynamics and the theoretical constructs of the multiverse underscores the need to question the extent to which our perceptions of (cosmic or media) diversity are colored by the principles of competition, exploitation, and individualism inherent in capitalism. Viewing the multiverse through a capitalist lens has illuminated the potential for an illusion of diversity—a mirage of infinite variety that, upon closer examination, reveals a limited set of underlying principles. This realization challenges us to reconsider the nature of diversity itself, both in the cosmos and within our socio-economic systems. It prompts us to ask whether our expectations of boundless uniqueness and variety in the multiverse are merely projections, rather than reflections of objective reality. *The Three-Body Problem* serves as a critical tool for examining the cyclic nature of our endeavors and the limitations of our imagination when conceptualizing alternate realities. It not only critiques the replication of capitalist ideologies across theoretical universes but also invites us to envision possibilities that transcend these constraints. Engaging with the mysteries of existence and the cosmos demands a radical departure from these frameworks, urging us to conceive of a universe where diversity is not just an extension of our worldly experiences but a genuine reflection of the boundless possibilities inherent in the fabric of reality. This endeavor requires not only a shift in our scientific and philosophical inquiries but also a transformation in the cultural narratives that shape our understanding of the multiverse.

References

Barad, Karen. 2007. *Meeting the Universe Halfway: Quantum Physics and the Entanglement of Matter and Meaning*. Durham: Duke University Press.

Dunne, Anthony, and Fiona Raby. 2013. *Speculative Everything: Design, Fiction, and Social Dreaming*. Cambridge, MA: The MIT Press.

Fisher, Mark. 2009. *Capitalist Realism: Is There No Alternative?* Winchester: Zero Books.

Frank, Thomas. 1997. *The Conquest of Cool: Business Culture, Counterculture, and the Rise of Hip Consumerism*. Chicago: University of Chicago Press.

Greene, Brian. 2011. *The Hidden Reality: Parallel Universes and the Deep Laws of the Cosmos*. New York: Alfred A. Knopf.

Harvey, David. 1989. *The Condition of Postmodernity: An Enquiry into the Origins of Cultural Change*. Oxford: Blackwell.

Jameson, Fredric. 1991. *Postmodernism, or, The Cultural Logic of Late Capitalism*. Durham: Duke University Press.

Jameson, Fredric. 2003. "Future City." *New Left Review* 21: 65–79.

Jameson, Fredric. 2005. *Archaeologies of the Future: The Desire Called Utopia and Other Science Fictions*. London: Verso.

Jenkins, Henry. 2006. *Convergence Culture: Where Old and New Media Collide*. New York: New York University Press.

Lenin, V. I. 1917. *Imperialism, the Highest Stage of Capitalism*. New York: International Publishers.

Liu, Cixin. 2014. *The Three-Body Problem*. Translated by Ken Liu. New York: Tor Books. Original work published 2008.

Rosenblum, Bruce, and Fred Kuttner. 2006. *Quantum Enigma: Physics Encounters Consciousness*. Oxford: Oxford University Press.

Unger, Roberto Mangabeira. 2014. *The Religion of the Future*. Cambridge, MA: Harvard University Press.

7

THE NON-BINARY MULTIVERSE POEM

Or Friday August 25th, 2023

Dusty Michael Perez

Or Friday August 25th, 2023

So my brother has COVID again,
   my nephew won't go outside at all,
my sister-in-law is busy working
   through pain, my mom is 92

and there's no time to list
   what's going on with her.
I got another bionic hip
   and feel less human though
walking well, though one leg
   feels shorter than the other.

But now I can walk without
   wincing. Parts of Hawaii
went up in flames, trans
   folks and their advocates
are being erased, I live in a state
   that considers this American,

DOI: 10.4324/9781003480846-9

what will unite these states,
        cool us down beyond one more
extra-extra large soda in a plastic
        cup destined to float in the land-
fill that unites us all? My mom

is writing her memoirs again. I imagine
        the heat from her hands as she types
and gives off magical ageless sparks
        I can't catch as I watch her, as
outside the epic temperature will be
        recorded much better than lives.

School starts soon. I shop online
        and prep for students expecting
real bonafide intelligence at the end.
        Not a simulacrum. Ugly word?

All I want to find is one word,
        no, one timeless sound, one
CD of voices that knew how
        to harmonize, The Shirelles,

The Crystals, LaBelle, The
        Marvellettes, who voice the
vinyl plasticized trajectory
        of girl groups turning, groove-
by-pressed group, into women

to gauge the unnoted thermostat
        inside the non-binary 'sir'
I answer to out and about
        running errands, but I can't run

at all, my lungs get this thing
        called air hunger, *Da Doo Run
Run Run, Da Doo Run Run Run,*
        so I sit and wheeze to
the news channel updates, then classic
        movies, sparkling water cans and pizza

boxes stacked up by the garbage can,
      programmed Bogie and Betty Bacall
then Betty Grable in *Mother Wore*
      Tights for at least the fourteenth time.

My mom has another bruise, a damn
      UTI. I can't breathe so well just
now, my doctor is out, but the nurse
      can see me next Friday. Who sees us
all without prior or new diagnoses
      to say this really isn't the way it will stay?

My brother may lend me money
      to help pay for my hip surgery.
Yesterday was my payday. Broke
      Already! Creditors, Amazon,

hips. I have no balance, minimum
      or otherwise. At 60 I'm living at
home again. Well, just for the summer.
      Life has a zillion syllabi everyone skims.

My mother's not supposed
      to be driving, good luck
with that Mom, she went out
      early and bought day-old croissants
and good soup but forgot the bacon

      and cream and jam. *Creole Lady*
Marmalade, but she asked for
      paper bags instead of plastic.
God and/or Goddess forbid, what

      kind of accidental collision could
keep her permanently in? There is no
      colliding small world big enough.
Danger / heartache dead ahead.
      We are not our cars, but almost,

reaching like mini steely launches
        through life. My car note is late.
Migrants are dying in heat too late.
        Do the only borders I care about
have personal skin to buffer them?
        Outside Floridians prepping for
a storm. Boy name or girl? It's best

to have equally-gendered natural
        named destruction, unless you're like me.
What is non-binary about the world? Yes.
        Will I still love all this tomorrow?

        Can I equally love the folks that hate with stickers,
        flags and signs what they can't see in the flesh?

Who buys the expectation that
        safety's just around the corner?
It's too hot to leave the house. We must.
        Does driving to the mailroom count?

Please Mr. Postman, look and see,
        if there's a package in your bag for me.
The geese line up to cross the street
        to automated honks. All stops.

They're following a boss-goose single-file.
        I meant to say *gander*, is there a third choice?
Nothing can replace their gorgeous gait,
        a sure-webbed, nonchalant swing. One web

in front of the other. Purpose. Direction.
        Nothing world-wide. I guess they're
looking for something green, or
        sustenance left. And aren't we all,

aligned as something inhuman takes the lead,
        balanced in the asphalt's daily rising heat
to trust that there will be a benevolent spark.
        Today's truest memoir can't be written yet.

Between awakening and sleep, between
     the flashed daymare that turns to well-lit dream,
the men I've loved come visiting. I'm supposed
     to be teaching, but I am learning, learning badly,

     well, or three times, or even never. Are we forgiven in
     the multiverse occurring as we make our morning beds?

Will I hear what that sounds like and feel it heal,
     surrounding me like a passed-down quilt? Will
the angry man I held too hard not hit me for it when
     I see him again? When will I know, much less feel,
more than three degrees, *When will I see you again/*
     When will we share precious moments—

Tomorrow the sun becomes the center of my
     universe again, as its surface
survives infinite small explosions. Or is
     the sun just firing at itself, over and over again,
mimicking, like a firearm vacuum, synaptic
     and humane connected targets? It's dusk

and I've yet to answer the multitudes
     vocalizing in my mind. My brother stays
home in his viral world and works remotely.
     I bring him Taco Bell and leave it at the door,

one more portal untaken today—is it only
     because it's forbidden for passage? In this
galaxy sometimes mortal love can be
     processed much faster than food. O blessed

     connectivity of girl group choruses
owning my phone. I hear you, girls,
     this girly-boy man-woman third thing
of unsung surfaces, much less depths. Please
     remind me to sing today a narration in glorious
tightening chords, like there's no tomorrow,
     harmonic, discordant or otherwise, all at once.

But how can I tell my life-triggered nephew
    that closure may come in a form he accepts
when his uniform body, outdoors or not, finds
    an infectious entrance of multiversed affirmation?

8

WHAT DOES IT MEAN TO BE ETERNALLY ALONE IN THE MULTIVERSE?

Adalberto Fernandes[1]

The concept of *multiverse*, with its literary roots in science fiction and distributed today by a powerful imagination industry, points toward a need to fight loneliness by reaching realities that are possibilities that cannot be constrained by what actually exists. That is, we try to reach "possible worlds" (Lewis 1986), which are parallel to ours because what we have in the present is not enough, namely a present where the others' company is useless. Against technophobic and romantic visions regarding nature, the multiverse inspires and produces new modes of encounters where the biological body ceases to be a limitation. This disembodied proximity allows the creation of relationships protected from the powerful presence of the organic body (smell, temperature, awkwardness, bodily social norms, etc.). It also reconfigures the way we produce meaningful encounters, in the same way as books have been always modes of producing friendly readers at a distance and long after the author's death (Sloterdijk 2009). This means that writing is the proto-multiverse where the hypothesis of the multiverse can encounter readers in different times and spaces—a literary multiverse. In sum, to be alone is the condition for imagining, and creating ways of, being accompanied in fictional devices such as the multiverse,

1 This chapter is financed by National Funds provided by FCT - Foundation for Science and Technology through project UIDB/04020/2020 with DOI 10.54499/UIDB/04020/2020 (https://doi.org/10.54499/UIDB/04020/2020).

DOI: 10.4324/9781003480846-10

especially when it is increasingly difficult to have meaningful encounters with others physically.

At the same time, loneliness is not only the cultural condition of multiverse imagination and industry. Paradoxically, the desire to produce a multiverse occurs in a networked world where being alone is increasingly difficult, if not completely impossible. CCTV cameras register our presence, smartphones track our localization and content consumption, apps build an individual archive of our data, etc. (Zuboff 2019). This means that if we want to fight loneliness by searching parallel immersive multiverse worlds, we are doing it with technologies that constantly remind us that we are not alone, given we are always being seen, tracked, analyzed, and so on. It is because we are connected, even when we are distant, that a cultural concept like multiverse could emerge as a world that already immerses us, that is, that never leaves us alone. To claim that the multiverse and the technologies that make it possible ruin an original loneliness of the ego, is to fail to address how our own individuality is already a collective work: we speak a collective language, our living conditions are granted by others, my memories come from the outside, and even when I talk silently with my own thoughts, I'm redoubling myself to be affected by me (Derrida 2011), creating a dialogue that was learned with the dialogues I had with others (Bakhtin 1981). In some sense, our ego is already a multiverse, an interconnected world where we are immersed with multiple selves without always the need to be physically with someone else.

In sum, when we embrace the multiverse, we are talking about a collective out-of-body experience that we have with ourselves when we are "alone." However, the impossibility of being alone and limited to my own body is the impossibility of being myself. According to Sartre's (1956) phenomenological approach, when the other is present I feel I cannot do everything I can with myself. The other reduces my space of possibilities given the impact of interactional norms or because the other occupies a space I want to use. It is when I am alone that I can do everything. The challenge ahead is: if the ego is a multiverse-ego, and if the body is a multiverse-body (i.e., an ego that is always immersed by others, and a body that always transcends its organic limits to be affected by others), is it possible to create a multiverse-loneliness—a world, or a nonworld, where the subject can be less, can be disconnected and protected from the networked world that fragments the ego and puts them in an uninterrupted connection with others? Is being alone a new form of resistance? Or is loneliness only a powerful fictional narrative that culturally pushes the joys and fears of immersive communicational technologies and imaginaries like the multiverse?

To approach these questions, it is productive to analyze, even if only exploratorily, the space, meaning, and narrative consequences loneliness has in blockbuster mainstream US movies that use the concept of multiverse (*The Marvels; Spider-man – Into the Spider-Verse; Doctor Strange in the Multiverse of Madness*). This is a way of doing a "cultural diagnosis" of powerful fictional dispositives that are shaping the culturally shared experience of multiverse-loneliness. A summary analysis of those movies reveals at least two crucial topics: (1) characters live in the present as if they were alone, due to their inability to use the present others to deal with the present crisis; (2) the constant multiplication of selves of the past and future that do not leave the present self alone. In the first case, the multiverse is used as an escape from the present in crisis, by altering the wrongdoings of the past or going to the future to find a solution to the present. In the second case, loneliness becomes impossible because past and future characters always invoke their present version, showing the impossibility of being alone in time.

A possible hypothesis of these exploratory topics is: the recourse to multiverse narratives appears to solve the loneliness that is produced in the present *physical* company. Once the present enters in crisis, time is freely manipulated to bring a more favorable state of affairs that the present lacks. The danger cannot be, thus, solved with the others who share the present in crisis because the presence of the others in the present is worthless. This is a loneliness that is felt even in the presence of others because their presence is useless to solve the present crisis. The multiverse solution to this loneliness in the face of crisis, namely, the duplication of oneself in past and/or future time, is a way of being accompanied by the same person, and not by a different one. While the multiverse shows no one has a chance of being completely alone from oneself because one is repeated three times, as present, past, and future versions and they always recognize each other. Instead of appearing as totally alien and unrecognizable, given the different effects of time on those three versions of the self, the multiverse-self is just producing another form of being alone in an accompanied form. The self is always with oneself. The reason is that the past and future selves are always recognized as part of the same self, and not as utterly strange, which means that time does not pass, it does not transform the self, making the multiverse a non-multiverse.

Multiverse is, according to the US movie industry cultural imaginary, a narrative device that uses aesthetical means to respond to and produce psychological effects of loneliness *with* the other, showing an incapacity or impossibility of being alone. That is, multiverse-loneliness is a product of being constantly in the *presence* of an other with whom we cannot establish a meaningful interaction in the present. As it was claimed above, when the present enters into crisis in the movies, the characters present at hand cannot come together to solve it, they need to go to find, or be found, by their present and/or past versions. They are physically together but the possibilities of togetherness are anulated. The crisis shows that sharing the present with others does not solve anything, so we need to go back or forward in time to find others with whom we can truly connect to come to a solution. The way multiverse is used to solve the crisis just shows that we are all alone in the present, the others do not make a difference in sharing the same time as us because the crisis can only be solved by abandoning the present. Going beyond overused claims of growing individualism, multiverse movies show that we are not individual enough because it is necessary to find others who are not those who lie close to us. They also show that loneliness stems from an incapacity to find solutions with the present other. At the same time, to be alone is something rare, even impossible in a multiverse logic, because my past and future selves are always ready to take over the present self when it enters in crisis. This means *that is not the present self or the present other that will solve loneliness*, a claim that is at all costs *averted and justified* by precisely returning to the company of others in the past and in the future who do not share my present.

For the sake of a future capacity to deal with this vicious circle, alternative aesthetical proposals to the world are needed to cope with the loneliness that assaults us from everywhere. Given the powerful image and logic of return, feedback, circles, and loops ingrained in multiverse narratives, it could be of great help to *return* to the Nietzschean proposal of the "eternal recurrence," which can be seen as a powerful antidote to the current multiverse fictions. Instead of developing resentful or nostalgic relationships with parallel worlds we cannot attain in the present, namely the golden age world of happy loneliness or meaningful togetherness, by inventing multiverse escapes from our present condition, we need, on the contrary, a Nietzschean "multiverse" that affirms all that has happened to us with joy, with a desire to accept as ours everything that happened to us in the past, it is happening in the present, and will happen in the future in the same way, be it good or bad (Nietzsche 1974).

The Nietzschean multiverse is a less a multiverse than a *uni-verse*, that is, a multiplicity of worlds that is happening in a *unitary* time because in the *eternal multiverse recurrence* what happens has already happened and it will happen again. We must, according to Nietzsche, affirm this recurrence of time with joy and make it our own, instead of being sad by the force of the repetition on us. One of the reasons is that that affirmation will also repeat itself in this eternal recurrence. In this way, we will be ready to deal with what the past made of us, and will continuously make in the future. Instead of rejecting as strange what has made what we are, we need to affirm *the eternal recurrence* as an opportunity for the active acceptance of what we cannot change and will be inevitably repeated in the future. According to Nietzsche, this piece of fiction possesses a profound impact on reality. Am I ready to embrace the notion that all of my past experiences will be replicated in the future? Would the notion of eternal recurrence be perceived as utterly horrifying? And what does that horror tell about my present life? Is my life not worthy of repetition? It means that with this fictional eternally repeated multiverse, a real question assaults our life: will I be ready to repeat everything I did forever? On the contrary, the current uses of the multiverse tend to convey the sense that it does not matter what is done in the present given that it can be solved or changed by abandoning it and restoring everything anew from the perspective of the past or the future.

The mainstream escapist version of the multiverse in US blockbusters blocks precisely these kind of questions by proposing multiverse aesthetical models of dealing with the psychological effects of time that construct individuals with no affirmative relation to time. The reason is that time serves to go back to repair the wrongs (negating what molded us), escape to the future when the present is bad (negating what is happening), or go back when the future seems problematic (negating what comes). This means that we become imprepared to change the present, accept the past, and face the future by failing to understand the power of what (repeatedly) makes us and cannot be undone, blocking the aesthetical and psychological means needed to deal with the challenging Nietzschean deterministic repetition of time.

Only a universe of eternally returning multiverses will allow an aesthetical management of the psychological effects of time repetition ready to face the lonely repetition of events. Lonely because it is not possible to go to the past to change it when the present enters into crisis, nor it is possible to reach for the future to anticipate the effects of the crisis in the present. When the present is always repeating the past and the present, we are stuck with ourselves. There is no other place to go to when we feel alone in this unparalleled universe. The multiverse actions of going back to the past or fast-forward to the future to find others is also an eternal escapist repetition, produced by dangerous multiverse narratives of US movies that reduce the aesthetical resources to strengthen our psychological capacity to deal with the present and the possibility that it is always repeating the same thing. Against an escapist multiverse, I propose a Nietzschean *eternal recurring uni-verse of multiverses* that is open to what happens to us in the present time, that affirms it, instead of hating the present with multiverse nostalgic and resentful relations with time, and makes ours the present that is made by the unchanging past and the future repeatedly. A repetition that includes the possibility that the proper multiverse is a thing from the past, that is being repeated in the present and recurring in the future, without the naïve use of multiverse narratives to escape the present or canceling its affirmative role in time. An *eternal recurring uni-verse of multiverses* to affirm and make ours the repetitions that make us on time.

References

Bakhtin, M.M. 1981. *The Dialogic Imagination: Four Essays*. Edited by Michael Holquist. Translated by Caryl Emerson. University of Texas Press Slavic Series. Austin: University of Texas Press.

Derrida, Jacques. 2011. *Voice and Phenomenon: Introduction to the Problem of the Sign in Husserl's Phenomenology*. Translated by Leonard Lawlor. Northwestern University Press.

Lewis, David. 1986. *On the Plurality of Worlds*. Wiley-Blackwell.

Nietzsche, Friedrich Wilhelm. 1974. *The Gay Science: With a Prelude in Rhymes and an Appendix of Songs*. Translated by Walter Kaufmann. Vintage Books edition. New York: Vintage Books.

Sartre, Jean-Paul. 1956. *Being and Nothingness: An Essay on Phenomenological Ontology*. Translated by Hazel Estella Barnes. New York: Philosophical Library.

Sloterdijk, Peter. 2009. 'Rules for the Human Zoo: A Response to the Letter on Humanism'. *Environment and Planning D: Society and Space* 27, no. 1: 12–28. https://doi.org/10.1068/dst3.

Zuboff, Shoshana. 2019. *The Age of Surveillance Capitalism: The Fight for a Human Future at the New Frontier of Power*. New York: PublicAffairs.

9

LUX IN TENEBRIS

Maria Faust

FIGURE 9.1 Nexus ~ Cosmos.

DOI: 10.4324/9781003480846-11

May the Universe Be Our Home

I sought solace in all this danger, and found the sun.

I hoped for knowledge in all this anger, and sorrows came undone.

I worked for ideals, less than mine, and fought for truth.

Simultaneous, and parallel, and war, and crime, a somber youth.

I asked another, one for each time.

Where sanity craved the northern hemisphere,

and the south, and neither to shine.

Which way to pursue, I ask – again and again,

one full of dreads, and weary and fearful,

or one, that helps to rise for diligence?

Each day – each night – each moment of now;

what are we here for?

-- I ask –

if not for the better,

if not for the integer,

yet, if not for the future of the momentarily how.

Home II

I lost a piece of heart, every place I went.

A thunder at the start, and peaceful in the end.

My sorrows came as wisdom, my shadows burnt in sand.

Each space brought fever, dawn - eternal dust.

All stars, the moon and yearning, made memories at last.

The pieces that I lost - came back so bright, complete.

My soul that always wanders - bore victory beyond defeat.

FIGURE 9.2 Cultura ~ Coexistere.

Ought To Be Overcome
Your notions of race, all stereotypes in space.
Your ideas of black and white,
shatter reality, stem from false pride.
The shades of grey, The Gods to whom we pray,
All equal, all unity, all flesh and bones.
The divisions you create, the borders you make,
through understanding overthrown.

Really Me
I am not solitary. In every we I find an us.
I am not necessary. The bonds we make, neither seperate nor less.
And inextricably bound, each move we make, in nature and in being.
Collectively we share, ideas and thoughts, each other's care.
We dive in seeing what has brought us here.
Without community, None of us were there.

Encounters
The truth in paradox, I ask to understand.
Black in all matters, and white, the other, they go hand in hand.
The dreams of class, shattered, since only sincerity can comprehend.
Wisdom beyond borders, a universe of orders,
and foolish those who limit themselves to one-dimensional ends.
I know and see - your eyes' long truth.
But forgiving and grace, humble days,
and consideration of coexistence belong to you.

FIGURE 9.3   Quo Vadis? ~ Lux in Tenebris.

Ophelia Ad Meliora
In all these sorrows dawns a treasure of another day,
The mere existence that we fail to measure vaults belongings which
opaquely veil.
Of what the haunted howling brought into the worldy sphere,
Came into being, each and every breath, beyond the gaze of another year.
We aim for this, we long for that, unveiled undone, beheld and shun,
Another dream is yet to come.
We ask for witness, we ask for truth, and dream mortality in every youth.
A dreadful universe that shone within,
A wealthy state, a brutal mind, a dreary leader, a kinships sin.
At last the first was yet to live, by nature`s promise fraud to miss.
We crave for dignity as we go, long for maturity in an age to grow.
The final stone is not yet set, apocalypse a promise of the ones unmet.
And hope in haze of every breath, the days to follow succumb death.

The Manifold Labyrinth
I wonder, I wonder, if this was all to the thunder.
I oblige, I oblige, the feeling was never just right.
And there's doom to this crisis, and failure to an end.
Was ever coincidence, just shallow of maturity's friend?
My depth in sorrows, this planet to come.
All concurrence shattered and the fruits of longing undone.
My dreams for perseverance, an artificial fruit to uphold.
When machines are fauxed into games, and humanity lost its essence in gold.
How do we refrain from the obstacle to belong?
And as we get closer, and as we preach,
The lessons to learn are the ones to teach.

10

TALES OF TWO MULTIVERSES

From Fan Service/Productivity to Shared
Inflationary Fantasies

Matt Hills

In this chapter, I want to explore a number of binaries that have circulated
around the concept of the multiverse. My focus will be on how the multiverse has been understood either textually (in relation to a range of pop-
cultural texts, but especially franchise movies) or—more unusually—as a
matter of audience creativity. Specifically, this text/audience binary has
been interpreted by critics and scholars as revolving around industry-fan
relations, whether via the multiverse as "fan service" in the text (Beaty
2016; Mooney 2021) or via the multiverse as "fan productivity" whereby
fans create variant versions of characters and worlds in their responses to
franchise fictions (De Kosnik 2016; Fazel and Geddes 2022). Associated
with this cleavage in scholarship, text-oriented/audience-oriented understandings of the multiverse also emerge out of, and feed into, several related
binaries, such as the multiverse being analyzed through a far longer history
(i.e., having related cultural concepts stretching back many centuries;
Rubenstein 2014 and Siegfried 2019) or a notably shorter history (usually
rendering the concept as a matter of pop-cultural franchise logics; see Proctor 2017 on Marvel multiverses; Martinez 2020 on the Spider-verse; and
Friedenthal 2019 and 2022 on DC and Marvel). The focus on text-based
readings has also incorporated what might be described as diegetic versus
extra-diegetic takes on the multiversal. However, I will argue that for each
binary—long history/short history, or wider cultural context versus nar-
rower franchise context; diegetic versus extra-diegetic; text versus audience—it is possible to see that the "multiverse" is securely utilized on one

DOI: 10.4324/9781003480846-12

side of the equation, whereas it remains rather more metaphorical on the opposed side. This indicates that the multiverse has a dominant set of logics in contemporary culture; it is part of the "franchise era" (Fleury, Hikari Hartzheim and Mamber 2019), a form of fan-targeted service based in texts and their diegetic representations. By contrast, emergent scholarly understandings of the multiverse in cultural rather than franchised contexts, as extra-diegetic, and as generated through the proliferation of fan fictions, involve more speculative extensions of the terminology.

My concern is not just with exploring these tales of two multiverses, though. By analyzing the associated binaries, I want to argue that the multiverse has become a multivocal term, appropriated by academics in order to contest its place as an industrial textual strategy. This raises the question of "why now?"—why has the multiverse become appealing as much to media/fan scholars as to those in charge of intellectual properties? When entire narrative "universes" aren't enough for fans and producers alike, then infinite possibilities, and do-overs might seem to speak (perhaps implicitly or even unconsciously) to an increasingly shared cultural awareness of markedly finite planetary resources (Highmore 2020).

In the closing part of the chapter, I will therefore turn my attention to a conjunctural analysis of the multiverse (Gilbert 2019; Grossberg 2019), considering the nature of its current appeal(s). At first glance, the multiversal offers a sense of highly individualized escapism or neoliberal wish-fulfillment, suggesting that we might all be living personally better lives in an alternative universe (Phillips 2013; Miller 2020; Sharzer 2022). This type of reading neglects the multiverse's vastly expansionist or inflationary tendencies, though—it doesn't just imply better or more consoling lives, or variant characters/outcomes, but also offers up a sense of the dazzlingly infinite which popular culture nevertheless struggles to narratively engage with. I will argue that the multiverse's conjunctural appeal relates to its *removal of limitation* at ecological scales far beyond the neoliberalized "trajectory of the self" (Giddens 1991, 70), making it potentially part of the "Anthropocene Unconscious" studied by Mark Bould (2021). First, however, I want to introduce a series of binaries through which the multiverse has been analyzed by cultural historians as well as academics in film and fan studies.

Dominant Versus Metaphorical Multiverses: Shorter/Longer Histories and Text-Centered/Audience-Centered Binaries

Multiple attempts have been made to map the concept of the multiverse, not just in terms of contemporary science and whether the multiverse is a properly "scientific" object at all, that is, whether it is capable of testing or observable evidence, but also as a matter of cultural history. In *Worlds*

Without End, Mary-Jane Rubenstein (2014) traces the term back to William James' work of 1895 (3–4), and the related notion of "cosmic multiplicity" is, in turn, traced back to Plato (18–19). Rubenstein suggests that the modern "multiverse" of Many Worlds theory, dating back to the 1950s, has its appeal in the fact that it provides a rationale for observable scientific constants (the fact that the universe seems, massively improbably, to support human life) without recourse to God or intentional "design"; if there are infinite other universes, then the mathematical constants of our universe form just one in an endless variety. In short, there is a "profoundly nontheistic (sometimes even antitheistic) motivation behind the turn to many-worlds scenarios" (17), even if the multiverse itself then ironically threatens to become a scientific article of faith.

And in *The Number of the Heavens*, Tom Siegfried (2019) similarly traces the multiverse not as "a uniquely modern idea, born of twentieth-century advances in physics and cosmology, but... rather an ancient idea with a rich... history" (271). Siegfried deconstructs the notion that science progresses in a linear fashion, arguing that there remain overlaps between medieval and modern approaches to the question of the multiverse, not least the issue of whether one should believe in something purely theoretical (273). Indeed, for Siegfried, a key historical moment is the 1277 decree, made by the Bishop of Paris, that enabled philosophers to consider, *contra* Aristotle, that God could have created many worlds.

But both these discussions of what might be thought of as *the longer history of the multiverse concept* fail to engage with its contemporary proliferation within popular culture, even though Rubenstein (2014) notes this "multiversal explosion" in passing (1). To read pop-cultural fascination off from scientific "nontheistic" or "antitheistic" motivations or developments seems unconvincing, failing to recognize the autonomy of media narratives with regard to scientific debates.

There is an under-developed alternative approach, however, which implies a sociological explanation of the pop-cultural appeal of multiverses, rather than harking back to specialist scientific disagreements. This is suggested by Anthony Giddens's (1991) *Modernity and Self-Identity*:

> Because of its reflexively mobilised — yet intrinsically erratic — dynamism, modern social activity has an essentially counterfactual character. In a post-traditional social universe, an indefinite range of potential courses of action (with their attendant risks) is at any given moment open to individuals and collectivities. Choosing among such alternatives is always an 'as if' matter, a question of selecting between 'possible worlds'. *Living in circumstances of modernity is best understood as a matter of the routine contemplation of counterfactuals.*
>
> *(28–29, my italics)*

This account, which is not about the popularity of multiverses but instead aims to characterize modern social life *tout court*, can nonetheless be read as arguing that the multiverse concept makes intuitive sense to people who experience their own modern lives as having "an essentially counterfactual character"—that is, "what if I took this course of action?," "what if I hadn't done X, or said Y?." Such everyday reflexivity, it can be suggested, means that in a "post-traditional social universe" (and it is striking that Giddens chooses *universe* as a term here, as if multiverse debates are ghosting across his own word choices) the science fiction or fantasy of represented multiverses can affectively echo peoples' lived experience. This is not because we believe in multiverses, or follow scientific debates, but because we are constantly engaged in kinds of counterfactual monitoring and imagination that feel like "multiversal" questions.

However, Giddens's arguments remain somewhat vague. When exactly are we supposed to have entered this "modern" stage of social life? Can we really generalize, in the spirit of grand theory, about social life as if it operates in the same way for everybody? And although Giddens's work might unintentionally open up an alternative longer history for the multiverse concept, where it is resonant with key conditions of (late) modernity, this work again fails to engage with the media, and pop-cultural narratives.

By contrast, film studies and fan studies have provided a media industry-focused *shorter history of the multiverse concept*, one largely unconcerned with the multiverse's philosophical-cultural history. Where the analyses of Rubenstein and Siegfried, and the allusive social theory of Giddens, give us ways of thinking metaphorically about multiverse-like cosmologies going back to the Ancient Greeks, or about multiverse-like modern social life, film/fan studies have engaged directly with prevalent, if not dominant, genre representations of the multiverse.

A key way of thinking about the multiverse in this context is to locate it not culturally but industrially, that is, as a matter of intellectual property franchising and franchise logics (see Davis's chapter in this volume). The "franchise era" is typically said to have become entrenched in the twenty-first century as a result of shifts toward the production of big-budget cinematic "universes" (Fleury, Hikari Hartzheim, and Mamber 2019, 12) and franchise adaptations (Meikle 2019). At the same time, franchise logics can be identified as "expansive" (Herbert 2017, 82), seeking to target as many different audience segments and identities as possible, and doing this not through single tent-pole movies but through "modular" proliferations of variant texts, where iterations of franchise characters and content can be designed for differentiated target audiences (Herbert 2017, 99). Derek Johnson (2013) has likewise analyzed media franchising as "a balance between sameness and difference, between source DNA and potential elaborations of it… [T]his potential for difference and variation proves…

important as a means of distinguishing... production identities in the use intellectual property resources shared across industrial networks" (107). Not coincidentally, Johnson introduces this discussion by applying ideas of franchise difference and variation to the 2009 animated movie *Turtles Forever*, where the villain of the piece, witnessing different versions of the *Teenage Mutant Ninja Turtles* crossing paths, concludes that "ours is but a single dimension in a multiverse of dimensions" (107). Here, the diegetic invocation of a "multiverse" becomes the textual device, even the "surprisingly cogent theoretical model" (107), for comprehending the franchise's recombination of elements as part of an anniversary celebration (see Ricardo Victoria-Uribe and Nazario Robles-Bastida's chapter in this volume).

Another example of the multiverse being diegetically used to position franchise variants occurs in J.J. Abrams' *Star Trek* (2009) where the film's new Kelvin timeline is distinguished from the "Prime" timeline of *Star Trek: The Original Series* (Hills 2015, 31). This means that Abrams' versions of Kirk and Spock are no longer the same characters as those played by William Shatner and Leonard Nimoy, and nor are they limited by previous continuity. But simultaneously, *The Original Series* is left in place as equally canonical—as the "Prime" universe, it cannot be superseded or wiped away by the later franchise iteration, and both are left to stand, diegetically, as different timelines of a *Trek* multiverse (see Geraghty's chapter in this volume). Such a diegetic strategy, of multiversal branched existence, has become relatively commonplace as a franchise logic permitting multiple historical versions of a character or world to non-hierarchically co-exist, appearing both in the animated film *Spider-man: Into the Spider-verse* (see Martínez 2020, 205) and again in the live-action *Spiderman: No Way Home* (2021) which combines three different iterations of the web-slinger's franchise history in the form of actors Tobey Maguire, Andrew Garfield, and Tom Holland.

Whereas variant "production communities" (Johnson 2013, 113) have mobilized the multiverse diegetically to set up their new IP iterations alongside established franchise continuities, the multiverse concept has also been extra-diegetically invoked by scholars. For example, in his analysis of the *Blade Runner* 5-disc Collector's Edition including Ridley Scott's *Final Cut*, Brooker (2009) argues that

By including the erasures and reworkings of Deckard's story through various editions, and presenting them all as potentially valid, the *Final Cut* celebrates what *Blade Runner* has become over the last 25 years: an unruly multiverse, a map of possible routes, a network of alternatives rather than a single narrative.

(90)

Yet there is no diegetic multiverse here: the term is utilized to characterize the extra-diegetic collection of different edits and versions of *Blade Runner*, as well as acknowledging its transmedia extensions. By displaying a kind of curatorial "transmedia memory" (Harvey 2015, 182) of different franchise versions over the years, *Blade Runner* presents its fans with a set of non-hierarchical narrative branches; the multiverse concept operates at the level of fan-consumer choice across narrative iterations rather than within *Blade Runner*'s story world.

Brooker isn't alone in academically applying the multiverse in this more metaphorical manner to capture likely fan experience rather than corporate diegetic moves. Carmelo Esterrich (2021) describes *Star Wars* by noting that the franchise "is not one universe. It is a multitude of them," with the fan-consumer being called upon to navigate "a multiverse" of licensed transmedia extensions as well as fanfiction (1). By rejecting the official industrial terminology of the "Expanded Universe" for *Star Wars* comics and books, as opposed to the films' centrality, Esterrich again seeks to reject a hierarchical franchise model, noting that his use of "a Star Wars multiverse" is not meant to reflect the way "contemporary science" has employed the term (3). Deploying the concept more metaphorically and allusively to reflect fans' pathways through a franchise—as well as including fan productivity in the form of fanfic and cosplay, etc.—opposes the notion of corporate or IP control over a multiverse, where this is diegetically used to wipe away messy continuity or to canonically map out multiversal options (or even numbered alternative worlds) akin to Grant Morrison's comic book series *Multiversity* (Friedenthal 2017, 104).

Both the longer/shorter history of the multiverse and its diegetic/extra-diegetic binary display one side of the opposition as dominant or more culturally secure, while its other term is quirkier and less agreed-upon culturally. Rubenstein and Siegfried's explorations of cosmic multiplicity—the multiverse *avant la lettre*, tracking back to Greek philosophy—are somewhat idiosyncratic, despite their value in highlighting patterns across cultural history; the more dominant take on the multiverse as it circulates outside current scientific debate is surely figured by how it has been denotatively represented in twenty-first century pop-cultural franchises. Similarly, for the diegetic/extra-diegetic binary, the scholarly use of the multiverse concept to characterize fans' consumption and navigation of franchise alternatives has had less purchase than the diegetic representation of blockbuster multiverses, and is left feeling metaphorical as opposed to the once again denotative versions of pop-cultural multiverses. Mass media representation, unsurprisingly, wins out culturally over specialized analysis.

The diegetic/extra-diegetic binary feeds into a wider splitting of the multiverse concept—that between *the multiverse as text-oriented or audience-oriented*. Fan studies have addressed this, as have film critics such as Darren Mooney. In his 2021 blog piece "Why the Multiverse is the Future of Shared Universe Storytelling," Mooney argues that textually "the multiverse is effectively a fig leaf that allows for the operation of industry machinery," with the continuity required by a shared universe such as the MCU capable of being loosened up, multiplied and re-imagined as IP rights-holders see fit. Variant versions of characters and premises can then be used to target different sections of the audience as well as different sections of a fandom. As Mooney observes: "a multiverse allows for the potential of multiple iterations of the *same* character—each with their own nostalgic fan base—to come together," as was the case for *Spiderman: No Way Home*. Mooney goes further in arguing that the diegetic, textual multiverse reflects a newer industrial norm of personalization and audience-targeting in the streaming era, where

> The shift towards the multiverse as an ordering principle seems to recognize this new reality. It is no longer enough to offer a single version of Batman that appeals to the largest possible audience; instead, the ideal is to create a multitude of Batmen, each of whom can appeal to a different section of the audience.

This industrial "fig leaf" amounts to a form of "fan service," where an "insider/outsider divide" is enabled among audiences. This is a divide between those who are into a specific version of a character or a specific pathway through a set of franchise alternatives and those who are affectively left outside the fan circle targeted by diegetic multiverse navigations (Beaty 2016, 322).

Countering such an emphasis on the multiversal as textual fan-targeting (or as an attempt to build coalitions between rival segments of a fandom), other scholars have instead stressed not how fans are industrially addressed (or even interpellated) by multiversal fictions, but rather how fan activity and productivity can produce multiversal pathways. Valerie Fazel and Louise Geddes (2022) argue that "all users who engage with the work contribute to a multifaceted, archontic development… that is best conceptualized as a multiverse" (5). Placing the multiversal on the side of fan experience, like Abigail De Kosnik (2016), Will Brooker (2009) and Carmelo Esterrich (2021) before them, but linking this more tightly to fans' production of fanfic rather than navigations of official (trans)media products, Fazel and Geddes pursue more inclusive "ideologies for the multiverse" (18). That is, they oppose any sense that only IP rights-holders, as supposedly official textual controllers, have authority over the creation of multiversal narrative/character pathways, viewing this opposition to any "limiting notion" of artificial authority as an "ontological" argument about the essence of pop-cultural and franchise multiverses. Fazel and Geddes' "archontic multiverse" (39) radically decenters and displaces any concept of 'canon.' Fans can begin from contingent affects and encounters with texts, situating their multiversal navigations within what amount to personalized frameworks (and hence this approach might also be argued to tacitly refract an era of media personalization and customization, albeit here any personalization is crafted by the fan-self rather than through industrial strategy). Placing the multiverse concept on the side of fan productivity rather than textual fan service is a powerful attempt to contest, if not reject, industrial and textual powers over the multiversal. In this model, fans are not "contained" by an industry-shaped multiverse, constrained only to selecting pathways through its pre-programmed alternatives, but can instead create infinitely new multiversal possibilities through the archontic addition of fanfic variants (De Kosnik 2016). And yet the cultural politics of this fan empowerment seem to use the multiverse concept in a novel, metaphorical, and less culturally secure way—as Fazel and Geddes recognize, their approach "promotes a new understanding of fandom" (5). We can thus suggest that the multiverse has itself, as a term, become multivalent and polyvocal, needing to be understood dialogically as a terrain of struggle between text-based and audience-based understandings when it is used to reflect on pop-cultural franchises, their fictions, and their fans. Industry-focused and text-oriented understandings may be more strongly established at present as dominant in associated binaries, but the multivocality of the "multiverse" means that its metaphorical and specialized language uses could yet take a fuller hold in ongoing struggles over the term's meanings.

This struggle over meaning also raises the question of why the "multiverse" has surfaced as a new key term in both pop-cultural franchising and fan studies, being fought over—and differentially claimed—by industry and sections of academia alike. In the closing part of this Chapter I want to address this "why now?" question by briefly adopting what will be defined as a conjunctural approach to the multiverse concept, as it has culturally circulated outside scientific debate and popular science.

Conjunctural Multiverses: The Shared Appeal of Inflationary Fantasy and World-Switching

Conjunctural analysis, a broadly Gramscian technique adopted by cultural studies, is usually applied to political movements and powers configuring in a specific cultural moment (see Grossberg 2018 and Clarke 2023 in US and UK contexts, respectively). But it also focuses on complex operations of culture within the slippery and hard-to-define specificities of a "conjuncture" (Gilbert 2019, 8–9 and 14; Grossberg 2019, 56). By usually restricting itself to national political contexts, however, conjunctural analysis has arguably failed to grapple with the vaster time scales of what's been termed the "Anthropocene" (Bould 2021), and hence has neglected issues of environmentalism and climate catastrophe (Highmore 2020). Analyzing the conjuncture (a phase of cultural-political forces) has meant seeking to map power relations in play. Also, it means mapping the specificities of the present, but in a radically contextualist manner that refuses to isolate out texts or topics, instead always asking "the question 'what does this have to do with everything else?' when examining any phenomenon, however minute" (Gilbert 2019, 5). My discussion of the multiverse concept as dialogic thus begins a tentative but related project for this term, as it has worked in recent popular franchises. The multiverse, perhaps unexpectedly, appears as a figure that at this moment, in this conjuncture of cultural-political forces, encapsulates power relations and struggles between fan creativity and the attempted industrial interpellation of fan audiences in an era of media personalization.

And yet, it is striking that the scientific/mathematical multiverse concept is far, far more expansive than even the most expansive of franchise logics (Herbert 2017), as well as being even more expansive than its representation outside franchising, for example, in *Everything Everywhere All At Once*, whose *mise-en-scene* cuts dizzyingly between many versions of its lead character Evelyn Quan Wang (Michelle Yeoh). Despite exploring absurdist universes where rocks are sentient, or where people have hotdogs for fingers, this narrative still reduces multiversal infinity to the emotional meanings of one mother-daughter relationship. Deployed in blockbuster and arthouse cinema cultures, the multiverse is made rapidly and narratively meaningful—it quickly becomes merely a sense of "this, not that"; this version of Spiderman and those other ones too; this version of our lead character, not that one. *Sliding Doors*-type moments of significant narrative choice are prioritized. But the multiverse, in both its scientific and pop-cultural circulations, has the potential to wholly overwhelm narrative coherence, especially if the model of an "Ultimate Multiverse" or "multiverse with a vengeance" (Rubenstein 2014, 6) is adopted where all possible worlds are thought to modally exist in an infinity of every possible conceivable variation. As such, the multiverse promises not just a neoliberal refashioning of the character/self as a better (or worse) incarnation, centering neoliberal ideologies of individualism at the same time as promising a compensatory or consoling, escapist view of desired alternative selves; a "late escapism" curiously of a piece with neoliberalism, perhaps (Sharzer 2022).

It also seems to hold out the sublime promise of spatiotemporality-without-limit, an absolute infinity that pop-cultural representations can only glimpse at. And this figuration arises at a point, and within a conjuncture, when the "planetary scale" of climate change has become largely politically accepted as well as socioculturally negotiated by economic interests (Gilbert 2019, 34), and where the "general climactic conditions on Planet Earth, and their particular manifestations in specific localities, …constitute crucial features of the contemporary conjuncture, which cannot be thought outside of the time-frame of the apparent onset of the 'Anthropocene' epoch" (Gilbert 2019, 35).

Mark Bould's (2021) *The Anthropocene Unconscious* gives us an important way to think through the spatiotemporality-without-limit of the popular franchise multiverse. In his analysis, and by drawing on the literary theory of Pierre Macherey, Bould argues that climate fiction ("cli-fi") should not be restricted to that which denotatively represents matters of climate change. Rather, the silences of texts can be read as a textual unconscious in which what is written out threatens to return—in this sense, Bould asks, "what happens when we stop assuming that the text is not about the anthropogenic biosphere crises engulfing us? What if all the stories we tell are stories about the Anthropocene? About climate change?" (17). Addressing sci-fi as well as cli-fi, Bould analyzes *Edge of Tomorrow* (2014) and its time-loop variations (a connotative version of multiversal logic) as figuring an impasse where "There is no way out" (117). Reading this as an unconscious analog for environmentalist despair, Bould ponders whether "impasse is actually transition?" (118), and whether the effects of imagined impasse and catastrophe might be transformable into motivations to act against climate catastrophe in the real world. But unlike *Edge of Tomorrow*, pop-cultural multiverses almost always promise some kind of liberation, some kind of radical change—they are, unusually, a generic space designed to permit massive alterations in character, scenario, and even basic narrative premise. Multiverses are *narratological machines of difference*, the very opposite of "impasse" (even if the infinity of modal possible worlds presupposed in some scientific versions of the multiverse would completely derail such narrative legibility and urgency). For example, in the TV adaptation of *Dark Matter* (2024), episode 3 suggests that the corporate actor responsible for accessing the multiverse has a specific rationale: "what if there was a world where cancer had been solved… Or climate change… What if we could bring that knowledge into our suffering world?" Otherwise, the vast possibilities of the multiversal are—at least initially—narratively reduced to a paranoid thriller where binary versions of a character, Jason Dessen (Joel Edgerton), swap across their two universes. But even within this narrative reduction, *Dark Matter* fleetingly starts to draw climate change into its diegesis, albeit via the centrist discourse of near-magical technological solutionism as a way out of our real-world environmental impasse.

Multiversal logics—both in text-centered and audience-centered approaches—hold out the possibility of *world-switching*, that is, fantastically and imaginatively swapping one universe for another rather than working to politically change our world. The finitude of natural resources can be evaded or sidestepped through such inflationary fantasies: there can always be more and different franchise variations; there can always be more and different fanfic variations. Even while the power struggles of industry-fan relations condense onto the franchise figure of the multiverse, there is arguably a powerfully Anthropocene unconscious to the appeal of the multiverse concept on both sides of our current conjuncture.

To conclude, I've considered some of the binaries that circulate around pop-cultural and franchise representations of the multiverse (longer/shorter history; diegetic/extra-diegetic; text/audience). In each case, I have argued that there are dominant and subordinated, or more metaphorical and less culturally secure, renderings of the multiverse. In the present moment, beyond the domains of scientific debate, it is industrial, diegetic, and textual senses of the multiverse concept that appear to dominate pop-culturally, over and above more specialized, scholarly readings of fan experience and productivity as multiversal. However, I ended the chapter by briefly introducing the idea of an "Anthropocene unconscious" to the inflationary, expansive fantasies of the multiverse in contemporary popular culture. This suggests that the multiverse concept, given its popularity and polyvalence, might do more than figure power struggles between industrial fan service and creative fan productivity. Contemporary proliferations of the multiverse and its spatiotemporality-without-limit may also speak—unconsciously, and through the substitutions or silences of "many worlds" textual denotation—to a repressed awareness of the ecological dangers facing our singular, finite, and all-too-limited world.

References

Beaty, Bart. 2016. "Superhero Fan Service: Audience Strategies in the Contemporary Interlinked Hollywood Blockbuster." *The Information Society* 32, no. 5: 318–325.

Bould, Mark. 2021. *The Anthropocene Unconscious: Climate Catastrophe Culture*. London: Verso.

Brooker, Will. 2009. "All Our Variant Futures: The Many Narratives of *Blade Runner: The Final Cut*." *Popular Communication* 7, no. 2: 79–91.

Clarke, John. 2023. *The Battle for Britain: Crises, Conflicts and the Conjuncture*. Bristol: Bristol University Press.

De Kosnik, Abigail. 2016. *Rogue Archives: Digital Cultural Memory and Media Fandom*. Cambridge: MIT Press.

Esterrich, Carmelo. 2021. *Star Wars Multiverse*. New Brunswick: Rutgers University Press.

Fazel, Valerie M. and Louise Geddes. 2022. *The Shakespeare Multiverse: Fandom as Literary Praxis*. New York: Routledge.

Fleury, James, Bryan Hikari Hartzheim, and Stephen Mamber. 2019. "Introduction: The Franchise Era." In *The Franchise Era: Managing Media in the Digital Economy*, edited by James Fleury, Bryan Hikari Hartzheim, and Stephen Mamber 1–28. Edinburgh: Edinburgh University Press.

Friedenthal, Andrew J. 2017. *Retcon Game: Retroactive Continuity and the Hyperlinking of America*. Jackson: University Press of Mississippi.

Friedenthal, Andrew J. 2019. *The World of DC Comics*. New York: Routledge.

Friedenthal, Andrew J. 2022. *The World of Marvel Comics*. New York: Routledge.

Giddens, Anthony. 1991. *Modernity and Self-Identity: Self and Society in the Late Modern Age*. Cambridge: Polity Press.

Gilbert, Jeremy. 2019. "This Conjuncture: For Stuart Hall." *New Formations* 96/97: 5–37.

Grossberg, Lawrence. 2018. *Under the Cover of Chaos: Trump and the Battle for the American Right*. London: Pluto Press.

Grossberg, Lawrence. 2019. "Cultural Studies in Search of a Method, Or Looking for Conjunctural Analysis." *New Formations* 96/97: 38–68.

Harvey, Colin. 2015. *Fantastic Transmedia: Narrative, Play and Memory Across Science Fiction and Fantasy Storyworlds*. Basingstoke: Palgrave Macmillan.

Herbert, Daniel. 2017. *Film Remakes and Franchises*. New Brunswick: Rutgers University Press.

Highmore, Ben. 2020. "Disjunctive Constellations: On Climate Change, Conjunctures and Cultural Studies." *New Formations* 102: 28–43.

Hills, Matt. 2015. "From 'Multiverse' to 'Abramsverse': *Blade Runner, Star Trek*, Multiplicity, and the Authorising of Cult/SF Worlds." In *Science Fiction Double Feature: The Science Fiction Film as Cult Text*, edited by J.P. Telotte and Gerald Duchovnay, 21–37. Liverpool: Liverpool University Press.

Johnson, Derek. 2013. *Media Franchising: Creative License and Collaboration in the Culture Industries*. New York: New York University Press.

Martínez, María Inmaculada Parra. 2020. "On comics, narratives and transmedia multiverses: re-envisioning the wall-crawler in *Spider-Man: Into the Spider-Verse*." *MHCJ* 11, no. 2: 201–220.

Meikle, Kyle. 2019. *Adaptations in the Franchise Era, 2001–16*. New York: Bloomsbury Academic.

Miller, Andrew H. 2020. *On Not Being Someone Else: Tales of Our Unled Lives*. Cambridge: Harvard University Press.

Mooney, Darren. 2021. "Why the Multiverse Is the Future of Shared Universe Storytelling." *The Escapist*, Feb 15th, https://www.escapistmagazine.com/why-the-multiverse-is-the-future-of-shared-universe-storytelling/, accessed 25/4/24.

Phillips, Adam. 2013. *Missing Out: In Praise of the Unlived Life*. New York: Farrar, Straus and Giroux.

Proctor, William. 2017. "Schrödinger's Cape: The Quantum Seriality of the Marvel Multiverse." In *Make Ours Marvel: Media Convergence and a Comics Universe*, edited by Matt Yockey, 319–345. Austin: University of Texas Press.

Rubenstein, Mary-Jane. 2014. *Worlds Without End: The Many Lives of the Multiverse*. New York: Columbia University Press.

Sharzer, Greg. 2022. *Late Escapism and Contemporary Neoliberalism: Alienation, Work and Utopia*. London and New York: Routledge.

Siegfried, Tom. 2019. *The Number of the Heavens: A History of the Multiverse and the Quest to Understand the Cosmos*. Cambridge: Harvard University Press.

PART II

Multiversal Constructs Across Media

11

IT'S A TRIP. IT'S GOT A FUNKY BEAT

Exploring the Sonic Multiverse through Hip Hop Sampling

Ian Sinnett

Please visit www.routledge.com/9781032770116 and click on Support Material to access the digital supplement for this chapter.

The poem "I Have a Dream about the Future," written by hip hop pioneer Afrika Bambaataa (1970) while still in high school, can be found deep in the archives of Cornell University's expansive hip hop collection, among the six hundred-plus square feet of material dedicated solely to him. It starts:

> I have a dream about the future. I dream that the world would be off the ground... Everybody will be able to live on different planets. That music will be different. Di-s[c]o-tex will be psychedelic. People be dancing in the air.

"I Have a Dream about the Future" portrays exactly what the title implies: Bambaataa's dreams of a futuristic utopia in which people fly around, can go anywhere they want, and immerse themselves in unique soundscapes and "psychedelic di-s[c]o-tex." The poem goes on to express the power of sonic relatability, concluding that "everybody will have a translator to speak other people['s] language ... and that maybe there might be peace in the world" (*Afrika Bambaataa Hip Hop Archive*). For Bambaataa, speaking other languages, traveling, and experiencing myriad cultures and ways of being is what can lead humanity toward peace. Although Bambaataa is not explicitly referring to "multiverses" in this poem, he provides a unique

DOI: 10.4324/9781003480846-14

example of early Afrofuturist thought that interrogates the ways in which interspatial travel and intersonic experiences may shape a humanistic harmony, a pattern of thought that I argue permeates throughout rap music and hip hop culture. Musically, this type of intersonic, transtemporal, cross-cultural traversal can be achieved through what I am calling the sonic multiverse.

In this chapter, I will be theorizing the sonic multiverse by investigating digital audio sampling (briefly defined as the act of using computerized technologies to cut and modify pieces of earlier audio recordings and placing them within a new composition) as it occurs in hip hop music. The sonic multiverse, I contend, describes a musical arrangement consisting of a complex and multifaceted assemblage of auditory components, each originating from a vast array of sources. I argue that the sonic multiverse bears striking similarities to how mediated multiverses have been portrayed throughout popular culture. For instance, in visual media, like film, the multiverse is shown to be a multitude of parallel universes—each with their own stories, timelines, circumstances, and characters. Yet, these multiversal narratives do not stand alone; they are all presented within a larger unified narrative, in context with each other, reshaping their respective events and timelines. In the film *Spider-Man: Across the Spider-Verse* (2023), for example, the film's titular Spiderman, Miles Morales, is subjected to a rip in dimensional barriers, exposing him to other iterations of the Spiderman narrative. These various Spiderpeople influence one another, reform each other's experiences, and shatter what was understood as each timeline's preconceived canon (see FitzWittemore's chapter in this volume). Similarly, the hip hop beat introduces, contrasts, and winds together multiple threads of sonic reference to construct a paradoxically unified and contradictory composition. This sonic multiverse, akin to the visual multiverse, is a form through which numerous discourses appear in tandem, consequently reconstructing their content and meanings by their proximal relational context.

While not solely confined to hip hop, the aesthetic form of the hip hop beat offers a novel means of both producing and consuming sonic multiverses. The sample-based composition (a musical piece that is made up primarily of digital samples) thrusts the listener into a multifarious web of sonic reference, revealing the contingency of narrative and discourse, and demonstrating that dominant narratives and meanings are subject to interpretation, reconstruction, and revision. Essentially, I argue that the formal qualities of the sonic multiverse may act as a means of shaping consciousness, that it is a modality of expression revealing the complexities of meaning and ways of being. It shows that the structures of society that we inherit are not unchangeable, nor are our futures fully predetermined.

This chapter contains a written portion and a digital accompaniment found on the publisher's website. This written section will provide an overview of the sonic multiverse as a conceptual theory. I will develop this theory by highlighting its connections to Black radical aesthetics, exploring how the sonic multiverse interacts with cultural memory, and displaying linkages to what Robin D.G. Kelley (2002) calls Black "freedom dreams." Afterwards, I put theory to practice in the digital supplement (https://prezi.com/view/kG0w0au2F3DyAoB73u99/) in which I break down and analyze the complex sonic multiverse of the Beastie Boys' song "B-Boy Bouillabaisse," from their sophomore album *Paul's Boutique* (1989). But first, I will provide a brief history of hip hop and digital sampling.

Hip Hop History, Black Radical Aesthetics, and the Advent of Digital Sampling

Although the origin point of any cultural form is often complicated and contested, hip hop's derivation has an almost universally agreed upon moment and location: the night of August 11, 1973, in the neighborhood of the South Bronx in New York City, at 1520 Sedgewick Avenue. This is where DJ Kool Herc (born Clive Campbell) pioneered the foundations of what we now call hip hop through his innovation of breakbeat deejaying. Herc, accustomed to deejaying in Brooklyn dance clubs, began this fateful set by doing what he usually did: playing hit dance songs in their entirety. However, realizing that the crowd only danced fervently during drum break moments of the funk and soul records he spun, he adjusted his deejaying style. "Instead of dancehall music," Joseph C. Ewoodzie (2017) explains,

> he played soul and funk records; and instead of playing songs in their entirety, he played the portion of the song with the most percussion... He went from the most danceable, high-energy peak of one song to that of another so the dancers would never feel a lull
>
> *(17)*

What Herc was doing, unknowingly, was setting up the foundation of an entire cultural form: the break.

What Herc was emphasizing are known as "break" moments of these songs—when most or all the melodic elements drop out, leaving just the percussive rhythm of the drumbeat. From here, the genre grew as other deejays further innovated on this breakbeat style, with the larger culture becoming more expansive, including breakdancing, emceeing (or rapping), and graffiti tagging (Chang 2005, Ewoodzie 2017, Rose 1994). According to cultural theorist Fred Moten (2003, 2017), however, the appeal of the break goes beyond hip hop. He argues that the break exists as a vital component of Black ontology, performance, and Black radical aesthetics. For Moten (2017), the break's continual appearance in a variety of Black expressive practices—from music to literature to visual art—indicates its embeddedness within Black experiences, originating from acts of resisting enslavement, and reappearing at particular moments and in different forms through the ongoing struggles with chattel slavery's "durational field" (*xii*). Or, in other words, as Christina Sharpe (2016) explains, the effects of the transatlantic slave trade are not confined to the past; they ripple forward, like the "wake" of a slave ship, perpetually shaping the present and eventual future. The break is a radical expression of freedom, a shriek that breaks time and domination, a revolutionary de/reconstructing of oneself from object-commodity (enslaved person) to speaking subject. The "speech sounds" that are articulated by the break, Moten (2003) claims, "[embody] the critique of value, of private property, of the sign" (12). Musically, the break manifests as "syncopation, performance, and the *anarchic organization of phonic substance,*" which "delineate an ontological field wherein black radicalism is set to work..." (85; emphasis mine). For Moten, the break is an essential aesthetic representation of and tool for Black resistance and activism. In hip hop, the break evolved alongside the evolution of sonic recording and reproduction technologies as producers took advantage of advancements in sampling technologies.

The introduction of digital sampling technologies to the wider consumer market in the 1980s reshaped rap music, ushering in what many dub as its "Golden Age," from the mid-1980s to the early 1990s. Seizing upon the capabilities of digital sampling technologies, deejays and producers were further transformed from bystander to composer, an evolution that Houston Baker (1991) explains as the creation of "a rap DJ who became a postmodern, ritual priest of sound rather than a passive spectator..." (220). Additionally, with these technologies, producers were not beholden to the breaks that were given to them. Rather, break moments could be created by the producer from tracks that may not have originally contained them. The break, Joseph Schloss (2004) argues, became "any expanse of music that is *thought of as a break* by a producer," (emphasis his) with the break being "brought into existence retroactively..." (36). Through this conceptualization, sampling is the active construction of new music with new meanings, assembled out of the audible remnants of the past. Digital sampling allows one to create unique soundscapes, to construct sonic multiverses out of original or fabricated breaks, creating and pasting together a variety of Moten-eque "shrieks" of resistance.

Dreaming the Sonic Multiverse

Importantly, these sonic remnants are not completely devoid of their original meanings. The samples arrive at the composition with resonances of their former significations. The prior content of the sample does not fully dissipate when cut from its source; it remains as flashes of cultural memory (Assmann 2011) that are recalled through their placement in the new composition. These meanings are then re-articulated (Hall 1985) by the compositional and social context into which the sample is placed. As I will depict in more detail in the chapter's digital component, taking multiple audio snippets from drastically different genres and spatiotemporal localities imbues each a different affective resonance, creating a tension between old and new, the familiar and the novel. Through the sonic multiverse, soundscapes are formed anew by seizing one's surrounding sonic material, using it to create new visions of the past, present, and future. These materials of the past, in which we are always immersed, act not only nostalgically, to remind us of what has happened. They act as building blocks with which to imagine and construct a different world.

I contend that the sonic multiverse functions as what Robin D.G. Kelley (2002) calls a "freedom dream." In his book *Freedom Dreams: The Black Radical Imagination*, he explains that "the map to a new world is in the imagination… [in] what we see in our third eyes rather than in the desolation that surrounds us" (2). Like Bambaataa's dreams of the future, Kelley explains that we cannot build toward a better, more equal, and egalitarian society without first imagining it. While the sonic multiverse can take many shapes, I argue that it fundamentally acts as a practice in imagination, in dreaming a radically different world, though one not entirely separate from the one in which we find ourselves. As Kelley thoughtfully puts it, "the most radical art is not protest art but works that take us to another place, envision a different way of seeing, perhaps a different way of feeling" (11). The sonic multiverse as I hope to show through my analysis of "B-Boy," can be that different place, that immaterial location in which we find ourselves seeing, hearing, and feeling differently.

Acknowledgment

A portion of the research for this chapter was aided by funding from the Center for Humanities Research at George Mason University.

References

Assmann, Jan. 2011. "Communicative and Cultural Memory." *Cultural Memories*, edited by Peter Meusburger et al., vol. 4, Springer Netherlands, pp. 15–27. doi: https://doi.org/10.1007/978-90-481-8945-8_2

Baker, Houston A. 1991. "Hybridity, the Rap Race, and Pedagogy for the 1990s." *Black Music Research Journal*, vol. 11, no. 2: 217. doi: https://doi.org/10.2307/779267.

Bambaataa, Afrika. 1970. *Hip Hop Archive, #8094*. Division of Rare and Manuscript Collections, Cornell University Library.

Beastie Boys. 1989. *Paul's Boutique. Capitol Records.*

Chang, Jeff. 2005. *Can't Stop, Won't Stop: A History of the Hip-Hop Generation.* New York, NY: St. Martin's Press.

Dos Santos, Joaquim and Justin K. Thompson. 2003. *Spider-Man: Across the Spider-Verse.* Sony Pictures. 2 hr., 20 min.

Ewoodzie, Joseph C. 2017. *Break Beats in the Bronx: Rediscovering Hip-Hop's Early Years.* Chapel Hill, NC: The University of North Carolina Press.

Hall, Stuart. 1985. "Signification, Representation, Ideology: Althusser and the Post-Structuralist Debates." *Critical Studies in Mass Communication* 2 (June): 91–114.

Kelley, Robin D.G. 2002. *Freedom Dreams: The Black Radical Imagination.* Boston, MA: Beacon Press.

Moten, Fred. 2003. *In the Break: The Aesthetics of the Black Radical Tradition.* Minneapolis, MN: University of Minnesota Press.

Moten, Fred. 2017. *Black and Blur.* Durham, NC: Duke University Press.

Rose, Tricia. 1994. *Black Noise: Rap Music and Black Culture in Contemporary America.* Hanover, CT: Wesleyan University Press.

Schloss, Joseph Glenn. 2004. *Making Beats: The Art of Sample-Based Hip-Hop.* Hanover, CT: Wesleyan University Press.

Sharpe, Christina Elizabeth. 2016. *In the Wake: On Blackness and Being.* Durham, NC: Duke University Press.

12

THE SHARED MULTIVERSAL POWER OF K-POP LORE VIDEOS AND VIDEOGRAPHIC CRITICISM

Louisa Ellen Stein

Please visit www.routledge.com/9781032770116 and click on Support Material to access the digital supplement for this chapter.

My video essay focuses on K-pop lore and how it deploys and exemplifies notions of the multiverse. In form, my video essay was inspired by fans' K-pop lore videos and by the practices of videographic criticism. In this supplementary written piece, meant to accompany my video essay, I consider both practices—K-pop lore fan videos and videographic criticism—and how we can understand their power through the concept of the multiverse.

Videographic criticism and fan-created lore videos both offer interpretative journeys for the creator and viewer through their layering of different rhetorical frames. Narration may be explanatory, or it may be personal, and a video clip may simultaneously be illustrative and poetic, making this a powerful and flexible form of critical creative expression. One's experience of a lore video may be as unique and changeable a journey as one's experience of the lore itself over time.

Defining Lore

Before we dive into understanding the form of K-pop lore videos, let me briefly define K-pop lore here in relation to the multiverse; it is my hope that my video essay takes on this task more fully.

K-pop lore is an English language term for **the stories that K-pop groups tell through their creative output.** Lore manifests across music

DOI: 10.4324/9781003480846-15

videos, trailer films, videos shown in concerts, interviews, choreography, costumes, and merchandise. As I argue in my video essay, lore often uses concepts of the multiverse to tell ongoing, flexible stories. Moreover, fans follow the "real life" narratives of a K-pop group's journey alongside the group's fictional storytelling. This multilayered web of storytelling means that K-pop itself is, in a sense, infused with the spirit of the multiverse.

Multiple creators shape a given K-pop lore, including producers, songwriters, lyricists, choreographers, marketers, merchandise designers, music video directors, cinematographers, and the K-pop group members themselves. All these creators contribute to the stories and themes that make up a group's lore. We could think of lore, up to a point at least, as a form of transmedia storytelling, where creators tell a coordinated story in multiple forms and shared through multiple distribution channels. (Jenkins 2007, 2011).

But where transmedia storytelling is primarily a producer-driven storytelling practice, seeking to tell a unified and controlled story, K-pop lore is much messier and more open-ended. Intended lore can shift rapidly and radically from album to album, and the production of lore doesn't stop with the official K-pop producers, for, in turn, fans seek out elements they understand as having lore potential, unpack and interpret them, and piece them together into larger narrative and thematic understandings. Thus, in the spirit of the alternative promises of multiverse, let me now offer an alternate definition of K-pop lore: **the long-form serial stories unpacked in an ongoing conversation between K-pop producers and fans, including the conversations amongst fans themselves** as they compare and expand on various interpretations.

K-pop Lore Videos, Videographic Criticism, and the Multiverse

Fans engage with lore across a wide range of spaces and interaction. Some of these are text-based, such as extended conversations in social media comments, where people unpack lore, offer their interpretations, and debate their differing understandings of specific details or larger story frameworks.[1] But much fan authorship on lore happens in video form. For

---

1 These fan-authored pieces of K-pop lore are really a form of what fans have in the past termed *meta*: fan analyses and theories offering larger perspectives on a given media text. (Neill Hoch 2022) These fan theories can be narrative, piecing together a larger story that is only doled originally out in fragments. A comparison could be made with fan theorization around the television series like *Lost* and *Sherlock* (Amo and García-Roca 2021) But fan meta can also focus on theme, unpacking key concepts in a given media text or even in fan engagement. Historically, the most common form for this type of meta fan theorization have been written analyses and, indeed, Archive of Our Own has a hashtag with an active history for "meta" where fans post analytic essays they have written. But one could argue that fan video from its earliest days was also potentially a form of fan meta, offering theoretical interpretations or narrative cohesion in sources that did not make such narrative linear narratives obvious. (Hofman 2018, Morimoto 2016, Stein 2021).

anglophone K-pop fandom, K-pop lore gets hashed out specifically in You-Tube videos where video authors present their interpretation of a group's lore, sometimes in shorter videos and sometimes in extended multiple-hour series. Such fan-made video essays explore how concepts connect with and across albums. Via editing, these videos map out extended unfolding stories, depicting them through compressed editing, with the addition of graphics and text. These videos then birth more textual engagement as viewers comment with their own interpretations of the lore under examination. In turn, other video essayists make additional videos offering their additional perspective or building on the arguments of the first videos.

These K-pop lore videos inspired the form of my own video essay, alongside the techniques and perspectives of the scholarly movement of videographic criticism. I believe we can usefully approach both of these creative practices—fan lore videos and videographic criticism (which I discuss more fully below)—from the perspective of the multiverse. Both in different ways depend upon the acceptance of multiple coexisting media truths and experiences.

K-pop Fan Lore Videos: Form and Function

K-pop fan video essays about a group's lore deploy a variety of stylistic, formal conventions, and rhetorical structures, put to a range of purposes. Videos may:

a   **Track lore in real time:** piece together the narrative bit by bit in installments as each music video, album, etc. is released.
b   **Analyze themes:** unpack themes as they have evolved through a group's work alongside the stories they tell, through explanatory audio narration or use of text.
c   **Summarize:** compress lore after the fact to offer a more cohesive, if simplified version of a group's lore.
d   **Juxtapose:** use multiscreen editing or intercutting to compare imagery from different videos.
e   **Personalize:** use voice over or text to offer the video maker's personal perspective or journey to understanding the lore.

Whether they're tackling themes or story line, in process or after the fact, with personal counterpoint or explanatory authority, lore fan video essays suggest a cohesiveness through a particular group's body of work. Lore videos often convey a degree of detail that could feel chaotic, but through editing choices offer a sense of legibility and coherence. Some provide overt interpretation, while others appear to be straightforward compilation of the storyline, but even these we can understand as interpretation through curation. Indeed, all these lore videos exist as powerful, curated, cumulations of moments, accented by accompanying narration to create interpretations of the lore in question.

Taken in collective, lore videos offer an understanding of a given group's lore shared across the collectivity of the fandom. This shared understanding likely includes contradictory elements or differing opinions, but all exist within what fans understand as an ongoing conversation and process of uncovering lore. Within individual videos or comments, fans put forth unified visions of lore, but the overall ethos of fan conversations within and around these videos allows for multiple interpretations. The collective goal is not a singular, finite understanding, but rather a process of multiple coexisting interpretations. In this way, K-pop lore videos exemplify the multiversal logics of fan engagement.

Videographic Criticism and the Multiverse

Videographic criticism—that is, media scholarship in intentional audiovisual form—offers a solution for the long-standing problem faced by media scholars: how to express the audio-visual experience of media through words only. With videographic criticism, scholars can express their arguments more directly through sound and moving image rather than needing to translate moments of film or television into textual representation. Moreover, videographic scholars can also potentially "show" rather than "tell" their arguments, using the tools of video editing not only to depict a given moment of media but to make an argument about it (Grant, Keathley, and Mittell 2019).

As media scholars have worked within the videographic form, some have found that the very process of working with the materiality of the media itself offers new insights. Catherine Grant describes her videographic work as "a work of material thinking... (that brings) to the surface of its production... new knowledge..." (Grant 2014) The video essay can reveal unexpected truths to the scholar in its very making.

However, such truths are by nature personalized, the product of fleeting moments of meaning-making on the part of the video's creator. Grant describes the unstable but rewarding process of discovery at work in making videographic criticism as a "cultural experience of not quite knowing for sure." This not-quite-knowing is not a bug of videographic criticism but an asset—videographic criticism can capture the powerful ephemerality of media experience. Indeed, one of videographic criticism's potential strengths is its ability to depict a scholar/creator's process, capturing the journey *to* knowing (something about) a media text. Depicting a video scholar's journey can contribute to a potent and expressive argument, guiding the viewer through the creator's interpretive journey and at the same time inviting them to come to their own interpretations.

This curiosity-driven, not-quite-knowing-for-sure project of video-graphic criticism feels multiversal at heart. Videographic criticism depends upon an assumption of multiplicity—the multiplicity of possible experiences of the original media and of the video essay on the part of the creator and viewers. Viewers watching a work of videographic criticism simultaneously experience the video's meta-analysis and the primary audiovisual sources reframed within the video. Thus, videographic criticism deploys poetics not only to make analytic interpretations but also to invite viewers to become active co-interpreters.

Conclusion

It is my hope that my video essay opens up moments of K-pop lore as I experienced them and invites you to accompany me retrospectively on my journey. My video essay offers a frame of interpretation but also invites you to draw your own conclusions. Plus, arguably my video offers an additional layer of lore, the story of my own discovery of K-pop lore.

Both videographic work and fan lore videos allow viewers to come to their own interpretations. In so doing, these forms reveal the power of the open-ended and the always-thereness of the audio-image; they invite viewers to take in at once the suggested analysis and the materiality of the primary source. This doubleness can support a reading offered by the video's author, but it can also potentially invite viewers to come up with their own alternate readings. The always-thereness of the audio-image, combined with the possibilities for multiple, shifting interpretations, allows for a dynamic, multiversal power, both in individual instances and in fandom collective.

References

Amo, José Manuel de, and Anastasio García-Roca. 2021. "Mechanisms for Interpretative Cooperation: Fan Theories in Virtual Communities." *Frontiers in Psychology* 12. https://doi.org/10.3389/fpsyg.2021.699976.

Grant, Catherine. 2014. "The Remix That Knew Too Much? On *Rebecca*, Retrospectatorship and the Making of *Rites of Passage*." *The Cine-Files: A Scholarly Journal of Cinema Studies*, Fall [Video and text]. ISSN 2156-9096. Online at: http://www.thecine-files.com/grant/

Grant, Catherine, Christian Keathley and Jason Mittell. 2019. *The Videographic Essay: Criticism in Sound and Image*. Montreal: cCboose/Rutgers University Press, 2nd ed. http://videographicessay.org/works/videographic-essay/scholarship-in-sound--image?path=contents

Hofmann, Melissa A. 2018. "Johnlock Meta and Authorial Intent in Sherlock Fandom: Affirmational or Transformational." *Transformative Works and Cultures* 28: 1–8.

Jenkins, Henry. 2007. "Transmedia Storytelling 101. Confessions of an Aca-Fan." 22 Mar. http://henryjenkins.org/2007/03/transmedia_storytlling_101.html.

Jenkins, Henry. 2011. "Transmedia 202: Further Reflections. Confessions of an Aca-Fan. 1 Aug. http://henryjenkins.org/2011/08/defining_transmedia_further_re.html.

Morimoto, Lori. 2016. "hannibal: a fanvid." [in] *Transition* 3, no. 4.

Neill Hoch, Indira. 2022. "Tumblr Meta-Fandom: Reflections and Repair." *AoIR Selected Papers of Internet Research*. (March) https://doi.org/10.5210/spir.v2022i0.13062.

Stein, Louisa. 2021. "No Limits: *The Untamed Fan Video, and Affective Repetition*." *In Media Res*, Feb 25. https://mediacommons.org/imr/content/%E2%80%9Cno-limits%E2%80%9D-untamed-fan-video-and-affective-repetition

13

THE MULTIVERSE OF MADNESS

Nev Fountain

The multiverse. It's such a thing now, isn't it? Everyone's got one. They are like air fryers. Marvel's got loads now. But before Marvel made multiverses fashionable *Doctor Who* was well ahead of them. *Doctor Who*'s multiverse derived from the early fragmentation of its spin-off media. Almost as soon at the TARDIS whirled away from Totter's Lane alternative universes were thrown up almost weekly, from comic strips, where the Doctor found himself with a new pair of grandchildren, to films, where even before Patrick Troughton sprang from the floor of the TARDIS the Doctor regenerated into a twinkly old man with a mustache, a corduroy coat, and a passing resemblance to Grand Moff Tarkin.

But let's talk Target books, the novelizations of the parent universe.

Target books were the pioneers in making their own pocket universes, places that were outside time and space. From the very first book written by David Whitaker in 1964 they thumbed their nose at canon, causality, linear storytelling, and the concept of beginnings, middles, and ends, and gave us experiences that you would never come across in any other franchise.

Example One: Marvel Has Spiderman, We Had Ian Chesterton

Here's an example to show how mad being a *Doctor Who* fan was. In 1979, I was ten. I bought *Doctor Who Weekly* magazine issue 1 which still was the most exciting thing that ever happened to me in my life, and it was in the

DOI: 10.4324/9781003480846-16

pages I saw a blurry photo of four time travelers rushing inside a police box to escape some angry cavemen. In that group was captioned a man called "Ian Chesterton" played by an actor called "William Russell."

This was, I was told, the Ian Chesterton who traveled with Doctor Who, which was news to me! I had already been familiar with Ian Chesterton for several years, but this man was a stranger. In fact, this stranger was the *third* Ian Chesterton I had come across. The second Ian Chesterton I met was the happy-go-lucky Ian Chesterton in the movie *Doctor Who and the Daleks*, the chocolate-squishing buffoon who gets scared by stock footage of Roman legionnaires and went on to present Record Breakers.

But the first Ian Chesterton I encountered was in the mists on Barnes Common, in the middle of a road accident. This was the Ian Chesterton in *Doctor Who in an Exciting Adventure with the Daleks* by David Whitaker. This was the original you might say. For me. A moody man, who was already angry before he even entered the TARDIS and the world of *Doctor Who*. He was cross because he'd torn his sports jacket on a nail. He was a still a teacher, granted, but he was restless, trying to leave the profession. He was going for a job in Donneby's in Reigate! He had a landlady! He lived in *rented accommodation* for God's sake.

To me, his whole backstory was somehow seedy and desperate. This wasn't the affable TV Ian, the jaunty well-fed role model from the children's Telly series with his Coal Hill tie and his down-with-the-kids love for John Smith and the Common Men, but a dangerous feral character who had sprung fully formed from the kitchen sink dramas of the 60s, from films like *Billy Liar* and *Look Back in Anger* and *Saturday Night and Sunday Morning*.

As I learned more about the prime universe Ian Chesterton, I could not reconcile him with this snippy, testy man who seemed to make Barbara burst into tears every five pages. I could not see him ever being played by William Russell. I saw this character being played by Albert Finney or Ralph Bates, or actually Ray Brooks. Why not Ray Brooks? The scratchy, squiggly illustration at the end of the book showed Ian with a huge shaggy head of hair, almost a quiff, that looked far more like Ray Brooks in *Kathy Come Home*, than a William Russell or Roy Castle.

And yes, let's talk about the end of that book. I read it at seven years old and I still remember the final line vividly. Doctor Who had given Ian and Barbara a choice; to stay on Skaro and rebuild a planet, or continue the mad crazy adventures with him. The final line was Ian's. He, of course, spoke for both him AND Barbara without consulting her, which was very like Richard Burton in *Look Back in Anger*.

He said to the Doctor: "we stay with you."

We stay with you.

Which was an incredible relief to me. I was relieved, and it seems silly now, with all that I know, I still feel relieved when I read those words, because of all the Ian Chestertons I have met, this one seemed ready to make that decision to stay. He seemed the one most likely to refuse to cling to the Doctor's skirts in the hope of making it back to London 1965. This one seemed capable of saying "sod you all, I've only got a landlady and a torn sports coat and a lousy teaching job I hate to go back to, I'm staying with the sexy blond people."

And every time I read it, I do feel relieved.

Because in my mind there might have been a part of the *Doctor Who* multiverse where he wouldn't have gone back to the TARDIS. Maybe he would have stayed on Skaro, and even persuaded Barbara and Susan to stay too. Perhaps the Doctor would have left Skaro, crazed with grief and loneliness, gone back to Earth, and made two android companions that he programmed to be his grandchildren, called John and Gillian? Who can say?

Even if this Ian Chesterton stayed in the TARDIS, I don't think this incarnation would have the same kind of adventures. They would be dirtier, nastier. *The Edge of Destruction* would be like the *Lord of the Flies*. *The Romans* would have had a lot more crucifixions. Women would have got hysterical and got slapped. And when Ian went back home, he would not have gone back to London 1965. He would be put back on Barnes Common. London 1977. He would save Marc Bolan from his car accident, and Marc Bolan would convince David Bowie to give up smoking and everything would be fine.

As you can see, Ian Chesterton is exactly like Spiderman. With Marvel, the Spiderverse has a grumpy Spiderman, a goofy naïve Spiderman, and a rather heroic Spiderman; the Chesterverse has exactly the same thing. I think it was a bit of a missed opportunity not to have William Russell, Roy Castle, and Ray Brooks all teaming up to defeat Old Mother.

Example Two: Marvel Has Agent Phil Coulson, We Had Captain Hawkins

If you don't know who agent Phil Coulson was, he was Nick Fury's right-hand man in the Avengers movie and a popular character. He died in "Avengers Assemble" and his death was the main reason the Avengers put their differences together to defeat Loki. His death had real significance in the story, so of course they brought him back. I don't know how they brought him back, and quite frankly I don't care. I assume it's something to do with the multiverse.

So let me tell you about Captain Hawkins, or in the Target multiverse, *Sergeant Hawkins*. For those of you not aware of his work, Hawkins was a soldier that helped the Brigadier fight reptile men in some caves in Dartmoor. He was dashing, practical, clever, loyal, and was basically all the best bits of Sergeant Benton and Captain Yates put together in the devastatingly handsome package of an in-his-prime Paul Darrow. Paul Darrow played the character of Captain Hawkins with beautiful understatement. Yes, you heard me. Beautiful Understatement and Paul Darrow. They are going in the same sentence, and you can't stop me.

For seven weeks, Brigadier Lethbridge-Stewart and Captain Sam Hawkins *were* the UNIT family. Hawkins was Agent Coulson to the Brigadier's Nick Fury (obviously, the Brigadier didn't have his eye patch quite yet, but that would come later). And then, in the last episode, he was killed abruptly and needlessly by the third eye of an angry reptile. It was so abrupt, it was a blink-and-you-miss-it death. And of course, with most *Doctor Who* fans of my age, I read the book first which had him survive, and then when I saw the Telly version that killed him, it was quite a shock to see him offed in such a casual way.

Thanks heavens for the Target book multiverse. When Malcolm Hulke novelized *Doctor Who and the Silurians* and it became the *Doctor Who and the Cave Monsters*, he obviously liked Captain Hawkins, or as he made him in the book Sergeant Hawkins. He was just as good as on the TV, if not even better. The Malcom Hulkaverse always found time to flesh out the characters, and Sergeant Hawkins absorbed storylines from other soldier characters, so he had even more of a presence in the story.

Best of all, this time, he did not die. He survived to the end. I spent decades after reading this book wondering why Sergeant Hawkins didn't return in later stories, until the release of the *Silurians* video cassette and then, sadly, I found out why.

As of this moment Sam Hawkins is alive in the Target multiverse. Where is he now? Did he end up scrawling cave paintings across the country like a Silurian-addled Banksy? Did he get seconded to another secret division of UNIT that also happens to be an anagram of *Doctor Who*? Did he lead "Cohort Dow" in daring raids on Silurian bases in Australia and Geneva and France?

Example Three: The DC Universe Had Batman, We Had Jo Grant

The DC universe loves doing Batman origin stories. Since 1989 they have done it four times and that's just the big live-action movies. But have they done two different origin stories in two years? That's what happened to Jo

Grant. In 1974 Jo arrived in the Target multiverse in *Doctor Who and the Doomsday Weapon*, a girl who came top of her class in spy school, who got into UNIT through her very powerful uncle, but got palmed off to the Doctor by the Brigadier. She demanded to be given something to do and got more than she bargained for. In this story, she is immediately plunged into danger and comes face to face for the first time with the mysterious evil Time Lord known as the Master.

In May 1975 they rebooted the character. In *Terror of the Autons*, she is a largely unqualified young woman who was picked by the Brigadier to annoy the Doctor so the Doctor could take his mind off Liz Shaw. But in this story, she is immediately plunged into danger and comes face to face for the first time with the mysterious evil Time Lord known as the Master.

Like Batman reboots, it does not really change Jo's character, but like Batman the origin is updated so it can fit the story. In *Autons*, she is a new companion trying to prove her worth. In *Doomsday Weapon* she is our identification figure, plunged into the unknown and trying to comprehend a strange alien planet.

In many ways I would argue Malcom Hulke's flexible treatment of Jo Grant's origin story—and indeed David Whitaker's flexible treatment of Ian Chesterton's origin story—is proved to be way ahead of its time. It's a precursor to another companion who comes much later, in 2012. Clara Oswald is a companion who nearly joins the 11th Doctor as a survivor of a starliner crash on a Dalek planet, then she nearly joins the 11th Doctor as a Victorian governess, and then she really joins the 11th Doctor as a twenty-first-century nanny and teacher.

Showrunner Steven Moffat created Clara, and we all know he was, and is, an avid Target book reader. There's that famous photo of Steven Moffat as a wee kiddie reading David Whitaker's *The Daleks*. Perhaps he's thinking "hum, multiple alternate introductions for a *Doctor Who* companion? Sounds rather fun!"

Example Four: Marvel Comics Has Their *What If* Series, *Doctor Who* Has Peri Brown

One of the newer introductions to the Marvel multiverse is a series of animated stories providing their characters with multiple different fates. This means they can kill characters, turn them evil, do anything they like to the poor bastards, and it doesn't matter.

How very unoriginal. Have they never heard of Perpugilliam Brown? A character who has had six fates and counting. Her two main television fates were a shock-surprise death and becoming the unwilling royalty of a backward civilization, both fates shared by that other female icon of the 80s, Princess Diana.

Even after John Nathan-Turner changed his mind mid-season and saved her from death by giving her a fate worse than death, the *Doctor Who* multiverse has worked tirelessly to fix her fate, I myself have been part of that huge project taking my trowel and adding a few bricks to that edifice, but it wouldn't be a messy bit of canon without the Target books having their say.

In *Mindwarp* writer Philip Martin decided that Peri's ultimate fate would be back on twentieth-century Earth where she would act as a manager for all-in wrestling champ Ycarnos, a fate that is as bizarre as it is fitting. And as the writer was the same guy who wrote the original script for *Mindwarp*, we have to accept it as canon, don't we?

There is something thrillingly subversive about original writer's mucking about with their own work within the Target universe, thumbing their noses at us. Philip Martin was one of the last Target writers to do this, and he joined a dishonorable tradition of Whitaker, Hulke, Dicks, Fisher, Saward, Hayles, Aaronovitch, and many others. We can only imagine what would have happened if Douglas Adams ever got his teeth into the *Pirate Planet* or *City of Death*. I guess we would have had to brace ourselves for a ride bumpier than the one experienced by Scaroth of the Jagaroth.

So there you have it, this is me talking about the multiverse to you, a bunch of avid Target readers. Or if you wait for the novelization, this is me not talking to you at all, or, me as the Antonine Killer, firing at a slave ship, or maybe I'm Linx, a microsecond from oblivion?

And as Target books are now continuing, I am sure there will be more opportunities for multiverse mischief, in books and all spin-off media Perhaps instead of David Tennant regenerating into Ncuti Gatwa he splits off into his own continuing regeneration, grows his own TARDIS and has his own adventures? Perhaps ALL the old Doctors carry beyond their regenerations in their own little TARDISs with their old companions?

Too much?

14

YOUNG FOREVER(?)

Looping Time and Fractured Futures in BTS's Bangtan Universe

Lauren R. O'Connor

April 11, Year 22. A known month and season, a slanted acknowledgment of age. This is the day Kim Seokjin returns again and again in the extended storytelling universe of Korean music superstars Bangtan Sonyeondan, also known as BTS. In addition to chart-topping albums, sold-out stadium tours, and visits to heads of state as official cultural ambassadors, BTS and their label, BigHit Entertainment (now HYBE), have crafted a sprawling multi-media narrative in which multiversality is a central theme. Kim Seokjin's infinite futures represent the dreams and possibilities of youth, though the story ultimately illustrates the bounded reality faced by modern adolescents.

The so-called Bangtan Universe (BU) is an alternate universe in and of itself, as its main characters are adolescent versions of the seven members of BTS: Kim Namjoon, Kim Seokjin, Min Yoongi, Jung Hoseok, Park Jimin, Kim Taehyung, and Jeon Jungkook.[1] The BU is also an example of what Henry Jenkins (2006) has termed "transmedia storytelling," or storytelling which "unfolds across multiple media platforms, with each new text making distinct and valuable contributions to the whole" (97–98). "Official" BU content exists in album art and jacket notes, music videos,

---

1 Throughout this essay, I will use the traditional Korean naming convention of family name first, given name second, for Korean people and fictional characters. The exception is for scholars whose work has been published in English with a western naming convention (i.e., Dal Yong Jin, whose family name is "Jin").

DOI: 10.4324/9781003480846-17

short films, prose novellas, and web comics; unofficial content abounds via fanfiction and fan art. However, the narrative itself is primarily and most concisely told via the SAVE ME webtoon.

The SAVE ME webtoon's main conceit—Seokjin's looping back in time to April 11, Year 22—references the idealistic notion that adolescent futures function much like multiverse stories. A young person theoretically has plenty of time to become whatever it is they would like to be, and they can imagine different versions of themselves in the future unencumbered by the weight of their own past. Yet, the story of the BU serves as a cautionary tale, illuminating how these societal beliefs about youth are often at odds with the actual opportunities available to them. In turn, the story fits neatly within BTS's broader critiques of wealth inequality, social stratification, and other issues preventing young people from shaping their own futures.

This essay will discuss both the form and content of the SAVE ME webtoon, in addition to referencing related works by BTS. I contend that the SAVE ME webtoon extends BTS's other artistic declarations about the predicament in which real-world adolescents—South Korean, American, and otherwise—find themselves in the early twenty-first century: they are the beneficiaries of theoretically infinite but materially inaccessible futures.

The Most Beautiful Moment Online

The BU was first formed within two of BTS's breakout hit album series, 2015's *The Most Beautiful Moment in Life* (abbreviated *HYYH*[2]) and their 2016 follow-up, *Wings*. Both series are concerned with the process of maturation, featuring song titles like "Begin," "Moving On," "First Love," and "Epilogue: Young Forever." *Wings* also engages heavily with the 1919 coming-of-age novel *Demian*, by Hermann Hesse (n.d.). Both albums were accompanied by a host of additional content, including extended music videos, non-music video short films, graphic lyric books, and more (see Stein's chapter in this volume). Although some of this content is more optimistic in tone, such as the "화양연화 on stage: prologue" film, much of it positions youth as a site of violence—sometimes on the part of the youth, but more often inflicted upon them. The BU is not a kind one; its adolescent protagonists suffer indignities as they attempt to navigate maturation.

---

2 *The Most Beautiful Moment in Life* is the English translation of the album's actual title, 花樣年華. In Korean, this is pronounced 화양연화, which is romanized as *hwa yang yeon hwa*. This romanization of the Korean pronunciation of the Chinese characters is where the common abbreviation *HYYH* comes from.

The SAVE ME webtoon was originally released in installments in early 2019, serving as a follow-up to the raft of content that built the universe during the *HYYH* and *Wings* eras. It is credited simply to BigHit Entertainment and LICO, a digital content production studio that is a subsidiary of the Korean corporation Naver. The word "webtoon" (a portmanteau of "web" and "cartoon") is both a specific brand and a more general term. The official "Webtoon" app, owned and operated by Naver, hosts the SAVE ME webtoon and thousands of others. (For the purposes of this essay, capital-W Webtoon will refer to Naver's site/app, while lower-case-w webtoon will refer to the form itself.)

As Brian Yecies and Ae-Gyung Shim (2021) write, the webtoon is a relatively new medium for the delivery of narrative-driven popular culture. Yecies and Shim invoke Jenkins as they situate the "webtooniverse" as "a revolutionary content-delivery method" and argue that webtoons "represent a convergence culture" (4–5). Although webtoons resemble a traditional comic in their blending of images and text, with words usually presented in bubbles to indicate a person's speech or thought, the webtoon reading experience deviates significantly from that of comic books. Comic books, both print and digital, often divide their pages into panels. There may or may not be strips of empty space, what Scott McCloud (1993) termed the "gutter," in between the panels. The books themselves are, of course, also divided into pages. Digital versions of print comics will often require a click or pressing a hotkey to move from one page to the next, and they are read horizontally, then vertically.

Webtoons, however, are meant to be read via continuous scrolling motion—they are meant to be read on a mobile phone (Jin 2023, 6). When a reader first opens a new Webtoon, a widget instructs the reader how to navigate the story: a little gray hand with index finger extended moves gently up the screen. Typically, stories are divided into episodes, which are akin to comic book issues. The SAVE ME webtoon, for instance, has 15 episodes and a prologue. Each episode begins at the "top" and the story continues vertically. Images are occasionally divided into panels with "gutters," but the size, shape, and color of these empty spaces vary widely; more often than in traditional comics, there is no gutter and images simply blend into one another. Narrative moments appear in image frames of different lengths, sometimes requiring the reader to shift the orientation of their phone to see the full scene.

The overall effect of reading a webtoon is thus smoother and more continuous than reading a comic, which has clear pauses and hard stops built into the medium; the SAVE ME webtoon's time loop narrative is therefore reinforced by the physical act and experience of reading a webtoon, a textbook example of Jenkins's claim that "[in] the ideal form of transmedia storytelling, each medium does what it does best" (98). Various media and technologies help to construct the BU, and the decision to place the time-looping piece of the narrative in a webtoon format creates a seamless marriage of form and function. The consistency of scrolling and the sensation of being brought "back" to the "top" of the story—where it began—with each new chapter simulates Seokjin's reversal of time. When Seokjin loops back to April 11, Year 22, the image readers see is that of a shattering pane of glass or mirror, as though the universe in which the webtoon is contained (the glass-screened phone) has been destroyed.

Crucially, the webtoon format also points to the primary intended audience for the BU. As any good ARMY (the nickname of BTS's fandom) will tell you, we are more than a group of screaming teenagers,[3] but webtoons are in fact largely targeted to and consumed by a young audience in Korea and abroad. In 2021, Naver Webtoon reported "over 72 million monthly active users (14 million in the US)" (Salkowitz 2021). Of those readers, "GenZ [sic] and younger Millennials make up 75% of Webtoon users worldwide, and 70% of the users in the US are under age 24" (Salkowitz 2021). A few years earlier, Korea's Yonhap News (2017; in Korean, cited in English by Jin 2023, 2) reported that the plurality of webtoon readers were in their teens (32.1%). This is not to say that the SAVE ME webtoon was exclusively read by young people, but simply to highlight that the webtoon's themes of recursive time and multiversality speak to a youthful audience populated by many who are, like the story's main characters, facing long and uncertain futures.

Boys Meet Evil

The BU is populated by seven main characters: Kim Seokjin, Min Yoongi, Jung Hoseok, Kim Namjoon, Park Jimin, Kim Taehyung, and Jeon Jungkook. Throughout the story, these teenage characters navigate grief, trauma, substance abuse, physical abuse, and suicidal ideation. The narrative that unfolds in the webtoon relies on the concept of overlapping existences: BTS' BU specifically mobilizes the idea of multiverse/time loops to explore the complicated nature of growing up.

The SAVE ME webtoon prologue introduces us to Seokjin and his friends, all of whom meet at school when they are punished for being late.

---

3 The author means to acknowledge the diversity of ARMY fandom, but would also like to go on record that they support and admire screaming teenagers. As David Holmes (2020) writes of BTS, "they are adored by screaming teenagers and we live in a society patriarchal enough to forget that screaming teenagers are nearly always right." No screaming teenagers were slandered in the making of this essay!

They are instructed to clean up an abandoned storeroom together, and the storeroom becomes their hangout, where Jungkook listens to Yoongi play the piano and Jimin and Hoseok dance to the music, while Taehyung and Namjoon laugh along. They enjoy many lovely days together, including a special trip to the beach. All of this is told in flashback, as the webtoon begins with Seokjin's return from studying abroad; he hasn't seen his friends in two years.

On April 11, Year 22, when Seokjin returns, he glimpses Jungkook walking to school and Namjoon working at a gas station. He does not speak to either of them, but on May 22, Year 22, he wakes up after having dreamed about all six of his friends. Searching for his friends leads him to a jail, where Namjoon is being held on assault charges. Seokjin learns from Namjoon that Jungkook and Yoongi are dead, and Hoseok is in a long-term hospitalization. Namjoon has not heard from the other two, but on his way out, Seokjin sees Taehyung being brought into the same jail for murdering his father. Later, readers learn that Jimin has been committed by his parents in a long-term stay at the same hospital as Hoseok.

Seokjin goes back to the beach where the seven friends once had their beautiful day and wonders how things might have been different: "If … if I had talked to Namjoon at the gas station that night, maybe we wouldn't have met in a jail." He looks at the sea and remembers, "There was a time when the seven of us were happy together, knowing we had each other … Where did it all go wrong?" A mysterious white cat appears and asks, "If you could turn back time, do you believe you can straighten out the errors and mistakes, and … save everyone?" (BigHit Entertainment and LICO 2019, Episode 1). Seokjin awakes the next morning feeling as if he had a strange dream, but in fact he has returned to April 11, Year 22, when the boys are separated but still alive and free. Over the 15 issues of the webtoon, Seokjin tries again and again to save them, and each time he fails he is sent back to April 11 to try once more.

Many media scholars have considered multiverse stories in which the narrative hinges on a specific moment in time. Allen Cameron (2008), in his exploration of modular narrative in film, borrows from Jorge Luis Borges in calling these stories "forking path narratives." He writes, "Forking-path narratives juxtapose alternative versions of a story, showing the possible outcomes that might result from small changes in a single event or group of events" (10). Karen Hellekson (2000), writing about alternative histories, calls such narratives "nexus stories," or "stories [that] occur at the moment of the break" (5). In the case of the SAVE ME webtoon, April 11, Year 22, is that break.

Such stories allow for both the time-traveling character(s) and the reader to see some of the potential differences that a small change in the past can make. For example, in one of Seokjin's early loops he is able to save Namjoon from being arrested and prevent Jungkook from ending his own life, only for all of them to get into a fatal car accident. Seokjin's continued return to April 11, Year 22, allows him to visit and shape multiple versions of his and his friends' futures. In episode six of the SAVE ME webtoon, Seojkin thinks, "The future keeps changing … because something from the past changed. It was me. I jumped and changed their fate. It was all because of me. I did this" (BigHit Entertainment and LICO 2019, Episode 6). Seokjin realizes that he is creating different existences—different universes— with each of his "jumps." This multiversality drives the plot of the SAVE ME webtoon, while also serving as a metaphor for the notion of malleable adolescent futures.

Twenty-First Century Youth

Recalling that the "home" of the BU is BTS's extended meditation on youth, adolescence emerges as a major theme in the SAVE ME webtoon. The story begins with the boys' school years, visually coded by school uniforms, desks, bookbags, and teachers. Though most of them have graduated or dropped out by April 11, Year 22, Jungkook, who is the youngest, remains a student. Some of them, including Jungkook, Taehyung, and Jimin, still live with or are under the legal guardianship of their parents. Hoseok works a typical teen gig at a burger joint, and Taehyung's preferred leisure activity is tagging in alleys. In a portion of the story told in the novella 花樣年華 The Notes 1 (BigHit Entertainment 2019), the boys visit the remnants of a supposedly magical rock on the beach that makes your dreams come true (32).[4] Their youth is visible and vital, and the impermanent nature of their jobs, homes, and relationships contributes to a sense of unformed-ness.

Historian Nancy Lesko (2012) notes that the idea of adolescence was defined and refined in the early twentieth century as a time of "becoming" (18). The framing of adolescents as "coming of age" or approaching some kind of threshold imbues them—their bodies, their social identities, their worldviews—with a sense of possibility. In this sense, their futures are their own tiny multiverses. At the same time, these possibilities are prescribed by social and political expectations. Lesko writes that descriptives like "'coming of age' and 'at the threshold' are also homiletic. These terms appear to give

---

4 The rock itself is destroyed, having been ground to gravel in service of a new luxury condo development. Though it is outside the scope of this essay, gentrification and the misery of late-stage capitalism is another recurring theme in the Bangtan Universe and BTS' larger artistic project. Interested parties are directed to songs like "Silver Spoon," "Am I Wrong," and "Paradise."

adolescence importance but really confer greater authority on the author of the homily" (2–3). Likewise, Marcel Danesi (1994) writes that "[the] social empowerment of the modern adolescent has been brought about by the world of adults" (5). In other words, it is adults and their institutions that are responsible for defining adolescence in the cultural consciousness. Lesko and Danesi's histories of the (western) concept of adolescence argue it is largely socially constructed, pointing to political, economic, and cultural factors that made adolescence "a place that people could endlessly worry about, a space that adults everywhere could watch carefully and that could be imagined to have many visible and invisible instabilities" (Lesko, 5). Danesi rather cheekily notes that "we really have no one to blame but ourselves" (5) for such concerns, acknowledging that modern adolescence has been, to some extent, designed to drive such worried attention.

The popularity of South Korean artistic output including television drama, webtoons, and (perhaps most tangibly) K-pop among young Americans reveals a shared sense of self and relationality between adolescents in both cultures. Although understandings of adolescence vary across cultures, adolescence is a similar source of scrutiny in South Korea, prompting research into how adolescence manifests alongside social convention, educational attainment, and personal development. Song Hana (2017) parallels Nancy Lesko's characterization of young people and their "becoming" in the *Korean Journal of Child Studies*, nothing that "each child has a great potential. Young children obtain lots of opportunity to improve various competence for a long time, and have a better life later on" (2). Midgette et al. (2016) have examined whether upbringing in Korean society, with its stronger emphasis on tradition, might alter the experience of adolescence described in the west. They note some differences in behavior along the axis of social convention, but they ultimately find that "an emphasis or greater value placed on conventions and traditions does not necessarily lead children and youth to develop a deeper structural understanding of the purpose of and affirmation of conventions." Instead, the adolescent experience of social integration closely mirrors that of youth in America. While the understanding of adolescence in South Korea and America are not identical, there is enough theoretical and practical overlap to characterize teenagers in both societies as experiencing a time of "becoming."

Despite similarities in societal understandings of adolescence, the timbre of this time of "becoming" is shaped by cultural convention and economic realities. In particular, the South Korean model of education and employment narrows the options of what a young person can become and how. Such social realities impose limitations on the multiverses of adolescents' futures; resistance to these realities and their limiting nature dominates the SAVE ME webtoon as well as BTS' other early musical releases.

Dream No More

Prior to the release of *HYYH*, the primary themes of Bangtan Sonyeon-dan's work vacillated between youthful romance and harsh criticism of the South Korean education system. Love songs like "Just One Day" and "Like" live alongside astute condemnations such as "No More Dream" and "N.O." The latter served as the title track for BTS's 2013 extended play, *O!RUL82?*, and it features the lyrics *"A nice house, a nice car, could such things really bring happiness? / In Seoul, to the SKY, would the parents really be happy?"*[5] "In Seoul" refers to attending university in Seoul, where most of South Korea's highly selective institutions are. "SKY" refers to three of the most prestigious universities in South Korea: Seoul National University, Korea University, and Yonsei University. BTS continue,

> 어른들은 내게 말하지
> *(The adults tell me)*
> 힘든 건 지금뿐이라고
> *(that the suffering lasts only momentarily now)*
> 조금 더 참으라고
> *(that I should endure it a little more)*
> 나중에 하라고
> *(that I should do what I want to do later)*
> Everybody say NO!
> 더는 나중이란 말로 안 돼
> *(The word "later" can't do anything anymore)*
> 더는 남의 꿈에 갇혀 살지 마
> *(Don't live your life by being trapped in someone else's dream)*
> *(Doolset Bangtan 2018)*

Film philosopher and BTS scholar Jiyoung Lee (2019) establishes a further connection between BTS' work and the society in which they emerged:

> Many young people around the world are on the same boat, coping with the uncertainty and instability of their futures … BTS' music can be seen as a medium that allows for the agonies and desires of young people around the world to resonate.
>
> *(17)*

---

5 Italicized lyrics are translated from Korean to English. Lyric translations provided by Doolset Bangtan at doolsetbangtan.wordpress.com.

The combined effect of songs like "N.O," "No More Dream," and "Intro: O!RUL8,2?" is a thoughtful criticism of a system in which young people are not given the opportunity to imagine their own futures—a singular, idealized path is already laid out for them, and deviation from that path is heavily discouraged even as the path itself is exceedingly narrow.

While educational attainment is valued in Western cultures, the college entrance admissions process in South Korea demands significantly more of children and teens. It is not uncommon for youth in South Korea to attend a regular school day, followed by a short break, and then three to four additional hours of schooling in the evening. As Patrick Hultberg et al. (2017) note, "Koreans spend a large share of their income on for-profit private tutoring academies called hagwon (학원)" (4). Lee Soojeong (2018) argues that many South Korean families participate "an infinite competition among them desiring for high status by being admitted to prestigious colleges" (126). Hultberg et al. invoke SKY universities when noting that "[in] South Korea, social hierarchy and status are paramount, and a degree from one of the top three elite tertiary institutions … provides an invaluable lifetime network and almost guarantees employment" in a stable field (3). However, entrance to these institutions remains incredibly competitive, and even upwards of 12 hours of school a day may not be enough to secure one of the limited spots in a new class. "In short, being a student in Korea, as in many other parts of the world, is expensive and private return on investment in some cases is minimal or equivalent to zero" (Hultberg et al. 2018, 14). As Jiyoung Lee writes, BTS are hyper-critical of this "survival of the fittest" mindset, "an idea enforced by older generations—an idea that restricts youth's ability to live, fly, decide, and dream, only to teach them how to live by established standards" (36). BTS identify this system as materially unsustainable, and, perhaps more importantly, argue it stifles imagination and robs young people of their futures.

Cultural conceptions of adolescence as a time of "becoming" mark it as separate from adulthood. In this formulation, reaching adulthood implies one ostensibly already "became" what one was going to be; adulthood is fixed in a way that adolescence is not. The present progressive nature of "becoming," coupled with more colloquial characterizations of youth as time when "the future is wide open" or one "can be anything [they] want to be," frames adolescence as a time when the future is particularly malleable. But what does this really mean in the lives of teenagers who feel pressured—culturally, economically, personally—to pursue a limited (and limiting) vision of success? The SAVE ME webtoon reveals this tension in the stories of teenagers whose malleable futures freeze rapidly into brittle presents, only to be shattered and started over again.

"If We're Together, We Can Smile"

So what becomes of adolescents who are squeezed out of their own futures, whose multiversal dreams are blunted by material realities? What do young readers of the SAVE ME webtoon have to look forward to? The intersection of adolescence, time travel, and multiversality within the SAVE ME webtoon tugs at threads trailing off the notion of youthful becoming, particularly the limited kind of becoming BTS inveighed against in "N.O."

Following their early work, *HYYH* takes a softer approach in its exploration of youth, but BTS have yet to abandon their skillful socioeconomic critique. The BU subtly de-natures and de-romanticizes growing up by putting youth and maturation into a political context. Indeed, the primary conflict driving the BU is how to navigate personal growth and relationships alongside external forces like poverty, trauma, and oppression. For example, Hoseok has lived most of his life in an orphanage perpetually under threat of having its funding cut. Namjoon has been the primary breadwinner of his family since he was in middle school, because his father is chronically ill and unable to work or access proper healthcare (花樣年華 *The Notes 1* 2019).

Seokjin's attempts to save his friends throughout the SAVE ME webtoon consistently fall short, because he is not merely engaging with isolated instances, but systematic issues that make accessing the singular, idealized path to maturity impossible. His time loops make this difficulty of a straightforward path to maturation legible, if in a fantastic fashion. The physical nature of a loop similarly resists the notion of linear, unidirectional growth. The characters of the BU cannot progress along the "proper" path to maturation, thrown off by economic pressures, mental health concerns, or a combination of traumas. Seokjin's looping resists cultural scripts that demand teenagers grow up a certain way, scripts that encourage competition with each other, and unidirectional focus on a specific set of achievements. Instead, as the SAVE ME webtoon unfolds, readers are treated to an alternative perspective rooted in connection and lateral growth.

Seokjin's failure to save his friends individually leads him, finally, to the realization that they must save each other. In episode 13, Seokjin thinks to himself, "Long after it became meaningless counting how many times I lived through April 11, the impossible happened" (BigHit Entertainment and LICO 2019, Episode 13). Taehyung shows up at his door first thing in the morning on April 11, and with his help, the seven friends reunite for a party at the storage container where Namjoon lives (illegally). "When I think I can't possibly fall any lower, you pull me back on my feet," Seokjin realizes. Although this loop similarly comes to ruin, the story ends with Seokjin remembering the white cat's final warning: "You won't make it out here alone, in this entangled destiny." He thinks to himself, "… no matter how many times I tried, I could never save all of them on my own," and looking at a picture of the seven of them he realizes how badly they need each other. The last words of the webtoon are Seokjin's thoughts, "Once again, just like before, together with you." The boys may be incapable of growing up, but they find vitality and purpose in growing out, toward each other.

Epilogue: Young Forever?

Whether or not Seokjin and his friends succeed in a collective effort to reimagine their futures is left ambiguous in both the webtoon and the notes. There is no simple ending here, merely the suggestion that perhaps through community and communal resistance, adolescents may create a version of their futures in which they can find moments of happiness. The "Euphoria: Theme of LOVE YOURSELF 起 Wonder" short film, released in 2018 as part of BTS's *Love Yourself* album series, implies at least a somewhat hopeful resolution for the characters of the BU.

Seokjin's ability to reverse time and create new futures aligns squarely with the understanding of adolescence as a time when the future is a multiverse of opportunity. However, his failure to keep all his friends alive in each universe cuts the notion of adolescence as a time of "becoming" down to the root. In the BU, as well as our own, strict educational institutions, increasing surveillance, and economic precarity tighten the loops, until teenagers are not able to envision becoming anything except that which can fit through.

Yet the Bangtan Universe remains an ongoing saga, alive and well in multiple media formats and woven into countless other stories, artistic works, and more via fan engagement. At the time of this writing, the BU is being adapted into a 12-episode televisual drama called "Youth" (Kim Soo Jin and Choi Woo Joo 2024). The founder and CEO of HYBE, Bang Sihyuk, recently alluded to a revival of the themes of *HYYH* upon BTS's return from mandatory military service in 2025, to celebrate the ten-year anniversary of the album's initial release (Bang Sihyuk 2023). It remains to be seen what BTS's real-world loop back in time to revisit *HYYH* will look like, but such an artistic direction evinces the resonance of the SAVE ME webtoon and its efforts to reclaim and make possible the multiversality of adolescent futures.

References

Bang, Sihyuk. 2023. "HYBE's Bang Si-Hyuk on Making Music for the Masses." Interviewed by So Hee Kim and Lucas Shaw. *Bloomberg*. October 12. https://www.bloomberg.com/news/videos/2023-10-12/hybe-s-bang-si-hyuk-on-making-music-for-the-masses-video

BigHit Entertainment. 2019. 花樣年華 *The Notes 1*. Seoul, South Korea: BigHit Entertainment.

BigHit Entertainment and LICO. 2019. SAVE ME. *Naver Webtoon*. https://www.webtoons.com/en/drama/bts-save-me/list?title_no=1514

BTS. 2013. *O!RUL8,2?* BigHit Entertainment, compact disc.

BTS. 2015a. 花樣年華 *Part 1*. Bighit Entertainment, compact disc.

BTS. 2015b. 花樣年華 *Part 2*. Bighit Entertainment, compact disc.

BTS. 2016. *Wings*. BigHit Entertainment, compact disc.

Cameron, Allen. 2008. *Modular Narratives in Contemporary Cinema*. New York: Palgrave MacMillan.

Danesi, Marcel. 1994. *Cool: The Signs and Meanings of Adolescence*. Toronto: University of Toronto Press.

Doolset Bangtan. 2018. Doolset Lyrics N.O. Wordpress. https://doolsetbangtan.wordpress.com/2018/12/15/n-o/

Hellekson, Karen. 2000. *The Alternate History: Refiguring Historical Time*. Kent: Kent State University Press.

Hesse, Hermann. n.d. *Demian: The Story of a Youth*. Penguin Classics Edition. Translated by Damion Searls. New York: Penguin Random House.

Holmes, David. 2020. "The Boundless Optimism of BTS." *Esquire*. November 22. https://www.esquire.com/entertainment/music/a34654383/bts-members-be-album-interview-2020/

Hultberg, Patrick, David Santandreu Calonge, and Seong-Hee Kim. 2017. "Educational Policy in South Korea: A Contemporary Model of Human Capital Accumulation?" *Cogent Economics & Finance* 5 no. 1: 1–16. https://doi.org/10.1080/23322039.2017.1389804

HYBE Labels. 2015. 화양연화 "On Stage: Prologue." *YouTube*. October 1. https://www.youtube.com/watch?v=Bt8648TNX1M&rco=1

HYBE Labels. 2018. "Euphoria: Theme of LOVE YOURSELF 起 Wonder." April 5. *YouTube*. https://www.youtube.com/watch?v=kX0vO4vlJuU&t=119s

Jenkins, Henry. 2006. *Convergence Culture: Where Old and New Media Collide*. New York: New York University Press.

Jin, Dal Yong. 2023. *Understanding Korean Webtoon Culture: Transmedia Storytelling, Digital Platforms, and Genres*. Cambridge: Harvard University Press.

Kim, Soo Jin and Choi Woo Joo (writers). 2024. *Youth*. Directed by Kim Jae Hong. Pre-production. https://mydramalist.com/62573-blue-sky. Accessed January 4, 2024.

Lee, Jiyoung. 2019. *BTS: Art Revolution*. Translated by Stella Kim, Myungji Chae, Chloe Jiye Won, and Shinwoo Lee. Melbourne: Parrhesia.

Lee, Soojeong. 2018. "Relationship among Education Fever, the College-admission Policy, and Shadow Education in South Korean." *KEDI Journal of Education Policy* 15 no. 2: 121–138. https://doi.org/10.22804/kjep.2018.15.2.007

Lesko, Nancy. 2012. *Act your Age!: A Cultural Construction of Adolescence* 2nd Edition. London: Routledge.

McCloud, Scott. 1993. *Understanding Comics: The Invisible Art*. New York: William Morrow.

Midgette, A., J. Y. Noh, I. J. Lee, and L. Nucci. 2016. "The Development of Korean Children's and Adolescents' Concepts of Social Convention." *Journal of Cross-Cultural Psychology* 47 no. 7: 918–928. https://doi.org/10.1177/0022022116655775

Salkowitz, Rob. 2021. "Webtoon CEO Sees Massive Growth and New Opportunities in U.S. Market." *Forbes*. November 2. https://www.forbes.com/sites/robsalkowitz/2021/11/02/webtoon-ceo-sees-massive-growth-and-new-opportunities-in-us-market/?sh=2adc1f3e7707 accessed 2loop/5/2024

Song, Hana. 2016. "Parental Influence on Children's Developmental Outcomes in Middle Childhood and Adolescence." *Korean Journal of Childhood Studies* 37 no. 5: 1–3. https://doi.org/10.5723/kjcs.2016.37.5.1

Yecies, Brian, and Ae-Gyung Shim. 2021. *South Korean's Webtooniverse and the Digital Comic Revolution*. London: Rowman & Littlefield.

15

"THE INFINITE KNOWLEDGE AND POWER OF THE MULTIVERSE"

Intersectionality as the Multiverse in *Everything Everywhere All at Once*

Shayna Maskell

The film *Everything Everywhere All at Once* (2022) has received critical acclaim for its inspiring storytelling and acting, becoming the most-awarded film in history (Du 2023). The film tells the story of an interdimensional battle between Asian American mother-daughter duo Evelyn (Michelle Yeoh) and Joy Wong (Stephanie Hsu), fighting to save—or destroy—existence. It is a story of family, of immigrants, of possibility, of nihilism, and of love. We are first introduced to Evelyn, the mother, wife, and laundromat owner, who is unhappily married to Waymond (Ke Huy Quan) and being audited by Internal Revenue Service (IRS) agent, Diedre (Jamie Lee Curtis), and disconnected from her daughter, Joy, whom she pretends does not have a girlfriend. It is in the elevator of the IRS where Evelyn is contacted by an alternate version of her husband, where she learns the multiuniverse is in peril and she must it from collapsing into a black hole created by an alternate version of her daughter, the parallel world-hopping villain Jobu.

Each version of the multiverse offers a different amalgam of Evelyn, posing nearly endless possibilities and alternative realities: Movie Star Evelyn, Hotdog-Fingers Evelyn, Chef Evelyn, Rock Evelyn, Dominatrix Evelyn. All of these Evelyns work to construct just one prong of the matrix of oppression, a sociocultural paradigm that understands different forms of oppression (e.g., class, race, sexuality, nationality) as interconnected (Hill Collins 2000). Multiverse Evelyns, and their single lens of identity, reflect the societal construction of personhood that flattens and

DOI: 10.4324/9781003480846-18

marginalizes individuals into one-dimensional, monolithic groups (i.e. Asian, immigrant, lesbian). Yet, primary Evelyn (the version we first meet) is able to verse-jump, to move fluidly between dimensions and Evelyns, resisting the typical construction of systems of oppression as parallel or separate. She is *all* Evelyns and thus *all* oppression(s), and it is only as this fully realized individual that she can save her daughter and all of the multiverse. Yet, Evelyn's fully realized self is complicated by Joy's intersectionality in the form of Jobu, who is all (marginalized) identities at once and reflects not only the constitutive nature of intersectionality and the *power* of this intersectional identity (as an all-knowing, all-powerful being) but also the complex emotional toll these interlocking systems of oppression have.

The Multiverse and/as Intersectionality

At its foundation, this film is a multiverse story, where Evelyn Wang's primary universe is just one in an infinite series of parallel universes, all of which comprise a different version of—or alternate reality for—Evelyn and her family. This multiverse is the theoretical and material setting of these separate but intersecting universes that the characters of the film jump to and from. Quantum theory postulates that it is nearly impossible to account for only one universe, with only one cosmological constant, that allows for the perfect conditions for which our universe to exist. Instead, if there are a bunch of different universes, all with different constants, we begin to understand that ours is simply a "Goldilocks" universe, where the constants are just right and are a result of the "random generation of universes throughout infinite time and space" (Rubenstein 2014, 207). Taken as a whole, these universes that comprise the multiverse are made—as the title of this film alludes to—everything that exists (all at once): energy, matter, time, physical laws, space. And while the multiverse is still scientific conjecture, it offers a space where philosophy, physics, religion, and science can speculate about the nature of not only the universe but also our place in it.

Intersectionality, as a theoretical model to understand power and our place in our own observable universe, is rooted in black feminist thought and argues that vectors of identity and difference (including race, gender, sexuality, and class) interlock in multiple systems of oppression, manifesting in institutional and social structures (Crenshaw 1991). As sociologist Patricia Hill Collins (2000) argues, "Intersectional paradigms remind us that oppression cannot be reduced to one fundamental type, and that oppressions work together in producing injustice" in what she terms a matrix of domination (18). These multiple systems of oppression create multiple, often reinforcing, forms of inequality, which often works to remarginalize specific aspects of identity. In this way, a black, queer, Muslim woman encounters different forms of inequity than a white, heterosexual, Christian woman. These multiplicities of sociopolitical locations, and the power structures that uphold and marginalize, are not a given. Instead, they are contingent, contextual, and relational. Such an acknowledgment demands recognition of the ways in which, say, feminism (focused primarily on white women) and racism (focused primarily on black men) can work at odds with one another and adjusts the sociopolitical paradigm to include interlocking and mutually constructive systems of power and marginalization. Thus, intersectionality works to question *how* institutions such as science, religion, politics, and economics creates inequities and naturalizes structures of power.

How then, should we understand intersectionality as the multiverse? I argue that the conceptual underpinnings of both posit not only a multiplicity of realities but also the relativity of those realities to one another. That is, just as theories of the multiverse necessitate an understanding of the one and the many through an order that is "constituted, dismantled, and renewed by an ever-roiling chaos" so too does intersectionality (Rubenstein 2014, 236). The "truth" of the multiverse is just as provisional as that of race, class, and gender. Both are attempts at conceptualizing the vastness (of the universe, of the self) and constructing an artificial order around and about it. Just as Alpha Waymond explains to Evelyn: "You underestimate how the smallest decisions can compound into significant differences over a lifetime. Every tiny decision creates another branching universe." In the same way, every vector of identity compounds to create a different matrix of domination. Both the multiverse and intersectionality attempt to disentangle ontology from truth.

This parallel is strengthened by the film's internal multiverse logic: multiple consciousnesses are from disparate universes but are retained by the individual. Indeed, the mechanics of interacting with these alternative universes in *Everything Everywhere All At Once*, called verse-jumping, is premised on allowing individuals to tap into not only memories and feelings of their alternate selves, but also their abilities and talents. This verse-jumping, the co-existence of a primary "self" that is always-already intersecting with and immutable from an endless number of other, different selves constructs a multiverse that is also-always about human existence and identity—how to bring meaning to the chaos of infinite realities.

The Evelyns

The film does not use time travel. There is no causal loop that characters use to change their lives and their outcomes. Instead, the characters themselves are the instrument of, and symbol for, the infinite multiverse. And Evelyn is the principal symbol and verse-jumper. In the primary universe, we understand Evelyn as harried, disconnected, and fragmented, both through the camera's quick movements, cuts, and disjointed sounds, as well as her storyline—being audited by the IRS, trying to maintain a failing laundromat, throwing her father, in from China, a birthday party. But as Evelyn visits alternate universes and alternate selves, we see different Evelyns: kung-fu Evelyn, hotdog-fingers Evelyn, rock Evelyn, and of course, primary Evelyn. Each of these Evelyns represents a different vector of identity in the matrix of domination, an embodiment and a possibility of what happens when a person is defined by one facet of their identity.

Primary Evelyn, as we first meet her, is miserable, embodying the hegemonic demands of contemporary (white) American womanhood. She is married to a man and in denial of her daughter's sexuality (even referring to Becky as a "he"), thereby reaffirming heteronormativity and the bifurcation of gender. Indeed, it is her (heterosexual) marriage that is naturalized in relation to the non-normative sexuality of her daughter. This heteronormativity is connected, as well, to Evelyn's race. She blames her misgendering of Becky on the nongendered Chinese word and denies their relationship in deference to her traditional Chinese father's homophobia. It is, then, Evelyn's relationship with her father that reinforces not only the patriarchal power structure of (hetero)sexuality and gender, but also the Americanized representation of an Asian woman. Socially constructed as "passive, weak, quiet, excessively submissive, and slavishly dutiful," Asian women are dually bound by both race and gender (Pyke and Johnson 2003, 36). Primary Evelyn is similarly compelled, constructed, and constrained by her womanhood (as a wife, as a mother, and as a daughter) and her Asian-ness (as a daughter, as a wife, and as a co-owner of a laundromat). We see this in her flashback, including her first memory, when, as a doctor presents her to her father, he says in Mandarin, "I'm sorry, it's a girl," to her marriage to Waymond and fleeing to America, which leads to her father disowning her, and then to her pregnancy and the birth of Joy. Her misery in these roles reveals the dangers of what Hill Collins (2000) calls "controlling images," whereby women of color are stigmatized and objectified by their Otherness. Despite adhering to Western hegemonic norms—owning a business (albeit a stereotypical one for Asian Americans), being in a heterosexual marriage, having children (Joy, decontextualized of her sexuality) —Evelyn is still a Chinese immigrant, and thus she is, in large part, defined as a discrete (if not subordinate) group apart from white women. These images control not only whiteness but also Evelyn's own sense of self and happiness. It is only on her verse-jumping quest, where she is able to access all the other versions of herself, that she is able to reflect on, question, and ultimately modify, her own identity.

Kung-Fu Evelyn

This Evelyn, one of the first we as the audience are introduced to, is premised on Evelyn's decision to *not* marry Waymond and *not* immigrate to America. As viewers, we witness the literal bifurcation of Evelyn's choices and thus life—in one, she goes to America, marries Waymond and comes primary Evelyn, the narrator of the movie. The other stays in China, is assaulted in an alleyway, recused by a woman who subsequently takes Evelyn under her wing and helps mold her into a kung-fu champion. Kung-Fu Evelyn is also premised on the seeming antitheses of Asian womanhood. While "women are granted little decision-making power and are not accorded an individual identity apart from their family role, which emphasizes their service to male members," kung-fu Evelyn is powerful as a warrior woman, matching the skill and grace of men in competition (Pyke and Johnson 2003, 38). Indeed, it is her womanhood that not only differentiates her from the long strain of male kung-fu masters and hypermasculinity martial arts conjures, but also *how* she fights. As kung-fu Evelyn learns from her master, "kung-fu is not just combat. Even this pinky can be kung-fu." As primary Evelyn uses this knowledge to fight off a horde of violent jumpers using only her pinky, it is clear that her strength and power does not come from muscle and masculinity but instead from the self-assuredness and comfort of making decisions for one's self. It is her womanhood, her physical connection to and understanding of her body that makes her formidable and indefatigable. And this strength was borne from her decision to *not* engage in the heteronormative demand of marriage and children. Importantly, that physicality and womanhood is discrete from her sexuality; kung-fu Evelyn is not married. And while she later crosses with once again with Waymond, her potency and capacity are *not* connected to her as a sexual or romantic being. As primary Evelyn tells Waymond (in Cantonese) after experiencing kung-fu Evelyn "I saw my life without you. I wish you could have seen it." And then, in English, she continues, "It was beautiful."

Yet, this womanhood is undeniably rooted in race. Her "martial virtue" is a "visceral experience of Chineseness" (Frank 2006, 202). Kung-fu Evelyn stayed in China, was taught in China, became a kung-fu movie star in China, and represents the cultural imaginary of not just womanhood but

Chinese womanhood. This racialized identity works to obfuscate or at least elide the national, regional, and local conflicts, generating an imaginary "Chineseness" that is rooted in a specific place and specific cultural imaginary of kung-fu. Such an identity is made apparent by a constant binary of East/West: the Western movie industry juxtaposed with the Eastern martial art; the Cantonese versus the English Evelyn speaks; a Chinese kwoon next to Los Angeles laundromat. Kung-fu Evelyn fully engages with both her womanhood, as a subversion of hegemonic constructions, and her race, as a reinforcement of her cultural identity. This duality is further reinforced by Michelle Yeoh, the actress who plays Evelyn. A Chinese movie star in her own right, she came to fame performing her own stunts in martial arts movies, crossing over to American fame with the blockbuster *Crouching Tiger, Hidden Dragon* (2000). The binaries and contradictions of race, gender, and nationality are emphasized and brought into the "real world" of the audience watching kung-fu Evelyn with interspersed archival footage of Yeoh at previous movie premieres. The tenuous line between real and artifice, Evelyn and Yeoh, Chinese and American, is undeniable.

Hotdog-Fingered Evelyn

In what is clearly the most bizarre and obvious iteration of Evelyn's journey to self-discovery and universe saving is the world where everyone boasts hotdogs for fingers. The hotdog finger reality is constructed, otherwise, as parallel to our world: only the fingers are different, described as "an evolutionary branch in the anatomy of the human race." But the real parallel to this universe is sexuality. In it, hotdog Evelyn is in a romantic relationship with Diedre, the IRS agent from primary Evelyn's universe. This relationship is signified by the fights the two have, their living together in an apartment, but also in a touching scene as the two women make their way back to one another. Primary Evelyn as hotdog finger Evelyn rejected her relationship with Diedre, shunning her and eventually causing her to pack her bags and leave. But, as the movie progress and primary Evelyn becomes comfortable in her multiplicity, in embodying what was foreign before, hotdog finger Evelyn tries to reconcile with Diedre, asking Diedre to play her a song on the piano. We see the two sit, side by side, as Diedre plays *Clair de lune* with her feet. This song, which repeats during a fight scene in primary Evelyn's universe, functions as a sign of love, conjuring the Verlaine poem on which it is based, "a vision of long-dead dancers in the moonlight dancing forever to a ghostly music" (Dawes 1969, 21). The emotional aspects of the song, including its minor modes, slow tempos, and lament line echo that of funeral marches and are associated with sadness and grief, while the use of harmonies and thick textures preserves the beauty (Ferguson 2011). This duality—of grief and love, of mother and daughter—poignantly mirrors the binaries mandated by hegemonic norms of (hetero)sexuality, while the hotdog fingers speak to the absurdity of being bound to such patriarchal and heteronormative demands. Hotdog-fingered Evelyn and her hotdog-fingered love, Diedre, symbolize the farcicality of expectations of and for romantic love, the preposterousness of heteronormativity.

Rock Evelyn

Rock Evelyn is nearly the last incarnation we see. Indeed, Evelyn is a rock in the alternate universe only after she is everything everywhere—an infinite number of Evelyns with an infinite number of ways of being and the nihilism of trying to make meaning conform to those multitudes. In this universe, which is unable to sustain human life—or, perhaps more to the point, a human life predicated on one singular identity—Evelyn and Joy exist only as a pair of large, smooth rocks overlooking the landscape of a canyon. And as Joy tells her mom, there is nothing to worry about in this universe: "Just be a rock." Existing—just existing—is all Joy and Evelyn have to do as rocks. Their "small and stupid" human selves do not matter. Just sitting, laying, being rocks, has no value in a universe without human life. It is, in philosophical terms, an existential conundrum. If there is no inherent meaning to life, what is its purpose? And it is against this vastness and stillness that Evelyn enacts a *different* ontology—a different way of being. The answer to the existential question, for her, is not nothingness (as it is for Jobu, as we will see in the next section) but in everythingness. That is, Evelyn verse-jumps from the nothingness of the rock universe to the primary universe, where she embraces herself as a fully intersectional human. She is everything and everyone, unraveling the ways in which power structures have divided and categorized identities and resisting against that. It is by being—not nothing but everything—that Evelyn saves herself, Joy, and the multiverse. But first, she must understand what she is fighting against. It is not just hegemonic norms that demand conformity and uniformity within singular conceptions of race, gender, and sexuality, it is also the oppression of those norms.

Jobu Tapaki: The Villain as Oppression

Jobu Tapaki is Joy after being unable to meet the near-impossible demands of her tiger mother Evelyn in the Alpha universe. She can inhabit every possible version of Joy in the multiverse, and, as Alpha Waymond describes her "sees every world at the same time...she's lost any sense of reality, any belief in objective truth." Her constant pursuit of finding a place for her identity—as a woman, as Asian American, as a lesbian—is marked by the villainous consequence of denying one's self: Jobu Tapaki and her quest for self-annihilation. Jobu embodies nihilism as the direct result of witnessing the same narratives, the same outcomes of the slightly different versions of herself in every single universe. Her omniscience makes her capable of knowing "what makes you tick and [on] what fragile branches your self-worth rests," and thus understands how hegemonic demands of race, gender, and sexuality structures a sense of meaning and purpose (or, in her case, hopelessness and nothingness). Jobu has seen that success in one universe is failure in another, that victory and loss are just cycles demanded by the hegemonic formulas of identity. It is the demands of society—to be a certain way, to conform to a certain concept of identity—that causes the never-ending suffering in each and every universe. This is epitomized in one of the first times we meet Jobu, as a pink-haired, Elvis Presley jumpsuit-wearing version of herself, walking with a strut and a leashed pet pig out of the elevator of the bottom floor of the IRS building. A cop, the epitome of authority, of white, cisgendered, heterosexual, masculinity, of power and punishment, tells her she can't be there. Jobu responds: "Is it that I can't be here or that I'm not allowed to be here? See, I can physically be here. But what you meant to say is you're not allowing me to be here." Possibilities are crushed by rigid demands. Her full non-normative, non-conforming self *can* and *does* exist in all universes (and at the same time). It *is* there. *She* is there. All of her. It is the artificial structures of power, the literal and figurative policing of bodies that does not allow her to be there. The violence against her existence is subverted and (absurdly) reconfigured when Jobu seizes the nightstick from the officer, turns it into a dildo, and proceeds to beat him to death with it.

The pinnacle of Jobu's nihilism, the tangible artifact of her despair, is the everything bagel. A singularity, a black hole, it is a metaphor for both the absurdity of life and the literal everything-ness of the multiverse that will be sucked into oblivion. As Jobu describes it, the bagel holds "all my hopes and dreams, my old report cards. Every breed of dog, every last personal ad on Craig's List. Sesame. Poppyseed. Salt." It is, as Jobu says, the everything of the multiverse and thus, "the truth." Her truth. And her truth is that the unending realities of a complex and unending self that cannot be recognized or validated ends in despair. In all universes throughout all time, Jobu has not found peace; she is constantly shattered into an infinite number of Joys, "never fully there. Just a lifetime of fractured moments, contradictions, and confusion." Yet, Jobu's search for her mother is not a typical villain arc—she is not trying to kill her mother or even the entire universe. Her goal in getting Evelyn isn't, as she tells her mother, "to destroy everything. It was to destroy myself." These multiple, fractured selves have disconnected Joy/Jobu from her corporeal and socioemotional self, and the end result can only be (self-)destruction. The rigidity of hegemonic norms of identity end in the destruction of self—whether than be the physical self or the socioemotional self. But before this self-demolition, Jobu makes one last effort to find her mother, to find the Evelyn who is "someone who could see what I see, feel what I feel." And it is this primary Evelyn, and all that she represents, that saves not only multiverse, but also her daughter.

Intersectionality as Hero

Primary Evelyn is the woman who can harness the knowledge, experiences, and powers of all multiverse Evelyns, reformulating her reality to her own liking. As she verse-jumps, she acquires the memories of the alternate Evelyns, knowledge of the different worlds that primary Evelyn can inhabit, but also the tangible skills from that alternate self, downloaded to her primary Evelyn stats. Kung-fu Evelyn gives her improved mobility and ingenuity; singer Evelyn gives her enhanced lung capacity and a lack of stage fright; blind Evelyn gives her enhanced hearing; chef Evelyn gives her dexterity. It is all of these abilities, all of these memories, all of these gifts that act as a metaphor for the different aspects of herself. And it is the use of *all* of these vectors of identity that make primary Evelyn so strong. The film attempts to capture this merging of selves using exceedingly fast cuts and changes in aspect ratio, showing the violent images of the multitudes of Evelyns and all her various forms of identities, finally using cuts so quickly that the Evelyns become indistinguishable: she is simply Evelyn, all of her possibilities and infinite variations. She *becomes* intersectional, as all aspects of her multiverse of identities combine to construct her current social reality. The collectivity becomes the singularity.

Importantly, however, it is not simply the *recognition* of primary Evelyn as intersectional, but *how* she uses this consciousness to fight to save the universe and Jobu/Joy. It is, as Waymond tells her in the kung-fu universe, kindness as/instead of fighting. Acceptance and love are strength, and Evelyn embraces all her selves as a way to fight the nothingness and chaos of the shattered self of Jobu/Joy. Kindness has been conceptualized by McIvor (2007) in three dimensions of behavior: (1) to oneself (as self-care and attitude; (2) to friends and family (as trust, compassion, and friendship); and (3) to community (as tolerance, responsibility, and service). Evelyn imagines otherwise, freeing herself from the rigid and dominant ways of thinking by imagining alternatives where identities grounded in social categories are mutually articulated and interlinked, accepted and validated. By recognizing the violence that is inextricable connected to the social structures that force, as Jobu tells her mother, the concept of "right" as a "tiny box, invented by people who are afraid" Evelyn subverts the dominant narrative. And she does so through kindness to herself, to her friends and family, and to her larger community.

Kindness-as-acceptance becomes the way that Evelyn defeats the nihilism of the everything bagel. With a googly-eye stuck to her forehead, echoing the third eye that in Buddhism represents the eye of consciousness, enlightenment that allows one to see beyond physical sight, Evelyn fights with love, joy, and tolerance rather than physical dominance. She dodges swords and punches, uniting two attackers in a passionate kiss instead; she turns a grenade into a perfume bottle, spraying an attacker with a scent that reminded him of his late wife and leaving him in sweet nostalgia; with her kung-fu skills, she fixes a neck impingement of another attacker, giving him physical peace; and stopping the scissor attack of yet another attacker, she places a ball gag in his mouth, satisfying his fetish for BDSM. Her verse-jumping allowed her to have compassion for her opponents' pain and turned that hurt into kindness.

This re-writing of narratives is most poignant and critical when it comes to the people in primary Evelyn's own life—her family. As she tells her daughter, after being in every single world and being every single Evelyn, there is "always something to love." It is *through* these differences—whether it be hotdog fingers, cat faces, sexuality, or gender—rather than *despite* them that we can understand the one's full identity and find love and acceptance. This means offering her IRS nemesis Diedre, a hug and understanding in the primary world, where the love they shared as mates in the hotdog universe offers empathy and understanding to her in this world. More essentially, this same-sex love of the hotdog world allows Evelyn to accept Joy's romantic relationship with Becky. As a part of this acceptance, Evelyn tells her father about the relationship, and in doing so throws off the shackles of her dad's disapproval. And her relationship with Waymond, on the brink of divorce in the primary world, is saved by her acceptance of who he is, and who they are together. The movie ends as it begins—with Evelyn and Waymond headed to the IRS for their audit. But this time, it is with the full understanding and acceptance of who they are—the messy, contradictory, disappointing, devoted, imperfect, evolving Evelyn and her family. It is the full knowledge of being—every possibility, every narrative, every version of herself—that allows Evelyn to be everything everywhere all at once and simultaneously with her family, at the IRS, reveling in the mundanity of the everyday.

References

Collins, Patricia Hill. 2000. *Black Feminist Thought: Knowledge, Consciousness, and the Politics of Empowerment*. New York: Routledge.

Du, Josh. 2023. "Everything Everywhere All at Once Passes Return of the King as Most-Awarded Movie Ever." *IGN*. https://www.ign.com/articles/everything-everywhere-all-at-once-return-of-the-king-most-awarded-movie. Retrieved April 25, 2024.

Crenshaw, Kimberlé. 1991. "Mapping the Margins: Intersectionality, Identity, and Violence Against Women of Color." *Stanford Law Review*, 43, no. 6: 1241–1300.

Dawes, Frank. 1969. *Debussy Piano Music*. London: British Broadcasting Corp.

Ferguson, Brent. 2011. "Moonlight in Movies: An Analytical Interpretation of Claude Debussy's "Clair de Lune" in Selected American Films."

Frank, Adam D. 2006. "Kung Fu Fantasies and Imagined Identities." In *Taijiquan and The Search for The Little Old Chinese Man: Understanding Identity through Martial Arts*, pp. 189–204. New York: Palgrave Macmillan US.

McIvor, Olivia. 2007. The Business of Kindness." *AMT Events* 24, no. 2: 80–82.

Pyke, Karen D., and Denise L. Johnson. 2003. "Asian American Women And Racialized Femininities: "Doing" Gender across Cultural Worlds." *Gender & Society* 17, no. 1: 33–53.

Rubenstein, Mary-Jane. 2014. *Worlds Without End: The Many Lives of the Multiverse*. Columbia University Press.

16

MARVEL'S MULTIVERSE OF MADNESS

The Implications of Alternative Realities, Variants, and Doppelgängers within the Marvel Cinematic Universe

Rob McLaughlin

The notion of variant characters appearing within comics has been a narrative device since the first appearance of Earth 2's Flash in *The Flash #123* (Sept. 1961, see also Davis's chapter in this volume). A plethora of contemporary examples of variants and alternative realities have come from both DC and Marvel, with *Age of Apocalypse*, *House of M*, and *Secret Wars* as core alternative dimensions that still impact current House of Ideas core titles. This integration of alternatives and variants has also translated into both the DC and Marvel cinematic universes, with (for example) *The Flash* providing a soft reboot to DC's cinematic endeavors via a 'Flashpoint.'

While in a diegetic context, this widening of the Marvel Cinematic Universe (MCU) to become the Marvel Cinematic Multiverse (MCM) drives the continuing release of the films within the franchise, outside of the diegesis of the films the expansion of the MCM also has a secondary significance in the utilization of the rights, intellectual property, and ownership of Marvels comic characters that Marvel do not outrightly own. In this chapter, I explore how the development of the MCM creates new opportunities for Marvel's continued industrial growth.

Origins of the Marvel Cinematic Universe/Multiverse

The notion of potentially limitless variants and iterations of characters appearing in "mainstream" continuity signifies that comic and comic

DOI: 10.4324/9781003480846-19

book films sit squarely within the realms of imagination and the fantastical and create world-building scenarios where anything can happen (Boillat 2022, 97). While the Marvel Comic Universe (also called the Prime or Sacred timeline) has, over the decades, expanded on the printed page, the MCU is now replicating its comic counterparts by both metaphorically and literally splintering its mainstream universe on screen (Suher and Tetik 2022).

In comics, the multiverse model has allowed publishers to experiment with new characters in prominent roles and different iterations of recognized superheroes, all without sacrificing the audience or continuity of their flagship titles (Frank 2016). Within the MCU, the studio has embraced the notion of variants, more subtly, as the transition to a cinematic medium has meant Marvel has to be more strategic in the narrative it has woven across the phases of its films and has to consider how what occurs in one medium could in turn have consequences in the others (Joffe 2019). The notion of alternative dimensions and the Multiverse in the MCU is initially referred to as far back as the original *Thor* (2011). Doctors Erik Selvig and Jane Foster posit the existence of a Multiverse via their descriptions of Asgard. They elaborate this notion further in *Thor: The Dark World* (2013), in which Selvig draws several diagrams and equations on a blackboard with the center of this multiversal representation showing the notation *616*. The importance of this number comes directly from the Marvel *Comic* Universe, where 616 is deemed the 'Prime' universe (first designated by comic writer Alan Moore in 1983) and in which the majority of the comic line inhabitants exist.

Mirror Mirror—Agents of Change

The concept of copies or duplicates occurring in an ongoing narrative often played out via supernatural means (Ruddell 2013), usually as a narrative device to bring a dark reflection of the protagonist into the diegesis. Within *Star Trek*, for example, the trope of evil doppelgängers that interact with their mainstream counterparts provided one of the most iconic presentations of alternative universes with the iconic episode "Mirror, Mirror" (1967; see also Geraghty's chapter in this volume). Viewers were presented with a parallel universe where mirror-image inhabitants copy the show's protagonists. By showing familiar traits, mannerisms and personalities subverted, "Mirror, Mirror" unnerves the viewer and reinforces Robin Wood's (1984) concept of "otherness," where the presentations of variants and doppelgängers are often framed as something split, divided, or repeated—the variant are as such familiar to the viewer and yet at the same time something outside, different, and unnerving.

The inclusion of an alternative or variant as a narrative catalyst is evident in the MCM's Phases 3 and 4, with two characters acting as a fulcrum to widen the MCU. The first of these catalysts is a variant of Loki—Sylvie, first seen in the Disney Plus series *Loki* (2021), who in the conclusion of the narrative, slays the series' main antagonist He Who Remains, a derivative of Kang the Conqueror. In the death of this character, Sylvie unleashes the Council of Kangs and an awareness of the Multiverse. In killing He Who Remains, Sylvie does what she thinks needs to be done, whereas Loki is unwilling to do such, and frames her as a monster. In this act of doubling, Sylvie becomes not only a physical but also a psychological threat and Loki sees a copy of himself pushing the boundaries of acceptability. *Loki* creates a dissolution of the self, as the character becomes a malevolent or darker version of themselves (Ruddell 2013). However, this iteration could also be seen as a better version of himself (Humann 2018, 89).

While Sylvie provides an example of a dark doppelganger, the other catalyst of the MCM is Doctor Strange. His act of transgression occurs when he assists Spiderman/Peter Parker from the 616/Sacred timeline in *Spiderman: No Way Home* (2021) and through this act opens the MCU to the Multiverse. This notion is explored in *Doctor Strange in the Multiverse of Madness* (2022), which embraces the multiverse in the film's first scene when a variant of Doctor Strange (sporting his comic book *Defenders* look) assists America Chavez's escape from an aggressive alternative reality. This opening not only provides an example of another variant being brought into the narrative, but also shows the notions of mistrust, antagonism, and the widening of the "familiar" turning into the (Doctor) "strange" (Baudrillard 1994), as this version of Doctor Strange is not as benevolent as his 616/Sacred timeline self.

This variant of Doctor Strange as a clones/replica also trigger debates of normativity (Halft 2014), which frame the representation of doppelgängers as deviant, monstrous, or unnerving by tapping into Freud's (1990) concept of the *uncanny*. For every benevolent meme of Spiderman pointing to another Spiderman, the concept of variants represents the idea of split, divided or mirrored self, a potential dark or evil reflection of a protagonist. In the representations of, for example, Sylvie, Doctor Strange as dark replicas (Nilsen 1998), the MCM narrative uses the notion of the duplicate to push the boundaries of expectation. As Halberstam (1995) notes, we "wear modern monsters like skin, they are us, they are on us and in us. Monstrosity… is replaced with a banality… because the enemy becomes harder to locate and looks more like the hero" (162–3). This concept instills fear and mistrust of a duplicate body image, as it questions the subjectivity of identity (Boss 1989, 98). Even the most benevolent character can become something sinister and nightmarish in the context of the narrative. This is evident to an even greater extent within *Doctor Strange in the Multiverse of Madness* (2022), which sees the Scarlet Witch, possessed by Darkhold, scour the Multiverse to find a counterpart she can possess and as such be reunited with her children.

What If...The Future of the Marvel Cinematic Multiverse

While the cinematic rights to *Iron Man*, *Captain America*, and *Thor* belong solely to Marvel Studios (which is owned by Disney), the ownership of other Marvel Comic University characters is more complex. Since the late 1990s, Sony has owned the rights to *Spiderman* and *Venom*, Universal owns the rights to *The Incredible Hulk*, and Fox owns the rights to the *X-Men* and the *Fantastic Four*. The continuation of a MCM requires building on existing properties/characters and a continued agreement with licensing partners to collaborate in future endeavors. Attempting to create a cohesive universe has become a difficult 'sell' to fans, as we've seen with Sony's *Mobius* (2022), *Madam Web* (2024), and *Kraven* (2024), as these satellite Spiderman titles all garnered poor reviews and box office returns. More success has been found by studios sharing characters' IP.

Success has also been found via a multiversal variant of the web-slinger in animated form with Miles Morales's *Spider-man: Into the Spiderverse* (2018; see also Whitmore's chapter in this volume). With the film's subsequent sequels embraced the multiverse to a much wider context, with appearances by Spider-Gwen, Spider-Ham, Jessica Drew, and Miguel O'Hara (the Spiderman of *Spider-Man 2099*). This iteration of Spiderman is aware of the multiverse and notes the events seen in *No Way Home* by referring to the MCU as "Earth-199999." More variants and derivatives could show in theory, including versions from the computer game *Spider-man* (2018), the Spiderman from *Spiderman and His Amazing Friends* cartoon show (1981) or even the *Theatrical Spiderman* from the 2011 *Spiderman Turn off the Dark* Broadway show.

The MCM also includes the Disney Plus animated *What If...?* (2021), which shows iterations and derivatives of alternative universes spun out from the existing MCU. The narration of the enigmatic Watcher in the first season of *What If...?* successfully presented more potential alternative dimensions but has not affected the Prime MCU sacred timeline... yet. However the merging, deletion, and cataclysmic repercussions of *What If...?*'s universal incursions is latent within the narrative, including characters like the dark iteration of Doctor Strange, the cast of the Time Variance Authority (TVA), America Chavez, Eddie Brock/Venom and Captain Monica Rambeau (who travels to the Fox X-Men universe) in *The Marvels* (2024). As of this writing, Marvel's embraces of a "Multiverse Saga" will culminate in the forthcoming *Secret Wars* (2025) film, which could see the shifting or indeed merging of comic universes or even the integration of the Fox, Sony, Universal, New Line, and even Toei-owned IP of Marvel properties into the 616/Scared Timeline or overall MCU narrative.

The notion of the intertwining of universes outside of the MCU also frames *Deadpool and Wolverine* (2024) which goes further in the exploration and alignment of the variant comic book universes in which these characters exist. The film presents not only in-universe character variants, such as Lady Deadpool and Dog-Pool, but also a plethora of prior and new variants, including a Zombified head of Deadpool called Head-Pool—which ties into a reality first explored in *What If… Zombies?!* The inclusion of a zombie-infested head could imply a potential inclusion of *Marvel Zombies* within the 616 universe and more horror-centric text like *Werewolf by Night* (2022). Historically, the moniker *Marvel Zombie* was used as term of endearment toward the comic fanbase by the editorial teams of the 1970s, written via the Bullpen Bulletins and other letters pages, editors notes and messages seen in the comics of *Marvel Zombies* by Robert Kirkman, Sean Phillips, Randy Gentile, and June Chung.

This use of characters, spaces, and contexts of other Marvel IP could tap into other licensed properties within a much wider sphere. As with the deal with Sony to bring prior Spiderman into the MCU, the possibilities of a shared IP MCM could in theory see iterations of the *New Mutants, Generation X*, and *Mutant X*, as well as the prior cinematic interpretations of Ghost Rider, Blade, and Elektra. An even more set of interesting propositions could be the inclusion of legacy properties such as Lou Ferrigno's 1970s *Incredible Hulk*, Toei 1970s Spiderman Takuya Yamashiro and his mecha Leopardon, Nicholas Hammond's 1970s *Spiderman*, and David Hasselhoff's *Nick Fury: Agent of S.H.I.E.L.D.* (1998).

This vast net of Marvel licensed properties could be cast even further and see the Multiverse embrace another *Doctor Strange* with Peter Hooton (1978), two *Daredevils*—the cinematic portrayal from Ben Affleck (2003) and the television incarnation from Rex Smith from *The Trial of the Incredible Hulk* (1989)—as well as three other *Punishers* (Dolph Lungren [1989], Thomas Jane [2004], and Ray Stevenson [2008]), and even three Captain Americas—Dick Purcell (1944), Reb Brown (1979) and the rubber eared Matt Salinger (1990). While it is unlikely that the Roger Corman-directed unreleased iteration of the *Fantastic Four* (1994) could be included, the possibilities of other Marvel licensed properties and a *What If...?* Multiverse might allow other possibilities.

Theoretically, the MCM could also incorporate other subsidiaries of Marvel licensing properties from its comic back catalog. During the late 1990s, Marvel bought Malibu comics, and while not the most recognized publisher, the line did produce some commendable and well-regarded comics including the vampiric *Rune* and the hyper-exaggerated Superman-inspired *Prime*. Some of the Malibu characters such as *Nightman* made the transition to television in a Glen A. Larson-backed series in 1997 and the Avengers like *Ultra-Force* whose animated exploits appeared briefly in 1995. Furthermore, a highly unlikely but fascinating proposition in tapping into prior publishing could in theory see some current DC characters appear in Marvel texts. During the mid-1990s Marvel had a relationship with Image during the *Heroes Reborn* saga (1996), which not only spawned the meme-centric Rob Liefeld Captain America but also saw a merging of the Wildstorm and Marvel Universe that saw members of the Wildstorm Universe (such as Grifter, Maul, Zealot, and Warblade) appear in conjunction with Marvel heroes. Now integrated into the DC Comic Universe, it might be a financial impossibility for these characters to appear within an MCM, but this embracing of variants, multiple universe, and doppelgängers by Marvel Studios will continue, no matter how far-fetched, tenuous, or driven by fan theories/fandom (Richter 2016),

Ultimately, the MCM proves that in this expanding multiversal model, the potential for characters, no matter how obscure or from another medium, provides endless possibilities.

References

Baudrillard, Jean. 1994 [1981]. *Simulacra and Simulation*. Translated by Sheila Faria Glaser. Chicago: University of Chicago Press.

Boillat, Alain. 2022. "*Cinema as a Worldbuilding Machine in the Digital Era: Essay on Multiverse Films and TV Series*." Indiana University Press.

Boss, Peter John. 1989. *Death, Disintegration of the Body and Subjectivity in the Contemporary Horror Film*. PhD diss., University of Warwick.

Frank, Kathryn M. 2016. "Everybody Wants to Rule the Multiverse: Latino Spider-Men in Marvel's Media Empire." In *Graphic Borders: Latino Comic Books Past, Present, and Future*, edited by Frederick Luis Aldama and Christopher González, 241–251. University of Texas Press.

Freud, Sigmund. 1990. "The Uncanny." In *The Penguin Freud Library, Vol. 14, Art and Literature*, edited by Albert Dickson, 336–76. London: Penguin.

Halberstam, Judith. 1995. *Skin Shows: Gothic Horror and the Technology of Monsters*. Duke University Press, North Carolina.

Halft, Stefan. 2014. "Clones as Human Monsters: Looking for Normality in the Age of Cloning." In *Monsters in Society: An Interdisciplinary Perspective*, edited by Andrewa S. Dauber, 1–12. Freeland: Brill.

Humann, Heather Duerre. 2018. *Another Me: The Doppelganger in 21st Century Fiction, Television and Film.* Jefferson City, MO: McFarland.

Joffe, Robyn. 2019. "Holding Out for a Hero (ine): An Examination of the Presentation and Treatment of Female Superheroes in Marvel Movies." *Panic at the Discourse: An Interdisciplinary Journal* 1, no. 1: 5–19.

Nilsen, Don LF. 1998. "Doppelgangers and Doubles in Literature: A Study in Tragicomic Incongruity." *Humor: International Journal of Humor Research*, 11, no. 2: 111–134.

Richter, Ádám. 2016. "The Marvel Cinematic Universe as a Transmedia Narrative." *Americana E-Journal of American Studies in Hungary* 12, no. 1. Accessed May 9, 2024. https://www.americanaejournal.hu/index.php/americanaejournal/article/view/45110/43761.

Ruddell, Caroline. 2013. *Besieged Ego: Doppelgangers and Split Identity Onscreen.* Edinburgh University Press.

Suher, Hasan Kemal, and Tuna Tetik. 2022. "The World Building in the Superhero Genre Through Movies and Video Games: The Interplay Between Marvel's Avengers and Marvel Cinematic Universe." *Games and Narrative: Theory and Practice*, edited by Barbaros Bostan, 155–170. Springer.

Wood, Robin. 1984. "An Introduction to the American Horror Film." In *Planks of Reason: Eessays on the Horror Film*, edited by Barry Keith Grant, 164–199. London: Scarecrow Press.

17

WHERE'S WANDA? COLLAGE AND COLLAPSE

Teresa Forde

WandaVision (Disney + 2021) explores the fragmented superhero/villain in Wanda Maximoff aka Scarlet Witch's retreat into grief and self-preservation. The attempt to avoid grief at her lost love, Vision, by reconstructing his image, is manifested in the world she has constructed. Wanda undertakes a variety of roles of women in American sitcoms from the 1950s onwards. *WandaVision* plays out versions of the ideal heteronormative couple and ensuing nuclear family, depicting Wanda and Vision's relationship, the birth of their two children, and attempts at happy domesticity. These collages, created by the author, are a response to Wanda's emotions, intentions, and revelations. In this work Wanda is in crisis, stressed from her inner and outer worlds and her fragmented psyche: the manipulation of Westview articulates her construction of a "world within a world" and is exemplar of her power within the multiverse.

In *WandaVision*, Wanda has taken the inhabitants of Westview hostage and suppressed their memories. Wanda's previous experiences of losing Vision and her family, attempts by the Avengers to control her and her brother Pietro by Hydra lead Wanda to galvanize her abilities to create a shroud in which to submerge herself. Wanda creates a digital field around Westview, referred to as the Hex, with its clear magical connotations. This boundary creates a local, insider position and sees the outside world as a threat. In effect, Wanda creates an insular space, a universe that she can control and manipulate: a palimpsest of the town of Westview. As Mary Beth Haralovich (1989) recognizes, the housewife was traditionally,

DOI: 10.4324/9781003480846-20

FIGURE 17.1   Teresa Forde 2024. *Psyche*. Paper Collage. 29.7 cm x 42 cm.

"promised psychic and social satisfaction for being contained within the private space of the home" (61). Wanda tries to establish this promised life for herself within her constructed world; yet this space is not without her exploitation of others and is, ultimately, unsustainable. The fabric of this world becomes finally torn when her adversary Agnes/Agatha Harkness reveals Wanda's manipulation and wields her own powers and Monica Riveau also penetrates this world on behalf of S.W.O.R.D. who seeks to monitor Westview.

The depiction of Wanda as a female superhero/villain who is led by her emotions is intriguing yet potentially problematic. For the most part, Wanda is the protagonist of her series, depicting what Jones (2022) argues is potentially a "narrative of emotion as weakness" (99), although he recognizes the ways in which "emotions are beginning to be seen as a source of strength rather than a weakness to be exploited" (201). Wanda, who can manipulate and traverse universes, is bound by her own grief, anger, guilt, and love. The ambivalence of Wanda's responsibility for this situation is key to the way in which we encounter her narrative. For Jones, Wanda takes this grief and shows us that "this is not weakness, but rather raw, unbridled, intensely powerful strength" (198) which establishes her power, if not her superheroinism.

FIGURE 17.2 Teresa Forde, 2024. *Find the One*. Paper Collage. 29.7 cm x 42 cm.

We can understand the fragmented experiences of Wanda's emotions as stages of grief that align with the development of the mother character within sitcoms, increasingly breaking out of stereotypical forms, only to momentarily relapse into another structure. For Jane Barnette (2022) the emotional turbulence forges a link with her propensity for magic, and "it is into this existential sadness that Wanda enacts her Hex" (50). The episodic and convoluted situation in Westview may be part of a wider recognition of relationships: as Matthew Hurley et al. (2011) suggest, "The love that lasts years is not an uninterrupted stream of loving, but rather a strong tendency, over those years, to repeatedly feel discrete instances of the emotion of love" (71). For Madison Kooba (2023), "foregrounding Wanda's struggles with mental health have made her an admirable character to many who see her drawing power from her emotions as a celebration of aspects of womanhood that have long been shamed by society" (3). However, the ambiguity of representation of emotions and the framing of Wanda in seeks a domestical ideal through manipulating her own desires. Wanda's response may be indicative of wider experiences of losing loved ones and the post-Blip shock, and she has experienced this return herself.

FIGURE 17.3   Teresa Forde, 2024. *Power*. Paper Collage. 29.7 cm x 42 cm.

In terms of Wanda's performances, we can consider Kathleen Rowe's (1990) account that "the parodic excesses and the comic conventions of the unruly woman provide a space to act out the dilemmas of femininity" (411). Such excesses call into question the construction of feminine behavior and the overarching "tropes of femininity" (412). The most social upheaval is created when Wanda and Agatha Harkness attempt to claim their "own" desires in a complex contestation within and beyond such tropes. Wanda's emotions are paramount as she transforms through her grief from one space to the next, jumping through time and processing her emotions in a largely solitary and stressful attempt to maintain everything she has created.

As Liesbet Van Zoonen (2013) recognizes in relation to the relationship between identity and identification: "the importance of the concept of authenticity points to a wider popular desire to identify 'real selves' that are true, single and consistent" (46). This is the consistency that Wanda seeks and is struggling to achieve but may not ultimately be possible if she remains isolated. In reflecting on the role of Wanda Maximoff, Elizabeth Olsen considers her desire:

> She needed a sense of self. We've watched her for such a long time wrestle with this idea of should she or shouldn't she be a superhero? What does it mean to her and everyone around her? She's lost because of it. This is the first time we've watched her have an acceptance of self. It's like a woman-coming-of-age story.

> *(Feinberg 2021, 18)*

FIGURE 17.4   Teresa Forde, 2024. *Darkhold*. Paper Collage. 29.7 cm x 42 cm.

It could be argued that this positive account of Wanda's situation is countered by the depiction of Wanda in *Doctor Strange in the Multiverse of Madness* (Raimi, 2022). In this film she is one-dimensionally portrayed as a version of Wanda who is confronting other lives that she craves. Wanda's use of the Darkhold leads her on a futile chase through the multiverse. In this version she is searching for herself as a mother until she realizes that she cannot fulfill this dream.

References

Barnette, Jane. 2022. "What is Wanda but Witches Persevering? Palimpsests of American Witches in WandaVision." *Theatre Journal* 74, no. 1: 41–57.

Feinberg, Scott. 2021. "'This Show Led Her To Become a Fully Realized, Autonomous Woman': WandaVision Star Elizabeth Olsen Reflects on Leading the First Marvel TV Series, which Gives the Actress and her Superhero Counterpart More Screen Time and More Agency in the MCU." *Hollywood Reporter* 427, no. 24:S18+ https://go-gale-com.ezproxy.derby.ac.uk/ps/i.do?p=ITOF&u=derby&id=GALE|A667860768&v=2.1&it=r&sid=bookmark-ITOF&asid=dbc61666. Accessed 10 Jan. 2024.

Haralovich, Mary Beth. 1989. "Sitcoms and Suburbs: Positioning the 1950s Homemaker." *Quarterly Review of Film and Video* 11, no. 1: 61–83.

Hurley, Matthew M., Dennett, Daniel C. and Adams, Reginald B. 2011. *Inside Jokes: Using Humor to Reverse-Engineer the Mind.* Cambridge, MA: MIT Press.

Jones, Benjamin. 2022. "The Evolving Portrayal of Female Emotions in the Marvel Cinematic Universe." *Journal of Feminist Family Therapy* 34, no. 1/2: 196–202.

Kooba, Madison M. 2023. "*A Cultural History of Anti-Feminism in Marvel's Scarlet Witch.*" PhD diss., University of Nevada.

Rowe, K. 1990. "Roseanne: Unruly Woman as Domestic Goddess." *Screen* 31, no. 4: 408–419.

Van Zoonen, Liesbet. 2013. "From Identity to Identification: Fixating the Fragmented Self." *Media, Culture & Society* 35, no. 1: 44–51

Film and Television

Doctor Strange in the Multiverse of Madness. 2022. Sam Raimi. Marvel Studios.
WandaVision. 2021. Disney.

18

EXPANDING THE SPIDER-VERSE

Infinite Possibility and Fan Plausibility

Lore FitzWhittemore

Unlike the *Spider-Man* trilogy from the 2000s and the *Spider-Man* films in the Marvel Cinematic Universe, *Spider-Man: Into the Spider-Verse* (2018) ushered in a new era of Spider-media, which shifted perspective from protagonist Peter Parker to showcase alternative possibilities for the individual behind the "Spider-Man" mask. The film focuses on the origin story of Miles Morales and foregrounds interactions between Spider-Heroes from different dimensions, indicative of a multiverse structure known as the "Spider-Verse." As the film navigates what it means for multiple characters to be a Spider-Hero, it thematically suggests through Miles' concluding narration: "Anyone can wear the [Spider-Man] mask. You can wear the mask. If you didn't know that before, I hope you do now."

Into the Spider-Verse was not the first narrative to suggest the moniker "Spider-Man" could be inhabited by a character other than Peter Parker; for instance, a *Spider-Verse* comic event in 2014 explored a similar multiversal narrative. This co-occurrence of characters emerged from comic production logics that lend themselves to multiplicity, including audience pleasure derived from exploring multiple iterations of the same characters and a global marketplace that can sustain simultaneously publishing multiple series by different authors with the same characters (Jenkins and Ford 2009; see Davis chapter in this volume). Despite the established role of multiplicity within comics, the sentiment suggested by *Into the Spider-Verse* that anyone could be Spider-Man, with the support of the multiversal

DOI: 10.4324/9781003480846-21

narrative to demonstrate the thematic significance of that statement, reso-nated with fan artists in a newfound way. This led to the rise of Spiderso-nas: representations of the self in some form as a Spider-Person.

The term *Spidersona* is an interplay between the words "Spider-Man" and "persona" and a continuation of fan practices to imagine a "self-insert" or iteration of the self (known as a "sona") into the story world (Cham 2024). While many artists interpreted Spidersonas as a challenge to represent their total selves with fidelity, others amplified a single aspect of their identity or interests within the context of the Spider-Man mythos (Broder 2024; Risk 2024). Spidersonas are indicative of and embrace the multiplicity of the Spider-Verse. Although multiplicity has historically been enmeshed within production logics managed by media industries, the Spider-Verse demonstrates multiverses can structure and sustain new col-laborationist interactions and create opportunities to reconfigure fan/pro-ducer boundaries to imagine alternative storytelling possibilities.

Theorizing Multiverse and Collaborationist Possibilities

Outside of comics, media franchises have often ascribed to a single conti-nuity and adhered to the production logics necessary to maintain that, including the "continuous production of culture from intellectual property resources shared across multiple sites of production" (Johnson 2013, 4). Fan communities have likewise produced their own creative works based upon these same intellectual property resources that they do not possess legal claim to. Media technologies, specifically those that have emerged since the 1990s, have further complicated boundaries between commercial culture created by media franchise producers and participatory culture driven by fan engagement. In response to this, Henry Jenkins (2006) iden-tified two characteristic responses of how media industries seek to manage these interactions from a corporate perspective: prohibitionists and col-laborationists. The prohibitionists seek "to regulate and criminalize many forms of fan participation," particularly unauthorized participation, whereas the collaborationists experiment "with new approaches that see fans as important collaborators in the production of content as grassroots intermediaries helping to promote the franchise" (134, 169).

Collaborationist strategies have previously been discussed primarily in relation to paratexts either authorized or restricted by corporate intent (Jenkins 2006; Scott 2008; Hills 2018). Such collaborationist cre-ations adhere to the established hyperdiegesis, or the "vast and detailed narrative space, only a fraction of which is ever directly seen or encoun-tered within the text, but which nevertheless appears to operate accord-ing to principles of logic and extension" (Hills 2003, 104). Hills identifies continuity as being crucial to the stability of a coherent hyperdiegesis. Multiverses provide opportunities for newfound collaborationist approaches—premised not on a singular, linear continuity but rather on the exponential expansion of hyperdiegesis. There is no longer a singular narrative space but a "plurality of worlds," to employ a phrase by Marie-Laure Ryan (2022), and consequentially the hyperdiegesis broadens to encompass a new sense of coherence spanning all continuities and possibilities implied by the multiverse.

This expanded hyperdiegesis has complicated classifications that traditionally structured discussions of authenticity between fans and producers. Factors such as inconsistencies caused by simultaneous production on multiple projects or exponential increases in story world information and uncertainty about who is authorized to answer clarifying questions have introduced new points of contention. Debates about canonicity exemplify one complication arising from a multiversal hyperdiegesis. "Canon" indicates an authorized, authorial, or acquired installment of the story world, whereas "fanon" refers to elements determined through fan community consensus (Whittemore 2024; Seymour 2018, 338). A singular story world with static, definitive continuity often ascribes to a binary of "canon" and "not canon." The emphasis on plurality within a multiverse, however, suggests a producer wanting to outright reject an element commonly embraced in fanon must instead disprove the existence of that element within every single possibility suggested by the multiverse. Fan/producer boundaries become even further complicated when, as will be discussed relative to Spidersonas below, fan creations inhabit a newfound space of possibility where they are validated and canonization is possible. Thus, multiverses possess the potential to structure new collaborationist interactions, defined by the inclusionary tendencies of hyperdiegetic plausibility rather than the exclusionary categorizations imposed by rigid structures of canonicity.

Spidersonas and New Collaborationist Interactions

Although Spidersonas were created prior to the release of *Spider-Man: Into the Spider-Verse*—including by Noah Molinaro and Rob Cham—there was a significant increase in the number of Spidersonas shared on social media platforms in the months following the theatrical release of *Into the Spider-Verse* in December 2018 (Wisneski 2024; Molinaro 2017; Cham 2018). The Spider-Verse and its expansion to encompass tens if not hundreds of thousands of Spidersonas indicates a new engagement with a reimagined collaborationist approach that reconfigures traditional fan/producer boundaries, as demonstrated through three types of interactions: the validation of Spidersonas through paratextual materials, the featuring of eighteen Spidersona artists in *Spider-Verse* (2019–2020), and the canonical elevation of the character of Sun-Spider.

Paratextual materials engaging with fans—like letter columns that emerged in the late 1950s—are an established practice within comic book culture, but Spidersonas diverge from the historical expectations (Stein 2013, 167). Prior to February 2009, Marvel Comics maintained an open submissions policy as a way to source new talent; their submissions guide emphasized only submitting work exclusively featuring preexisting Marvel characters (CBR Staff 2009; "Submissions Guide" 2001; Quesada 2005). Furthermore, fan art contests coinciding with theatrical releases of movies from the Marvel Cinematic Universe meticulously specify that only characters appearing in the corresponding film are eligible for use ("Marvel's Avengers: Age of Ultron Fan Art Contest Official Rules" 2015; "Marvel Studios' Doctor Strange in the Multiverse of Madness Fan Art Contest | Official Rules" 2022). Official channels for engaging with fan works, including the artistic creations of aspiring professionals, have often been restricted to existing characters for legal purposes. This approach of trying to distance an authorized production from fan works to avoid litigation and accidental intermingling is well-documented in fan studies and has been articulated by industry professionals like Amber Benson and Sara Gamble (Benson 2013, 385; Zubernis and Larsen 2012, 182).

Official producers, however, publicly acknowledged the proliferation of Spidersonas—and the fact that many of these fan-designed characters were created with the intention of existing in a universe separate from the primary *Spider-Verse* characters allowed for creative separation and narrative distance. The plurality of representation actualized through Spidersonas was cited as inspiration for *Spider-Man: Across the Spider-Verse* (2023):

> In their quest to create new and diverse Spider-Heroes for the movie, the creative team was clearly inspired by the Spidersona movement that happened online when the first movie was released. 'All kinds of people really got inspired to see themselves as a Spider-Person,' recalls writer and exec producer Chris Miller. 'We got to see designs of Spider-People of every background, shape, ability and interest. [...]'
> *(Zahed 2022, 120)*

Additionally, Sony Pictures Entertainment recognized the promotional potential of Spidersonas and partnered with Vermillio to create a Spider-Verse AI engine "trained exclusively on original, authenticated content from the Spider-Verse films" that generated the user's "new Spidersona avatar set in the Spider-Verse" ("Issue II" 2023; "Get Started" 2023). The multiversal element, combined with the plurality of how many artists participated in creating Spidersonas, has enabled new collaborationist practices that acknowledge and reference fan creativity; producers can accredit inspiration to fan creations rather than attempting to maintain a degree of separation for legal purposes.

In addition to paratextual references to Spidersonas, Marvel Comics featured eighteen fan-created Spidersona designs in the back matter of *Spider-Verse* (2019–2020), a comic run launched ten months after the theatrical debut of *Into the Spider-Verse*. Several individuals employed at the Marvel Comics offices noticed the increase in Spidersonas on social media platforms like Twitter following the release of *Into the Spider-Verse* (Wisneski 2024). Nick Lowe, the editor for the *Spider-Verse* comic series, advocated for the idea to showcase Spidersonas to editor-in-chief C.B. Cebulski; the showcase was approved, and several employees—including Nick Lowe and assistant editors Kathleen Wisneski and Martin Biro—were tasked with selecting which Spidersonas were going to be featured (Wisneski 2024). Rather than duplicating the preexisting creative works that artists had shared on social media platforms, the Marvel Comics creative team commissioned original pieces of each of the Spidersonas (MacKay 2019a, 19).

Diverging from collaborationist approaches that reinforce established fan/producer boundaries—such as engaging with video game fan communities to consider player engagement or providing materials for fan productions with the intent of promotion—the collaborationist interactions relating to Spidersonas obscure professional boundaries delineating "fan" and "producer" (Jenkins 2006, 159–68; Scott 2008, 223). The featured artists were required to sign a work-for-hire contract that transferred ownership rights of their Spidersonas to Marvel Comics in perpetuity, allowing Marvel the "right to register copyrights, trademarks, and patents based on the creative design" in exchange for a one-time commission payment rather than a royalties-based structure (Bolk 2024; Poole 2024; Styles 2024; Broder 2024); artists also signed a non-disclosure agreement to keep the project a surprise until it was formally announced (Poole 2024; Bolk 2024). While this was not the first time that artistic imaginings have been acquired by Marvel Comics using a work-for-hire agreement—for instance, Marvel purchased the design for the black Spider-Man suit that became the foundation for a symbiote from Randy Schueller for $220 in 1982—Spidersonas are notable for the fact that Marvel Comics acquired more than a costume: they now had the rights to eighteen new characters (Cronin 2007; DeFalco and Frenz 2005). Kevin Bolk, creator of Spider-Manly, verbalized the way this collaborative interaction complicated fan/producer boundaries, acknowledging he is "a fan as much as I am a professional" and contextualizing his participation in this Spidersona showcase within his fifteen-year-long professional cartoonist career (Bolk 2024).

The fan/producer distinguishment was further complicated by the fact that the character of Sun-Spider, created by Dayn Broder and featured in *Spider-Verse* #3, was narrativized beyond the initial *Spider-Verse* run and became a protagonist in *Edge of Spider-Verse* #4 (MacKay 2019b; Franklin 2022). Even though Broder transferred their ownership rights of Sun-Spider, Marvel Comics reached out to Broder to further collaborate on the development of the character beyond visual design and appearance, and Broder worked alongside editor Nick Lowe, assistant editor Kaeden McGahey, and writer Tee Franklin to establish the character's personality and background (Broder 2024). Sun-Spider has also appeared in multiple issues of *Spider-Man: End of the Spider-Verse* (2022–2023). Furthermore, Broder was also surprised to discover upon the public release of the *Spider-Man: Across the Spider-Verse* promotional poster that their character was featured on it—and was even more surprised to learn Sun-Spider has a brief speaking part in the sequel film itself (Broder 2024). The possibilities for these new collaborationist interactions are well-exemplified by the fact that a Spidersona created after the release of *Into the Spider-Verse* could be included on-screen in *Across the Spider-Verse*.

Infinite Possibility and Fan Plausibility Through Spider-Verse Hyperdiegesis

The multiverse structure and hyperdiegesis inherent to the Spider-Verse catalyzed the rapid creation of Spidersonas and new modalities of collaborationist interactions; they also support new modes of meaningful participatory engagement that imagine new possibilities for the Spider-Verse. The *Spider-Verse* films have been acknowledged for their foregrounding of characters from underrepresented and marginalized communities, aptly surmised by Mav, a Spidersona artist:

> What they've done with Spider-Man, specifically with Miles Morales, where so many different people are being represented, people feel seen and they feel like, hey, I really could be Spider-Man in some universe, somewhere across time and space. Maybe I could be Spider-Man. And it just encouraged them to [create Spidersonas].
>
> *(2024)*

Spidersonas thus also indicate additional opportunities for more inclusive representation beyond what is currently present within the films, creating narrative spaces where Indigenous, neurodiverse, nonbinary, and non-Western characters can embody the role of a Spider-Person (Allen 2024; Blue 2023; Mav 2024). Many artists represent aspects of their identity through their Spidersonas that are not commonly represented in media, and that reimagining of what a Spider-Person could be was the manifestation of the infinite possibility embedded into multiverse structures.

Kathleen Wisneski—one of the assistant editors on the *Spider-Verse* series that published the eighteen Spidersonas—described the Spidersona showcase as a way to acknowledge the identities and positionalities of individuals who have not historically been given recognition within the comics industry, calling it a "wealth of perspective that we aren't able to get so easily" (2024). In my interview with Wisneski, she acknowledged Sun-Spider in particular, stating the character embodied

> a concept that maybe we never would have thought to do on our own or it could have taken a while. I'm glad that there was an artist who was able to push people in that direction, and it was a good design.
> *(Wisneski 2024)*

Dayn Broder represented their own experiences with their genetic disability, Ehlers-Danlos syndrome, and imagined the necessary adaptive technology through Sun-Spider. The collaborationist method of integrating Sun-Spider into the *Spider-Verse* comics and films has created a space for a positive media portrayal of a disabled individual that was previously absent (Broder 2024).

The collaborationist inclusion of Sun-Spider in *Across the Spider-Verse* and the multiverse structure more broadly has complicated traditional notions of canonicity. Several Spidersona artists interviewed made statements indicating they believe the multiverse structure of the Spider-Verse extends some form of canonicity to all Spidersonas:

- "I really like the fact that Spider-Verse, the way Spider-Verse works is that all of these characters could be canon in some way" (Risk 2024).
- "It's canon because the Spider-Verse can be anything. So, you know, so many different things are in the Spider-Verse. So like, you can do whatever you want and it would be canon" (Laila 2024).
- "Any creation that you feel can like put on the suit and be a hero, basically it can be canon" (Roman 2024).

Rather than traditional notions of "canon" signaling the authorization of certain creations and the exclusion of others, the multiverse suggests an inclusive hyperdiegetic plausibility—meaning that infinite possibility extends a sense of canonical possibility to fan creations within the Spider-Verse hyperdiegesis. This plausibility is a renegotiation between fans and producers with acknowledgment from industry professionals. In addition to this plausibility being affirmed through Sun-Spider's appearance in *Across the Spider-Verse*, specifically mentioned in two of my interviews, Wisneski also acknowledged: "In a way, all of the Spidersonas are canon kind of, in a way that me as Captain America isn't" (Roman 2024; Allen 2024; Wisneski 2024).

Conclusion

Infinite possibility and indicated plausibility interrelated with the multiverse allow for experimental exploration of creativity and, in this instance, the radical reimagining of cultural icons. The Spider-Verse cultivated a newfound sense of belonging that unifies fans, producers, and any individuals lingering somewhere between due to the blurring of traditional boundaries; it generates a positive feeling of plausibility, excitement, and investment on behalf of the artists without overlooking the intellectual property rights of Marvel and traditional production logics. Because of this, Spidersonas indicate a new modality for navigating collaborationist approaches with multiverses and, more specifically, the potential for multiverses that are more inclusive and indicative of imaginative possibilities beyond stories confined to a singular story world.

Acknowledgments

Thank you to Kathleen Wisneski and the Spidersona artists whose contributions shaped this project. To learn more about the artists and watch a free documentary about their Spidersonas in their own words, please visit expandingthespiderverse.github.io.

References

Allen, Tate. 2024. *Spidersona interview about Spider-Man Choctaw*. Interview by Lore Whittemore. Zoom interview.
Benson, Amber. 2013. "Blurring the Lines." In *Fic: Why Fanfiction Is Taking Over the World*, edited by Anne Jamison. Dallas, Texas: Smart Pop.
Blue. 2023. *Spidersona interview about Spider-Fool*. Interview by Lore Whittemore. Zoom interview.
Bolk, Kevin (K-Bo). 2024. *Spidersona interview about Spider-Manly*. Interview by Lore Whittemore. Zoom interview.
Broder, Dayn. 2024. Spidersona interview about Sun-Spider. Zoom interview.
CBR Staff. 2009. "Marvel Announces New Submissions Policy." *CBR*. February 27, 2009. https://www.cbr.com/marvel-announces-new-submissions-policy/.
Cham, Rob. 2024. Spidersona interview about Spider-Cham. Interview by Lore Whittemore. Zoom interview.

Cham, Rob (@robcham). 2018. "listen, there are multiple universes in Spider-Man Into The Spider-verse and honestly that means infinite universes and in one of them you could be a Spider-man. And so I think we should all draw ourselves as Spider-people. our #Spidersona if you will." Tweet. *Twitter*, October 2, 2018. https://twitter.com/robcham/status/1047151099159822336.

Cronin, Brian. 2007. "Randy Schueller's Brush With Comic History." CBR.cc Presents Comics Should Be Good. May 16, 2007. https://web.archive.org/web/20070518022241/http://goodcomics.comicbookresources.com/2007/05/16/randy-schuellers-brush-with-comic-history/.

DeFalco, Tom, and Ron Frenz. 2005. Black and White and Read All Over: The Spider-Man Extreme Makeover. Interview by Dan Johnson. 2005. "Back Issue #12." Marvel Comics, October.

Franklin, Tee. 2022. "Edge of Spider-Verse #4." *Marvel Comics*, September.

"Get Started | Spider Society." 2023. https://gb.spidersociety.com.

Hills, Matt. 2003. *Fan Cultures*. 1st ed. Hoboken: Taylor and Francis.

Hills, Matt. 2018. "Fandom." In *The Routledge Companion to Imaginary Worlds*, edited by Mark J. P. Wolf, 274–80. Routledge Companions. New York London: Routledge, Taylor & Francis Group.

"Issue II | The 'i' in Generative AI: Where Creators Are Using Gen AI." 2023. Vermillio. https://vermill.io/q2-the-i-in-generative-ai/.

Jenkins, Henry. 2006. *Convergence Culture: Where Old and New Media Collide*. New York: New York University Press.

Jenkins, Henry, and Sam Ford. 2009. "Managing Multiplicity in Superhero Comics: An Interview with Henry Jenkins." In *Third Person: Authoring and Exploring Vast Narratives*, edited by Pat Harrigan and Noah Wardrip-Fruin. Cambridge, Mass: MIT Press.

Johnson, Derek. 2013. *Media Franchising: Creative License and Collaboration in the Culture Industries*. New York: NYU Press.

Laila. 2024. Spidersona interview about Spider-Skates.. Interview by Lore Whittemore. Zoom interview.

MacKay, Jed. 2019a. "Spider-Verse #1." *Marvel Comics*, October.

MacKay, Jed. 2019b. "Spider-Verse #3." *Marvel Comics*, December.

"Marvel Studios' Doctor Strange in the Multiverse of Madness Fan Art Contest | Official Rules." 2022. *Marvel.Com*. April 5, 2022. https://web.archive.org/web/20220405005024/https://www.marvel.com/doctorstrangecontest.

"Marvel's Avengers: Age of Ultron Fan Art Contest Official Rules." 2015. Tumblr. Marvel Entertainment Official Tumblr. https://marvelentertainmentfanart.tumblr.com/rules.

Mav. 2024. Spidersona interview about Spider-Martian. Interview by Lore Whittemore. Zoom interview.

Molinaro, Noah (@gojigochi). 2017. "The piece I drew for my and @SiobhanAislinn art trade. She had me draw her "spider-sona" again! #arttrade #spiderman #marvel". Tweet. *Twitter*, October 3, 2017. https://twitter.com/gojigochi/status/915320313658986496.

Poole, Steve. 2024. Spidersona interview about Spider-Ramen. Interview by Lore Whittemore. Zoom interview.

Quesada, Joe "Senor Swanky." 2005. "Submissions Guide." *Marvel.com*. August 26, 2005. https://web.archive.org/web/20050826222056/http://www.marvel.com:80/about/submissions_guide/submissions.html.

Risk. 2024. Spidersona interview about Spider-Man China. Interview by Lore Whittemore. Zoom interview.

Roman. 2024. Spidersona interview about Disco-Spider and Inca-Spider. Interview by Lore Whittemore. Zoom interview.

Ryan, Marie-Laure. 2022. *A New Anatomy of Storyworlds: What Is, What If, As If. Theory and Interpretation of Narrative*. Columbus: The Ohio State University Press.

Scott, Suzanne. 2008. "Authorized Resistance: Is Fan Production Frakked?" In *Cylons in America: Critical Studies in Battlestar Galactica*, edited by Tiffany Potter and C.W. Marshall, 210–23. New York: Continuum.

Seymour, Jessica, ed. 2018. "Racebending and Prosumer Fanart Practices in Harry Potter Fandom." In *A Companion to Media Fandom and Fan Studies*, 333–47. Wiley Blackwell Companions in Cultural Studies. Hoboken, NJ: Wiley Blackwell.

Stein, Daniel. 2013. "Superhero Comics and the Authorizing Functions of the Comic Book Paratext." In *From Comic Strips to Graphic Novels: Contributions to the Theory and History of Graphic Narrative*, edited by Daniel Stein and Jan-Noël Thon, 155–89. Narratologia 37. Berlin Boston: De Gruyter.

Styles, Rachael "Sheeps." 2024. *Spidersona interview about Spider-Wool*. Interview by Lore Whittemore. Zoom interview.

"Submissions Guide." 2001. Marvel.com. August 2, 2001. https://web.archive.org/web/20010802120106/http://www.marvel.com/about/submissions_guide/submissions.html.

Whittemore, Lore. 2024. "'Canon' as an Emerging Meaning-Making Process." In *Society for Cinema and Media Studies Annual Conference*. Boston, MA.

Wisneski, Kathleen. 2024. *Spidersona interview about Spider-Verse (2019–2020)*. Interview by Lore Whittemore. Zoom interview.

Zahed, Ramin. 2022. *Spider-Man: Across the Spider-Verse: The Art of the Movie*. New York: Abrams.

Zubernis, Lynn, and Katherine Larsen. 2012. *Fandom at the Crossroads: Celebration, Shame and Fan/Producer Relationships*. Newcastle upon Tyne, UK: Cambridge Scholars.

19

TURTLEVERSE!

The Teenage Mutant Ninja Turtles Multiverse in Media as a Tool to Explore Diverse Genres

Ricardo Victoria-Uribe and Nazario Robles-Bastida

The multiverse as a concept has gone from an explanation in philosophy and physics that seeks to elucidate the possibilities of the universe we live in—and to reconcile the strange realm of quantum physics with the reality we experience—to a tool widely used in storytelling, in particular science fiction and to a lesser degree fantasy. As exemplified by the Marvel Cinematic Universe (MCU), the multiverse is used these days to create epic narratives where the fate of the universe is at stake. Nonetheless, the multiverse as a narrative tool goes beyond the creation of this type of epics; it also allows storytellers to explore diverse genres by putting their core cast of characters through different narrative styles, be they adventure, horror, romance, etc. The multiverse offers an opportunity to muse about how a character—and by extension the audience—might react under different circumstances to those of the main continuity, without derailing the canon narrative. In these stories, characters are often forced to confront themselves, not only at a spiritual level, but also at a physical level, by encountering their alter egos from different universes. Franchises such as *Star Trek*, with the Mirror and the Kelvin universes opposed to the main timeline, *Sliders*, a trip into alternate history, the MCU, and the Spiderverse, to mention just a few, have allowed media producers to hold a mirror, sometimes a dark one, to their characters and stories to create new narratives that more often than not, exist in a different genre than their original counterpart.

DOI: 10.4324/9781003480846-22

The aim of this chapter is to explore how the multiverse explores different genres in storytelling, through diverse iterations of the same core cast, though a less-studied case study: *The Teenage Mutant Ninja Turtles* (TMNT) franchise. Currently a complex multiverse, the worlds of TMNT has been built on a core cast of characters where each iteration has had a different storytelling focus, that, while still maintaining similar narrative ideas and origins for the Turtles, offers different textual possibilities, as they often explore different genres. While the TMNT rarely interact with their other selves, each version is considered canon within this multiverse, taking the same narrative core beats that distinguish the Turtles from other franchises but introducing enough differences in terms of storytelling and genre to enable the grown of an ever expanding multiverse, with each of this versions geared to a different audience, a fact that has allowed the franchise to remain fresh after 40 years.

Multiverse in Theory

If a universe can be explained as a group of elements sharing the same spacetime, a multiverse is a group of universes that might have or not the same elements, with subtle or major changes, that might share a same point of origin, a bifurcation point, but will have different endpoints. It is the idea that out there, in some extradimensional plane, there exists a different universe from us, where a distinct version of us has a different life, be it in minor or in major aspects. There are two different concepts of multiverse. One comes from the inflationary cosmology, the theory most people know as the Big Bang, where different regions of spacetime, self-contained as pocket universes or extra dimensions, have their own rules but remain part of the same universe, we just cannot observe them (Carroll 2023). The other comes from quantum physics trying to reconcile its postulates with the world we experience every day. This interpretation is known as "many worlds" and explained in crude terms, means that every decision or event taking place in our universe, can have two (or more) outcomes, and said outcomes become their own reality, expanding into different versions of the universe, going from barely discernible changes to those were life as we know it cannot exist (Carroll 2023). This latter version is the most well-known by audiences, thanks to science fiction stories presented in books, movies, TV shows, and videogames (i.e., *Star Trek*, the MCU, *Sliders*, to mention a few). However, conceptually speaking, the concept of multiverse is not new when it comes to narrative genres. It is not even a new concept in terms of scientific theory or cultural beliefs. One of the earliest appearances of the concept can be traced to religion.

Diverse religious beliefs around the world mention different planes of existence, multilayered cosmologies, and previous worlds created and destroyed. For example, Hinduism, through the Puranic texts of the *Bhagavata Purana*, mentions countless universes, enveloped by other universes, like an onion (Alonso-Serrano and Jannes 2019) In terms of media, while there are earlier mentions to this idea in H.G. Wells books the origin of the use of the concept can be traced to the science fiction pulp magazines and the story *Sidewise in Time* by Murray Leinster, first published in 1934 in *Astounding Stories* (McMillan 2022). Since then, the idea of a multiverse as a tool for writers that allows exploring diverse narratives and genres with the same cast, or alternative versions of the same characters brought up through different circumstances, has been a staple of comics and science fiction literature, that has expanded gradually to other media.

Multiverse in Media and Genre Exploration

The notion of genre has experienced important transformations since its inception in ancient Greece. As Jenkins asserts in conversation with Ito and Boyd (Jenkins, Ito, and Boyd 2016), scholarship on television and film genres no longer conceptualizes genres "as rigid formulas or sets of fixed textual features" (65). Instead, we must think of genres as possessing fluid boundaries that are often influenced by the *interpretive strategies* used by readers to decode texts. From this perspective, reading—and we would argue also writing—a genre text is a question of both social and literacy background (Jenkins, Ito, and Boyd 2016). Genres are then, not simply inscribed within a text but emerge at the point where the text encounters the reader. It is at this moment when textual boundaries are established and a piece of media becomes *horror*, *science fiction*, etc.

Of course, this does not mean that a text is just a disconnected series of narrative images, textual features, and vignettes that can be used by the readers to define genre boundaries in any way they desire. Contrary to a model of audiences as individual consumers of media that appropriate texts "to suit specific and often incommensurable needs" (Cartmell et al. 1997, 3), every text possesses a certain degree of interpretative closure. In his discussion of the film *Casablanca*, Umberto Eco (1990) explains that while texts are always open to interpretation and appropriation by audience members, they also must display certain textual features, some specific quality, to become a cult object. He goes on to explain how through these textual features, a film, book, etc., becomes "a sort of textual syllabus, a living example of living textuality" (199). This living textuality is precisely the genre into which a given text belongs, as a function of certain textual features that allow readers to classify it as *this* (horror, for example) and not *that* (a Western). That this classification results from the *genre literacy* of the reader must always be kept in mind as to avoid falling back into the idea of genre-as-rigid-formula.

Essential to the idea of living textuality and to the possibility of classifying a text into a genre, is Eco's notion of "intertextual archetype." He defines this concept as "a preestablished and frequently reappearing narrative situation, cited or in some way recycled by innumerable other texts and provoking in the addressee a sort of intense emotion accompanied by the vague feeling of a déjà vu" (Eco 1990, 200). The intertextual archetype connects the text not only with other similar texts but also with the readers, inscribing a specific piece of media into a particular genre which boundaries are defined by the interpretive strategies of audience members. As explained by Eco (1990), this process also brings forward intense emotions on the part of this same audience members, a fact that helps to explain the strong attachment that some of them develop toward specific texts and genres.

Enter the multiverse. This narrative tool allows both writers and readers to increase the living textuality of the franchise in which they take part. By being able to craft alternative iterations of their characters, that exist in a different universe to the original one, creators can incorporate to their work new "intertextual archetypes" that allow their texts to "speak" with a new set of texts, located in a completely different genre. Fans of the franchise, for their part, can experience these new archetypes in a new space that at once protects their beloved original and offers new insights and emotions. What would happen if Iron Man became a zombie and Winter Soldier had to kill him? What would Kirk do if he were given command of the USS Enterprise at a younger age to his original counterpart? These are the types of questions that both writers and readers can explore within the multiverse; the living textuality of the franchise is expanded in this way, without the risk of altering the original narrative too much.

Multiversal Turtles

When Kevin Eastman and Peter Laird came up with the concept of a turtle holding a nunchaku back in the 80s, little did they know that a behemoth franchise and an expansive multiverse would spawn from that sketch. TMNT is now a ubiquitous franchise including comics, movies, animation, live-action series, videogames, toys, books, concerts, and crossovers with other franchises, be it Power Rangers, Batman, Usagi Yojimbo, Street

Fighter, Ghostbuster, or Stranger Things. Each version of the Turtles is different to the others, in tone, narrative, and even genre, but the core cast of Leonardo, Donatello, Michelangelo, and Raphael, plus Master Splinter and April O'Neil remain the same, with each interaction exploring themes such as family, honor, heroism, love, violence, identity, loneliness, teenagerhood, legacy, and redemption.

Interestingly, all these versions of TMNT are canon. Well before the MCU introduced the concept of multiverse to a wider part of their audience (or at least the non-comic part) in the current century, the Ninja Turtles did it in 2009 as part of the twenty-fifth anniversary of the franchise. The tool used for that was the animated movie *Turtles Forever* (Burdine and Goldfine 2009) which served as a coda or epilogue for the 2003 animated version of the Turtles, considered as grittier and more faithful to the Mirage comics, while still being toned down for TV broadcasting. In this movie, a multiverse was established, by joining two different cartoon iterations of the four Turtles (1987 and 2003) against their respective villains that were trying to get rid of the Prime Turtles—the Mirage version of the original, more violent Turtles—to erase countless TMNT universes. This movie firmly established that every version of the Turtles that has ever existed is part of the same multiverse, as it showcased all of them in several screens of the multidimensional machine feature in the film. At the end, three versions of the Turtles, including the Prime Turtles, save the day, defeating the villains and restoring the multiverse. The movie ends with the final vignettes of the original Mirage comic, with the Turtles striking hard and fading away into the night. A further crossover between the 2012 and the 1987 versions of TMNT cemented this multiverse, which includes the live-action movies, the sentai shows, the 2007 CGI animated movie, comics, videogames, etc.

These varied iterations of the TMNT have allowed different creative teams to explore different genres, among them: Noir, Martial Arts, Comedy, Science Fiction, Fantasy, Horror, Musicals, and Sentai. With each iteration, the world of TMNT expands not only in narrative terms, but also on account of certain textual features and intertextual archetypes that characterize diverse genres. In the different TMNT stories, you can find critters from other dimensions, aliens, mystic sages, multiverse tournaments, ghosts, sentient robots, immortals, reincarnation and New York City. As a result of the creative freedom granted by this multiverse to writers and artists, the world (or worlds) of the Turtles no longer has clear boundaries between different genres and resists any simple classification. Different elements, meanings, and plot developments offer readers a unique textual experience that, very much like its mutant protagonists, is everchanging, transformative, and complex.

Exploration of Genre through Chelonian Multiverses

A discussion on the TMNT multiverse must begin with the original version of the Turtles, the Mirage Studios version. As discussed in the previous section, this is one of the most violent versions of the TMNT. Interestingly, this original version was already an experiment on intertextuality, given that its first issue was heavily inspired by the more mature Daredevil comic created by Frank Miller for Marvel. In an example of the use of intertextual archetypes, Eastman and Laird spoofed several elements from this comic book: Matt Murdock aka the vigilante hero Daredevil, was trained by talented martial artist "Stick" to fight the group of evil ninjas known as "The Hand." In the TMNT comic, the latter two became the mutated martial artist rat "Splinter" and the evil ninjas "The Foot," respectively (Bisges 2008).

This original issue of TMNT is a self-contained history of revenge in modern New York City, which ends with the death of Shredder at the hands of the Turtles. In line with its origins as a spoof of Daredevil, it uses textual features closely aligned with the superhero genre. However, the success of this first comic after its publication in 1984, led to the transformation of this one-shot into a comic book series. Eventually, the increasing popularity of Ninja Turtles led to a cartoon spin-off (Bisges 2008). This cartoon, released in 1987, toned down the grittier aspects of the original comic, added a lot of humor, visual gags, and a more dynamic and colorful element to the story. To date, this version constitutes the most fondly remembered and well-known iteration of the Turtles, having influenced a whole generation of fans that grew up with them during the 1980s. Having inspired most TMNT videogames, the lasting influence of TMNT 87 can be seen also in the fact that, as discussed in the previous section, they took part in the first multiversal adventure of the Turtles, *Turtles Forever*.

The textual differences between the Mirage and the '87 version of TMNT, due to a change in intended audience (from adults to children) and to a change of medium (comic book to cartoon) already illustrate a genre change for the Ninja Turtles. Moving from the superhero genre to the comedy adventure genre that characterizes many American cartoons in the 1980s, TMNT began to expand its living textuality by playing with intertextual archetypes from different genres. At this point, keeping the narrative of each version separate from the other was an easy task, given that they existed in different media without much crossover in terms of audience. Things became more complicated once a live-action movie was greenlit.

Teenage Mutant Ninja Turtles (Barron 1990) capitalized on the success of both the comic book and the cartoon. In another example of intertextuality, the movie takes a few elements from the cartoon to appeal to a wider audience but adapts the first arcs of the more violent Mirage comics, leaving most of its grittiness intact. Taking place mostly at night, the movie constructs a noir urban world, complete with a zealous journalist (April O'Neil) and a mysterious criminal gang that threatens New York City and its citizens. The violence, while not overt and bloody, is not simulated either. Contrary to the cartoon, the Turtles do not take on robots, but very human ninjas. The death of Shredder, while not too graphic, is implied to be gruesome as he gets crushed by a trash compactor. At the end, while retaining elements from its predecessors, this film stresses noir, and martial arts textual features, once again exploring the boundaries between genres.

It should be noted that the multiverse has not yet been introduced in the world of TMNT at this point. However, the exploration of genres and intertextual archetypes that happens through these different iterations of the franchise, already announces its genesis. As discussed in the section above, the TMNT multiverse was not introduced until 2009 with *Turtles Forever*. We have already discussed two of the turtle teams that take part in this film, the Mirage and '87 versions. The other main team in the movie, the 2003 animated version of the Turtles, is also relevant to this discussion of genre and the multiverse. In one of the arcs for this version of TMNT, the four Turtles join the Battle Nexus, a multidimensional martial arts tournament previously won by Splinter and Shredder. They cross paths in this tournament with Usagi Yojimbo (created by Stan Sakai) possibly the character that has had more crossovers with the Turtles. The arc follows the traditional martial arts tournament structure that characterizes several shonen anime, with a mystery side plot that reaches its climax after taking over the tournament arc.

The martial arts tournament has been a staple of shonen anime for a long time. From this perspective, the intertextual archetypes used by the 2003 animated version of the Turtles cites and interacts with texts created outside of North America, a development perhaps not surprising given the Japanese influences present in the original comic. This Japanese influence is also clear one of the most recent iterations of the world of TMNT which has taken fans of the Turtles by storm: *The Last Ronin* (Eastman, Laird and Waltz 2020).

Released in 2020 by IDW publishing, this comic book miniseries takes place in an alternative future where all Turtles except for one have been killed. The surviving turtle must traverse a dystopian world to avenge the death of his brothers. Full of violence and futurist elements, *The Last Ronin* once again creates a new narrative universe that in this case, uses intertextual archetypes associated with cyberpunk and dystopian fiction to explore themes of loss, death, and loneliness. The degree to which these themes are explored within this story would be impossible to achieve in other TMNT narratives, as in most universes, the brothers remain alive and together. The fact that *The Last Ronin* was released after the inception of the TMNT multiverse is also significative as it exists in its own official universe, aptly called the "Roninverse." With both a videogame and a film in development, *The Last Ronin* illustrates the myriad of narrative possibilities of the TMNT multiverse.

To end this discussion on genre and the TMNT, it is interesting to mention a recent addition to the narrative worlds of TMNT, the film *Teenage Mutant Ninja Turtles: Mutant Mayhem* (Rowe 2023). The most recent iteration of the Turtles explores topics such as isolation, family, forming part of a group, doing things beyond a parent's wish or expectation, and growing up. Contrary to the dark tone of *The Last Ronin*, *Mutant Mayhem* is a hopeful movie that endeavors to make the Turtles into actual teenagers with its corresponding problems and attitudes, to the point that, for the first time in their publication history, the film ends with them attending a regular high school, finally interacting with human teenagers. Using intertextual archetypes that correspond to the coming-of-age genre, the film once again presents new interpretations of the Turtles and their world.

Conclusion

The diversity of creators, comic imprints, media companies, etc. that have participated in the creation of the TMNT multiverse is significant. As such we have focused our discussion regarding this case study on some examples that we found illustrative in terms of genre and multiverse. By no means does our analysis include all TMNT versions and media texts, as there are many other versions left outside due to space constraints. Nonetheless, our examples shed light on the complex nature of this narrative multiverse and how different iterations of TMNT use different textual features and intertextual archetypes to explore a variety of genres. Either as mutant warriors in a superhero world, as lone ninjas seeking revenge for their brothers or as teenagers going to high school, the TMNT continue to change and evolve, pushing the boundaries of genre through the multiverse.

References

Alonso-Serrano, Ana, and Gil Jannes. 2019. "Conceptual Challenges on the Road to the Multiverse." *Universe (Basel)* 5 (10): 212.

Barron, Steve, dir. 1990. *Teenage Mutant Ninja Turtles*. New Line Cinema.

Bisges, John. 2008. "Turtle Power!: How Four Mutant Teenagers Nuked the Entertainment Industry." *The Journal of Popular Culture* 41 (6): 918–33.

Burdine, Roy, and Lloyd Goldfine, dirs. 2009. *Turtles Forever*. The CW4Kids.

Carroll, Sean. 2023. "Are Many Worlds and the Multiverse the Same Idea?" *Discover Magazine*, July 2023.

Cartmell, Deborah, Heidi Kaye, Imelda Whelehan, and I.Q. Hunter. 1997. *Trash Aesthetics: Popular Culture and Its Audience*. London: Pluto Press.

Eastman, Kevin, Peter Laird and Tom Waltz. 2020. *The Last Ronin*. San Diego: IDW Publishing.

Eco, Umberto. 1990. *Travels in Hyperreality*. San Diego: Harcourt Brace.

Jenkins, Henry, Mizuko Ito, and Danah Boyd. 2016. *Participatory culture in a networked era*. Cambridge: Polity Press.

McMillan, Graeme. 2022. "A brief history of the multiverse - the term, and its proliferation at DC, Marvel, and well... everywhere." *Popverse*, June 2022.

Rowe, Jeff, dir. 2023. *Teenage Mutant Ninja Turtles: Mutant Mayhem*. Paramount Pictures.

PART III

Multiversal Constructs Across Literature and Games

20

THE COMIC BOOK MULTIVERSE

Creative Playground / Corporate Plaything

Blair Davis

The multiverse is best known by mass audiences from recent Marvel and DC Comics films like *Spider-Man: No Way Home*, *Spider-Man: Into the Spider-Verse*, *Doctor Strange in the Multiverse of Madness*, and *The Flash*. But this recent resurgence of the multiverse in popular culture has a long history, one that dates back to the early 1960s. The 1961 story "Flash of Two Worlds!" from DC's *The Flash* #123 marked the beginning of the comic book multiverse, which has remained a constant narrative trope ever since. From *The Flash*, *Justice League of America* and *The Avengers* in the 1960s and 1970s, to major company-wide crossover events like DC's *Crisis on Infinite Earths* and *Zero Hour* in the 1980s and 1990s, the multiverse was used across dozens of titles to tell stories with an ever-widening interdimensional scope. By the early 2000s, multiversal storytelling became even more commonplace as events like *Infinite Crisis*, *Flashpoint*, *Spider-Verse*, *Secret Wars*, and *Dark Crisis* occurred almost annually at both Marvel and DC.

Disney and Warner Bros. eventually used the concept of the multiverse as a way to create multiple versions of their numerous characters while developing new phases of their cinematic superhero franchises. Along with creating a larger, multi-universe mythos for their heroes' adventures, the multiverse also allowed studios to build narrative connections with prior iterations of these characters' filmic adaptations which hadn't been thus far considered part of the current franchise's continuity. But by the time

DOI: 10.4324/9781003480846-24

Hollywood started using this formula (with its dual benefits, both commercial and creative), comic book publishers had already used the multiverse for similar aims for decades. The multiverse allowed previous incarnations of characters whose titles had previously been canceled and/ or rebooted to be tied more directly to newer versions, thereby connecting older series and storylines to current ones.

Such efforts are not done purely for creative reasons. Along with the creation of a bigger sandbox for their characters to play in, the multiverse has larger corporate implications. By building new narrative connections between previous iterations of characters, publishers have revitalized previously outdated entries in their back catalogs. In this way, the multiverse is both a narrative playground for creators to build stories with as well as a corporate plaything for corporations to wield whenever their products (and profits) need revitalizing.

Multiple Earths Emerge

At the core of multiversal storytelling is the idea that an established narrative world does not exist in isolation within time and space. Rather, that world exists in alternate forms and permeations via a multitude of universes. The multiverse therefore consists of different worlds in parallel universes in which a single character exists in related (but still distinctive) variations. DC Comics made it a regular selling feature of their stories by building a larger narrative continuity around the idea, beginning in the 1960s.

In 1961's *The Flash* #123, the titular super-speedster Barry Allen meets his counterpart Jay Garrick from an alternate earth in the story "Flash of Two Worlds." Both men are versions of the heroic Flash in their respective worlds who meet when Allen breaches the barrier between universes: "The way I see it, I vibrated so fast -- I tore a gap in the vibratory shields separating our worlds!" explains Allen. Writer Gardner Fox faced two challenges in making this story plausible to readers—one a narrative hurdle, the other an editorial complication. The narrative challenge involved creating a pseudo-scientific rationale for how multiple earths might exist. Physicists had not yet developed plausible concepts for the potential existence of parallel universes by 1961, with developments such as string theory arriving as "a universal theory of microphysics" by 1974 (Dawid 2013, 10). String theory, explains Richard Dawid (a physicist and a philosopher), allowed for a "multiverse scenario of infinite inflation," with the multiverse serving as "a huge, maybe infinite number of universes in an inflationary background space" within the cosmos (90; 149). But Fox has Barry

Allen, a police scientist, propose a theoretical coexistence of worlds based on differing vibrations:

> My theory is that both earths were created at the same time in two quite similar universes! They vibrate differently -- which keeps them apart! Life, customs - even languages – evolved on your earth almost exactly as they did on my earth! Destiny must have decreed there'd be a Flash... on each earth!" he explains.
>
> *(Fox 1961, Flash 123, 8–9)*

Barry Allen's debut as the Flash came five years earlier in 1956's *Showcase* #4, written by Robert Kanigher. Here we find him reading an old issue of *Flash Comics* starring Jay Garrick, a series that had actually been published throughout the 1940s. "I wonder what it would really be like - to be the fastest man alive on earth? Well... I'll never know – The Flash was just a character some writer dreamed up!" (Kanigher 1956, 3) This, then, is the second hurdle Fox faced in having two versions of the same character meet. Barry's debut as the rebranded Flash came only seven years after the Garrick-led series ended in 1949, and many comics fans would surely have recalled those earlier adventures. To clarify this conundrum without adding any new backstory to explain Garrick's recent inactivity, *Showcase* #4 repositioned the original Flash as "just a character" in a comic book rather than a preexisting hero in Allen's narrative world.

The challenge came in reconciling how the newer stories about Barry Allen handled what narrative theorists term the "diegesis," or the world of the story. When DC debuted their new Flash in 1956, they didn't want the character to inhabit the same diegesis as Garrick in a way that might cause readers to wonder why there were two different versions of the character, so they framed the original speedster's adventures as mere stories that existed in Allen's narrative world. In this way, Garrick's adventures became a minor "intradiegetic" detail, allowing the older tales from *Flash Comics* to still exist "inside the story" for readers in a way that connected the past to the present with minimal explication (Shen 2005, 107). Media scholar Andrew J. Friedenthal (2019) writes in *The World of DC Comics* of how Kanigher's "metatextual twist" of making Garrick's adventures mere stories that Allen had read in his youth "was likely more of an in-joke, a knowing nod for fans old enough to remember the Golden Age Flash, it would ultimately prove to be the first glimmering of the DC Multiverse" (22).

But the multiverse was indeed still only a potential glimmer in Barry Allen's eye in 1956. When Fox (who penned Garrick's original tales from the 1940s, co-creating the character in *Flash Comics* #1) wrote *The Flash*

#123, he had an opportunity to reestablish Garrick's 1940s adventures as a more direct part of DC superhero continuity once more with the introduction of multiple earths. Fox had been writing Allen's adventures as the Flash in the pages of *Justice League America* (and before that, in three issues of *The Brave and the Bold*) since 1960, so had already written several stories about the new Flash before using the multiverse as a framework to reinvent how readers understood DC's Golden Age tales. Fox even inserted himself into the story—as an unconscious multiversal moderator: "A writer named Gardner Fox wrote about your adventures – which he claimed came to him in dreams! Obviously when Fox was asleep, his mind was 'tuned in' on your vibratory earth!" says Allen. When Barry tells him that the series was canceled in 1949, Garrick confirms that to be the same year he "retired" from crimefighting (Fox and Infantino 1961, 10).

In positioning himself as the fulcrum between the story worlds of the two eras of Flash comics, Fox not only reclaimed a prior decade's worth of stories that had been written out of the publisher's narrative continuity by Kanigher, he also opened up new worlds of storytelling possibilities by introducing parallel earths as an ongoing concept at DC. Most comic book historians consider Barry Allen's 1956 debut as the Flash as the beginning of the "Silver Age" of comics, replacing "Golden Age" predecessors like Garrick (see Schoel 2019). Yet as significant as that transition was for superhero comics, the arrival of the multiverse in *The Flash* #123 added a new level of narrative continuity—and complexity. Not only did this new approach solidify the fact that a new era of comic book storytelling was well underway, it also established key tropes and themes that DC would return to, reinvent, and rebrand in each and every decade that followed.

Alternate realities were relatively uncommon in comic books until the 1960s. In 1953, Kanigher wrote Wonder Woman's meeting with a parallel version of herself from a twin earth in "Wonder Woman's Invisible Twin" from *Wonder Woman* #59, but the concept of parallel earths wasn't brought back in the series thereafter. By the end of the 1950s, variations on existing heroes like Batman and Superman typically came in the form of what would eventually be termed "imaginary stories." They weren't considered part of a character's regular narrative continuity, but were usually contextualized as a dream, a computer simulation or other practical means. In 1959's *Batman* #127, the tale "The Second Life of Batman" finds the hero witness what his life would be like if his parents had not been murdered, thanks to a computer's help. Similarly, in *Superman* #132, a computer shows us what the Man of Steel's life would have been like had Krypton not exploded.

The following year saw *Superman's Girl Friend, Lois Lane* #19 feature what it termed the "First of an Imaginary Series" in the story "Mr. and

Mrs. Clark (Superman) Kent!," in which the pair are wed. Readers, we are told, had been requesting

> an imaginary story showing how things would be if Lois Lane learned Clark Kent were Superman and married him! This is only the first of many such tales which could very well happen in the future lives if Lois Lane and Superman, but perhaps never will!.
>
> *(Siegel 1960, 23)*

Such stories were playfully speculative, regularly pondering the romantic entanglements of their characters in a way that later generations of readers would do via the "shipping" practices of fanfiction which "hypothesize an unfolding relationship between the series' protagonists" in various pop culture texts (Scodari and Felder 2000, 238).

But these imaginary stories weren't referred to again in later issues. The traditional norms of comic book storytelling in this era emphasized self-contained narratives within a single issue rather than serializing tales across multiple issues—a strategy which allowed readers who hadn't read earlier issues to enjoy a story on its own merits without feeling lost. But the introduction of the multiverse helped to change those norms. When Jay Garrick returned six issues later in 1962's *The Flash* #129, Fox takes only two panels to recap the concept of parallel worlds, in textual captions which announce:

> "For as *Flash* readers know, there is another earth – almost an exact duplicate of our own – where the Flash is not Barry Allen but an older man named Jay Garrick... Yes, reader, in this other world the Flash is Jay Garrick, who was made super-swift when he knocked over a retort filled with hard water and inhaled the fumes..."
>
> *(Fox and Infantino 1962, 2)*

This direct appeal to the reader assumes and/or encourages a certain familiarity with previous storylines, much as it stresses an ongoing narrative continuity for its characters. By the time that Garrick and Allen teamed up for a third adventure in June 1963's *The Flash* #137, even less time is spent establishing the multiversal premise of parallel worlds and dual speedsters. As Allen gazes at a map, he muses to himself, "I'm the only person in my world who's ever seen *this* map of Earth! This is the Earth of my good friend Jay (Flash) Garrick, where there's a Keystone City in the place of Central City where I live!" (Fox and Infantino 1963, 3) As these adventures on multiple earths continued each year in the pages of *The Flash*, the multiverse premise soon became an established and regular fixture of DC

Comics stories across many of its series in the 1960s. Not only did this strategy give the publisher a new form of spectacle to market to comics fans, it also revitalized DC's brand at the same time as Marvel Comics' new superhero line grew in popularity throughout the decade.

Just two months after Garrick's third encounter with Allen in *The Flash* #137, DC expanded the scope of their multiversal team-ups in 1963's *Justice League of America* #21 and #22, their premier superhero team consisting of beloved characters like Superman, Batman and Wonder Woman and the Flash. Here, the Justice League of America (JLA) teamed up with their Earth-Two counterparts, the Justice Society of America (JSA). Made up of Golden Age heroes like Doctor Fate, Hour Man, Black Canary and the Atom, the JSA originally appeared in DC's *All Star Comics* between 1940 and 1951. The two supergroups teamed up for a two-part adventure (with #21's story title "Crisis on Earth-One!" followed by "Crisis on Earth-Two!" in the next issue) that saw each traverse the vibrational barrier between worlds to visit their earthly counterparts and thwart a group of supervillains that have captured both Flashes.

The two-part story cemented the reclamation of DC's Golden Age narrative continuity, permanently folding it into the publisher's ongoing Silver Age storytelling by offering a creative rationale for the JSA's absence since 1951 (one that is kinder to our heroes than the industrial logic of having their series canceled due to low sales). The JSA's time away is likened to an early retirement, with Fox explaining how the group has reunited "after more than a decade of inactivity..." The writer adds: "True, there are a few grey hairs showing – and their faces are lined with the passage of time – but their mighty powers are only slightly dimmed...". (Fox and Sekowsky 1961, 3). By bringing the heroes out of retirement, DC not only created a multi-decade continuity for their expanded roster of superheroes, but also developed a new type of product to sell readers: epic-length, multi-issue narratives that enticed/required readers to pick up the next issue to see how things resolved. In subsequent decades, such stories grew ever-larger in scope as the corporate logic of franchising affected comic book publishing trends.

From Two to Three to Infinite Earths

At the end of *Justice League of America* #22, one of the thwarted villains posits the existence of a third Earth ("I've got it! There is an Earth-One and an Earth-Two! Somewhere there must be an Earth-Three!"; Fox and Sekowsky 1963a, b, 22). Sure enough, the next time the JLA and JSA teamed up in 1964's *Justice League of America* #29 and #30, the heroes battle a threat from Earth-Three that, we're told, "marks the beginning of a spectacular interdimensional struggle that shakes the very structure of three worlds!" (Fox and Sekowsky 1964, 1). Such interdimensional crises became an annual tradition in the pages of *Justice League of America* throughout the 1960s and 1970s, as well as in issues of *The Flash* and *Green Lantern*.

Marvel Comics also ventured into multiversal storytelling by the end of the 1960s, introducing a new superhero group within the pages of *The Avengers* that would soon become known as the Squadron Supreme—heroes from another dimension whose powers and costumes somewhat resembled those of the Justice League (meant as a winking imitation of rival publisher DC's leading lineup of superheroes).[1] The Squadron Supreme returned in multiple issues of *The Avengers* between 1975 and 1976 before gaining their own titular twelve-issue mini-series in 1985.

This period also saw regular multi-earth reunions of the JSA and JLA each year in *Justice League of America* through 1984. These team-ups used the term "crisis" as a recurring label (and marketing tagline), in keeping with the earlier story titles like "Crisis on Earth-One." One such adventure opens with the description, "Each year for the past several years, the members of these two societies have met to confront a new, seemingly annual crisis... These meetings have led to friendships... and rivalries... and a deep sense of comradeship..." (Conway and Dillin 1979, 2). Such rhetoric suggests ongoing growth in characterization, with the serialized format of DC's comic books allowing an ever-building narrative continuity—much like at Marvel in the pages of ongoing series like *The Avengers*.

By the mid-1980s, multiple decades worth of stories were well entrenched as a backstory for these multiversal meetings. As 1979's *Justice League of America* #171 explained at the outset of a new JLA/JSA crossover: "There are as many possible worlds there are decisions to be made in a single day, for each decision is a choice, and for every choice there are at least two alternatives..." (Conway and Dillin 1979, 2). But DC soon became concerned that their ever-widening array of parallel earths might be causing potential readers too many "decisions" in their purchasing choices. They were losing readers to Marvel, which launched new a strategy in the early 1980s to gain a wider audience: special events in which dozens of heroes joined forces, like 1982's *Marvel Super Heroes Contest of Champions* and 1984's *Marvel Super Heroes Secret Wars*. Both series saw the publisher's most popular superheroes confront new cosmic threats together—but without the need to explain an increasingly complex, decades-long history

---

1 The characters that eventually became known as the Squadron Supreme in the February 1971 issue of *The Avengers* (#85) were a revamped version of the villainous Squadron Sinister, which was first seen in *The Avengers* #69 (October 1969).

of multiple worlds to their audiences. Many DC insiders worried that by drawing on what was by then over forty years' worth of storytelling in their alternate-earth-adventures, new readers increasingly saw this as a hindrance rather than a selling point.

DC's solution was 1985's *Crisis on Infinite Earths*, a twelve-issue miniseries that served as both a significant event in its own right as well as the flashpoint for a revamping of numerous series moving forward (with long-running titles like *Superman*, *Wonder Woman*, *The Flash*, and *Justice League of America* coming to an end, replaced by new series with the sales benefit of new number-one issues). Legal scholar Benjamin Authers (2012) notes how *Crisis on Infinite Earths* was spurred by both creative and corporate factors, with the series being "propelled both by the valuing of a complex narrative that draws on multiple aspects of DC's publishing history, and by a commercial desire to constrain that complexity." Several of DC's creators and editors, he says, believed the publisher's narrative continuity had become so "byzantine" by the 1980s that it was "discouraging new readers," leading to what writer Marv Wolfman called the need to "straighten out and simplify" things in the hopes of increasing sales (66).

Crisis on Infinite Earths was the apex of DC's multiverse: a major event that tied in to nearly every ongoing DC series between 1985 and 1986. In turn, it drew readers' attention to the grandeur and spectacle of multiversal storytelling in a way that the publisher had never done before, bombarding comics fans with an inescapable storyline that stretched across dozens of titles. The irony is that if DC really did want to make it easier for readers to navigate their narrative continuity, they did so in form of the most editorially complex cross-over event the comic book industry had ever seen.

Crisis on Infinite Earths proved consequential in its legacy. Major characters like Supergirl, the Earth-2 Superman, and Barry Allen were killed off, seemingly. Long-running titles such as *Superman*, *Wonder Woman*, *Justice League of America*, and *The Flash* were canceled or revamped, with new series, narrative continuities and even new characters replacing them. The relaunched versions of *Superman* and *Wonder Woman* jettisoned their prior story histories, chronicling the seemingly early-days of their heroes anew as a way of enticing readers looking for a fresh start with a new product line.

Such rebranding exercises were a symptom of a new corporate logic among comics publishers that took hold by the mid-1980s. Comics scholar Sean Guynes (2019) notes how "*Crisis* exemplified the very process of collaboration between creators and corporate stakeholders that enlivened and fueled franchising in the era of media conglomeration," using the multiverse to create an extreme example of intra-industrial franchising in the midst of the 1980s mediascape, since the story captured all of DC's IP [intellectual property] into one series that narrated the internal relations of as many properties to one another as possible.

(184)

By the 1990s, crossover storylines were commonplace at most major comic book publishers. At Marvel, numerous series intersected for such events as "The Infinity Gauntlet," "Onslaught," and "Maximum Carnage" (which had its own associated video game).

Marvel mostly stayed out of the multiverse, however, having already established their series *What If…?* as a forum for alternate histories and incarnations of their heroes. DC followed suit, creating a separate imprint called Elseworlds in 1989 to tell stories about alternate versions of their characters, with the stated caveat that readers should understand such tales as the equivalent of from the "imaginary stories" of the 1950s and 1960s, much like in *Superman's Girlfriend, Lois Lane*. Such stories begin by informing us, "In Elseworlds. heroes are taken from their usual settings and put into strange times and places – some that have existed, or might have existed, and others that can't, couldn't or shouldn't exist." (*Elseworlds: Batman*, 2016) In this way, creators could explore alternate possibilities for characters without such stories becoming part of DC's narrative canon.

But as much as they sought to simplify their narrative continuity and to keep separate multiple variations of popular characters, DC couldn't seem to help dipping multiple toes into the multiverse pool throughout the 1990s. In 1994, *Zero Hour: Crisis in Time* attempted to resolve what writer/artist Dan Jurgens (2019) described as "continuity bugs [that] were popping up throughout the DC Universe." At the same time, franchise logic dictated that the series could "provide a platform for each individual DC title to cross over with the story," by using alternate timelines as a way for the publisher's heroes to resolve yet another crisis (7–8).

That same year also saw a separate multiversal crossover in *Worlds Collide*, which saw series from various Superman-related titles like *Superman: The Man of Steel*, *Steel* and *Superboy* tie in with series from DC's Milestone Media imprint, which featured series about heroes of color like *Static*, *Icon*, *Hardware*, and *Blood Syndicate*. While Milestone's titles had initially introduced their own narrative world in 1993, it was little more than a year before DC merged the two universes with what the cover tagline of *Worlds Collide* #1 called a "Rift Between Worlds." (*Worlds Collide* 1994, front cover)

By bringing these worlds together, DC sought to expand readership for their Milestone series by getting Superman fans to purchase all of the crossover event's tie-in issues. Whereas multiversal stories were self-contained within titles like *The Flash* and *Justice League of America* in the 1960s and 1970s, in later decades the multiverse was predominantly used as a way to anchor large, multi-title storylines that were just as much about bolstering franchises as they were about narrative world-building.

The 2000s: The Ultimate Flashpoint Convergence of the Infinite Final Dark Crisis in the House of Secret Spider-verse Wars

The early twenty-first century saw the strategy used on a near-annual basis at both DC and Marvel. Both publishers realized that they could sell more copies of a given character's series if it was rebooted with a new number-one issue. Such reboots often took the form of a new narrative continuity, which frequently involved alternate timelines and/or universes.

In keeping with DC's designations of Earth-One, Earth-Two, etc., Marvel also began expanding the number of narrative worlds by giving them unique numbers. In 2000, Marvel launched a new imprint called "Ultimate Marvel" in which new incarnations of familiar heroes were given a fresh start in what became known as Earth-1610. While the traditional Marvel universe is set on Earth-616 (with its multi-decade continuity), readers could see the Earth-1610 Peter Parker gain his powers for the first time in the pages of *Ultimate Spider-Man*. ("Earth 616" n.d.).

The two universes remained separated at first, but rifts and crises and infinite possibilities awaited in later years—culminating in their eventual merger with the 2015 event *Secret Wars*. Besides pure spectacle, the biggest reason for this fusion was to allow popular character Miles Morales (the Earth-1610 version of Spider-Man, having replaced Peter Parker a few years prior), to be a fully connected part of Marvel's dominant continuity. As such, Marvel shone a spotlight on the character that aided in various licensing efforts ahead of his starring role in the 2018 animated film *Spider-Man: Into the Spider-Verse*, which saw webslingers from numerous universes unite. Marvel introduced the concept of the Spider-Verse in comics starting in 2014 as a way of bringing together all of the different incarnations of Spider-Man that existed in prior decades, such as Spider-Ham, Spider-Man 2099, Spider-Man Noir, and Spider-Girl, along with the debut of new variations like Spider-Punk. If earlier decades saw comic book creators and editors shudder at the thought of "continuity bugs," by the 2010s comics' continuity was swarming with multiple versions of popular characters on a near-constant basis.

At both Marvel and DC, 2005 was the year in which the multiverse became a modern-day foundation of twenty-first-century comic book storytelling thanks to the success of two massive cross-over events—Marvel's *House of M* and DC's *Infinite Crisis*. *House of M* saw an alternate world in which mutants control society. While order was eventually restored and universes separated anew at Marvel, DC's *Infinite Crisis* saw the multiverse crisis once again placed at the forefront of the publisher's storytelling agenda, led by writer Geoff Johns. DC then followed *Infinite Crisis* with numerous events: 2008's *Final Crisis*, 2011's *Flashpoint*, 2014's *Future's End*, 2014's *The Multiversity*, 2015's *Convergence*, 2017's *Dark Nights: Metal*, 2020's *Dark Nights: Death Metal*, 2021's *Future State*, 2021's *Infinite Frontier*, 2022's *Dark Crisis on Infinite Earths*, and 2022's *Flashpoint Beyond*. Each one involved parallel worlds and/or timelines as they chronicled the multiversal potentials, portents, and probabilities of nearly every DC character, team, and era of narrative continuity (from the Golden Age and the Silver Age to the 1980s reboots and beyond).

It was a lot to keep up with. For decades now, DC has seen a constant push and pull between building continuity and simplifying it as they try to serve two different sets of readers—longtime fans and new generations. Some stories delved deeper into the history of the multiverse like *Infinite Crisis* (unveiled the year after DC's final batch of Elseworlds stories were released in 2004, which seems uncoincidental). Others rebooted earlier histories like *Flashpoint* (and the subsequent creation of "The New 52" franchise in 2011). While each storyline attempted to address certain "continuity bugs" and establish new possibilities, one thing remained a constant throughout regardless of any specific story details: the multiverse *is* the message (as we can only assume that a parallel-Earth Marshall McLuhan once said). DC has embraced the multiverse for over sixty years now, and regardless of whether any particular story undoes or resurrects a particular timeline, universe, or plot point, the most important factor for fans to keep in mind year after year is that all of these stories have been told, are remembered by many readers and might potentially be referenced again in the future by any given creator depending on the editorial and corporate whims and needs of DC Comics.

Conclusion: The Infinite Frontier of Comics

In July of 2020, as his new series *Dark Nights: Death Metal* was released, comics writer Scott Snyder (2021) took a stand on what the multiverse means for readers. Much as the notion that we were "all in this together" took hold during the early days of the COVID-19 pandemic, so too did Snyder see comic book publishers, creators and fans as a united force.

"For better or worse, we're in it together, as an industry, and as fans... and that means embracing things that you might be uncomfortable with in different ways," he says about the need for fans to take the good with the bad when it comes to the history of DC's tales. "We're trying to unify all of the DC stories," he says, even the ones we might not "agree with," adding that he wants "to take everything and say 'all of the stories are part of the DCU." In this way, he argues, creators and fans alike can "embrace all of it, and make continuity celebratory, conclusive, more united, and connected." (Arrant 2020) Snyder soon summed up this sentiment in a pithier way: "it all happened, it all matters." (Norman 2020) Snyder in turn used the story title "It All Matters" for the first issue of his series *Dark Nights: Death Metal* as a way of entrenching the idea further.

And if it all matters, then it can all be marketed. Reprinted editions of graphic novels have surged in recent years, with more sales coming from traditional book sellers in the wake of Penguin Random House becoming Marvel's distributor since 2021. The multiverse allows older storylines and characters that were once forgotten to gain a fresh relevance to new audiences, thanks to storylines like *Infinite Crisis*, *The Multiversity* and *Secret Wars*, along with films like *Spider-Man: Into the Spider-Verse*. Not only do multiversal storylines target fans seeking new forms of spectacle, they can also drive increased sales of older material, like new reprints of older comics titles.

In keeping with Snyder's credo that every DC story matters, *Dark Nights: Death Metal* concluded with the revelation that an "infinite web of multiverses" has formed into an "omniverse" which will serve as an "infinite frontier" for the entire history of the publisher's tales. If multiverses can be reset, rebooted, and rebranded at will by a publisher, then Snyder's efforts serve as an attempt to preserve the relevance of both current and past stories to future generations of comics fans with the creation of an omniverse that can henceforth be used to make any given story "matter" again.

In this way, the multiverse has the potential not only to reinvent franchises for new generations but also to maintain or reignite interest among older fans by way of nostalgic appeal to earlier eras. By merging the original, decades-old versions of characters with modern iterations, comics publishers can appeal to multiple demographics simultaneously. And while it only takes one corporate decision to replace a given narrative continuity with a brand new one, future creators can potentially use Synder's notion of the omniverse to once again reboot and rebrand previously discarded storylines and continuities. Snyder ends *Dark Nights: Death Metal* #7 with a proclamation: "May we see that everything we do, every story – mine and yours and theirs – is ours… and it all matters. Always." (Snyder and Capullo 2021, n.p.) From the Golden Age to the present, the multiverse has long been used as a way of making it all matter to comic book fans.

References

Arrant, Chris. 2020. "Scott Snyder: 'Everything Matters' in Death Metal, Including Wally West and His Controversial Past." *Newsarama*, July 14, 2020. https://www.gamesradar.com/scott-snyder-everything-matters-in-death-metal-including-wally-west-and-his-controversial-past/.

Authers, Benjamin. 2012. "'What Had Been Many Became One': Continuity, the Common Law and Crisis on Infinite Earths." *Law, Text, Culture* 16, 65–92.

Conway, Gerry, writer, and Dick Dillin, penciller. 1979. *Justice League of America* 171. New York: DC Comics, October.

Dawid, Richard. 2013. *String Theory and the Scientific Method*. Cambridge: Cambridge University Press.

Earth 616. n.d. *Marvel Database*. https://marvel.fandom.com/wiki/Earth-616.

Elseworlds: Batman, Vol. 2. 2016. New York: DC Comics, n.p.

Fox, Gardner, writer, and Carmine Infantino, penciller. 1961. *The Flash* 123. New York: DC Comics, September 1961.

Fox, Gardner, writer, and Carmine Infantino, penciller. 1962. *The Flash* 129. New York: DC Comics, June.

Fox, Gardner, writer, and Carmine Infantino, penciller. 1963. *The Flash* 137. New York: DC Comics, June.

Fox, Garnder, and Mike Sekowsky. 1963a. *Justice League of America* 21. New York: DC Comics, August.

Fox, Garnder, and Mike Sekowsky. 1963b. *Justice League of America* 22. New York: DC Comics, September.

Fox, Garnder, and Mike Sekowsky. 1964. *Justice League of America* 29. New York: DC Comics, August.

Friedenthal, Andrew J. 2019. *The World of DC Comics*. New York: Routledge.

Jurgens, Dan. 2019. "Foreword." *Zero Hour: Crisis in Time 25th Anniversary Omnibus*. Burbank: DC Comics, 7–9.

Kanigher, Robert, writer, and Carmine Infantino, penciller. 1956. *Showcase* 4. New York: DC Comics, September–October 1956.

Norman, Bug. 2020. "DC's Future Multiverse May Already Be Changing." *Screen Rant*, December 20. https://screenrant.com/dc-future-multiverse-changing-comics/.

Schoel, William. 2019. *The Silver Age of Comics*. Albany, GA: BearManor Media.

Scodari, Christine, and Jenna L. Felder. 2000. "Creating a Pocket Universe: 'Shippers,' Fan Fiction, and *The X-Files* Online." *Communication Studies* 51, no. 1: 238–257.

Sean, Guynes. 2019. "Worlds Will Live, Worlds Will Die: Crisis on Infinite Earths and the Anxieties and Calamities of the Comic-Book Event." *Inks: The Journal of the Comics Studies Society* 3, no. 2: 171–190.

Shen, Dan. 2005. "Diegesis." *Routledge Encyclopedia of Narrative Theory*, edited by David Herman, Manfred Jahn and Marie-Laure Ryan, 107–8. London: Routledge.

Siegel, Jerry, writer, and Kurt Schaffenberger, penciller. 1960. *Superman's Girl Friend, Lois Lane*, 19. New York: DC Comics, August.

Snyder, Scott, writer, and Greg Capullo, penciller. 2021. *Dark Nights: Death Metal*, 7. March 2021, Burbank: DC Comics, March 2021.

Worlds Collide 1. 1994. New York: DC Comics, July.

21

MULTIVERSALITY IN CHILDREN'S PICTURE BOOKS

Portals to Many Worlds

Ondine Park

Scientific theories and discourses of the multiverse easily, and in some ways by necessity, slide into cosmological concerns about the origins of our universe and how its origin might be related to the possibility, probability, characteristics, and potential origins and development of the multiverse. Such considerations evoke vastness and violence—the (depending on the theory) violent eruption of our universe and/or the multiverse or (again, depending on the theory) its violent destruction. Universes may bubble up only to collapse, unable to endure because of their particular combination of characteristics. The multiverse may be iterations of computer simulations or of mathematical objects, which render the familiar world and its eventful and everyday life as mere, possibly meaningless, simulations. In science fiction and fantasy representations of the multiverse, such as in *Rick and Morty*, *Adventure Time*, *Spiderman into the Multiverse*, or *Dark*, it is not unusual for the fact of the multiverse to become known because of tragic or dire circumstances. And, in part because of this, the knowledge of the multiverse and the capacity to access other worlds pose grim cataclysmic threat to one or more worlds and thus imposes immense responsibility on those humans (or other beings) who know and can traverse the many worlds.

In contrast to these epic, violent, and danger-posing versions of the multiverse, in children's picture books, a different and friendlier sense of multiversality is represented, even if not precisely articulated as *the multiverse*. Although children's picture books, particularly ones for younger children,

DOI: 10.4324/9781003480846-25

do not feature the multiverse in the explicitly scientific or even science-fiction conception, I suggest that they do frequently represent what I might describe as *multiversality*. By multiversality, I mean a taken-for-granted orientation to the nature of physical and temporal reality wherein the world in which each story unfolds is understood to be only one self-contained world, or universe, of potentially many, possibly an infinite number of, such simultaneous, parallel universes. And, these other universes may be quite similar to each other or quite different in a variety of superficial or fundamental ways. In works such as *The Adventure of Beekle: The Unimaginary Friend, A Child of Books, The Red Book, The Three Pigs*, and *Windows*, multiversality is presented not as a dangerous, cosmic risk, but as a believable, accessible possibility. It is illustrative of the world-creating, expansive possibilities inherent in stories. Reading multiversality and its depictions in children's picture books renders sensible the complex ways in which many worlds are enmeshed with, curled in, bubble up from, or otherwise emerge out of the familiar and ordinary. In particular, attunement to multiversality suggests possibilities of recognizing and operating in relation to incongruous logics and imaginaries without dismissing, submerging, or minimizing any at the expense of other, particularly dominant, ones. In this way, multiplicities of temporo-spatialities can be taken as equally real or potential, even if not necessarily easy to access. In daily life, children may feel and, indeed, have little control or authority in dictating how things should be understood, related to, or undertaken. They may find the social meanings, conditions, and norms into which they are always in the process of being socialized to be both mysterious and arbitrary. Yet, in reading multiversal picture books, the child has an opportunity to re-envision (or possibly reaffirm) their world as replete with many equally legible and valuable ways of making sense of, and being in, the world. And, in this way, such stories may offer the child recognition and validation for their own idiosyncratic ways of storying and inhabiting the world, different from the grown-up stories and world to which they are otherwise always adapting. Multiversality in children's books, then, allow children to imagine and play with the possibilities of reality as being full of creative potency through the liveliness of imagination.

Picture Books

In the beginning paratext of *A Child of Books*, written by Oliver Jeffers and illustrated by Sam Winston, the following epigraph, attributed to Muriel Rukeyser, accompanies the author's and illustrator's dedications: "The universe is made of stories, not of atoms." This notion of stories creating a universe and comprising whole worlds is a common motif in children's literature. It can be taken figuratively and evocatively. But I suggest it can also often be taken literally, rendering books as the keepers of, or entry-points into, stories and, therefore, universes. In this way, therefore, stories are also windows on, portals to, or even the place of other worlds. As described in the opening lines of the picture book *A Dragon in a Wagon*, every book is a "word window," and so the book-filled shelves of a library offer "Rows upon rows of Word Windows. ... [E]very time you open one, a new adventure has begun." Although *A Dragon in a Wagon* doesn't significantly explore the possibilities of its opening image, the richer and more complex engagement with these themes in the picture books that I will be discussing in this chapter creatively and innovatively render perceptible the surprising and delightful power of imagination and stories (in words, pictures, and books) to provide a glimpse into alternative realities or to create worlds.

Picture books include pictures that are at least as important as, if not sometimes more than, the written narrative. In picture books, the words and pictures work together to create a story (if both are present—although sometimes there is no textual narrative). Because both elements convey an important but different aspect of the story, they can work cooperatively, complementing each other, or they can compete with, interrupt, or even contradict each other (Park 2011, 175; Shulevitz 1996; Johnston and Mangat 2003). In some cases, a number of different styles of the textual and/or visual elements may be present, creating multiple layers of playfulness and possibility. Just as the social world can be mysterious for kids with properties or knowledges often seeming to appear through utterly irrational, random, or fantastical irruptions, children's picture books, like the ones I consider below, often also re-draw, perforate, or obliterate the otherwise seemingly strict line between the everyday and the fantastic, the taken-for-granted and mysterious, what can be expected and what is astounding. The fantastical and creative worlds imagined and explored in children's picture books help to articulate the mutual constitution and deep inter-embeddedness of the real and imaginary, fact and fiction, worlds storied into creation by words, and those imaged through pictures.

Many Logics, Beings, and Becomings

In picture books, especially those intended for very young, possibly pre-literate, children, the status of fantastical beings, places, and activities are often presented as amazing but ordinary while ordinary ones might be unexpectedly marvelous. If these beings, places, or activities are presented as real, no matter how unlikely or improbable, then the reader is invited to accept they are real in some way without trying to pin down what their actual ontological status is. Thus, for example, in *The Adventure of Beekle:*

The Unimaginary Friend by Dan Santat, Beekle is explicitly described as an imaginary friend who has not yet been imagined by a "real child." He waits years, playing on a "far away island" with other yet-unimagined imaginary friends, all waiting for their friends-to-be to imagine them. Thus, Beekle is an imaginary friend who has an existence independent of and prior to the friendship that will eventually come to make him an imaginary friend. As such, Beekle is simultaneously imaginary and unimaginary, unreal and real, born a friend yet one who never has had a friend. Beekle finally "does the unimaginable" and, rather than continue to wait to be made into an imaginary friend, strikes out on his own to go find his friend in the "real world." Eventually, Beekle and his friend Alice find each other. Through this journey, Beekle is transformed from an imaginary friend-to-be and becomes an unimaginary friend. Together, through their friendship, Beekle and Alice come to discover that the real world "began to feel a little less strange." Yet, in making the world less strange, they are empowered together to do "the unimaginable," echoing the *unimaginable* ontological-crossing journey Beekle made from the realm of imaginary friends-in-waiting to the real world. Thus, through the adventure of Beekle and Alice's friendship, the world is simultaneously made, without contradiction, both more and less familiar, and both less and more "real" and "imaginary." In this way, this picture book contains multiple different, even discordant logics and incompatible states of being that are, nevertheless, co-constitutive and co-present.

Opening Portals

Where *Beekle*, through its notion of unimaginariness creates multiversality by simultaneously holding together many different logics of spatialities, beings, and becomings, the sense of the multiversal in *The Three Pigs* by David Weisner is rendered by connecting but not reductively integrating many incommensurate stories. Here, the creation of new narratives by introducing dissonance between words and pictures open portals into and across many worlds.

The Three Pigs begins with the familiar story of the three pigs, each of whom builds a house. The first pig builds his house out of straw, the second out of sticks, and the third out of brick. In this familiar version of the story, the wolf implores each pig in succession to come out of his house and, when each pig refuses, threatens to, and then in fact does, blow down the house, eating each of the two pigs before being foiled by the third pig. This familiar version of the story, conveyed in narrative text and complementary pictures, is initially contained within orderly, sequential, framed rectangles. In Wesiner's picture book version of the nursery story, however, when the wolf blows down the first pig's straw house, he also blows the pig out of the familiar story. As the pig falls out past the frame of the story into the white space between frames, the pig remarks in a newly introduced comic book-style speech balloon that the wolf has blown him "right out of the story." The wolf remains contained in the story unfolding within the frame; but, with the pig missing, an increasing dissonance takes place within the frames between the narrative text, which continues telling the familiar story, and the pictures, which reflect the unfolding of a differentiating timeline of events. Thus, where the text in the panel reads that he "ate the pig up," the picture shows the wolf perplexed, unable to find the pig amid the debris contained within the frame. As the story continues, the straw-house pig appears in the white space between panels to implore the stick-house pig, who is still inside his own story panel, to join him "out here" in the blank space outside of the framed panels where it's safe. Before the wolf blows down the stick house, the stick-house pig exits the frame. Again, there is a misalignment within the story panel between the text narrating the wolf eating the pig and the picture showing a baffled wolf. The two pigs go on to lead the brick-house pig out of his story panel into the blank, white space, which is now expansive, while the story panels take on the appearance and materiality of posters that can be moved out of their regular, squared organization, and can be pushed and twisted.

While the pigs three-dimensionally leap about, walk on top of, or climb between the narrative panels that lay stacked on top of each other like a pile of papers the wolf remains two-dimensionally within the panel (himself folded or bent as his panel is folded or bent). The pigs make a paper airplane of the panel the wolf is in, climb on top of it, and fly through a number of blank pages before they crash land. One pig, inspecting the space, appears as if he is looking directly out toward the reader, noting "I think … someone's out there" while the other two pigs call him to help them with a new scene. In the next spread, the pigs enter into the nursery rhyme in which a cat plays a fiddle and a cow jumps over the moon. The style of illustration in this spread is flatter and softer and the pigs can be seen transforming into this style as they enter the world of the nursery rhyme, and returning to their more realist, three-dimensional appearance as they leave the scene. The cat with the fiddle also leaves the scene and similarly changes in drawing style. They enter again into the blank space, now filled like an art gallery lined with rows of poster-like panels depicting a plethora of different stories. The pigs then enter into a brown-line drawn set of sepia-tone pages. They rescue a dragon about to be slain by a knight, by showing the dragon how to escape the story.

Back in the gallery-like blank space, the three pigs, the cat with the fiddle, and the dragon happen upon the panel containing the third pig's brick house. The pigs decide to return home. To do this, they re-organize the panels of their story into the original narrative order. In the panel, after the wolf attempts to blow down the brick house, the dragon, along with the pigs and cat, come out of the front door, shocking the terrified wolf backward and jumbling up the words and letters of the narrative. The group collects up the loose letters and, on the last page, arrange these letters to read "And they all lived happily ever aft[er]."

In the summary provided on the copyright page, the three pigs are able to evade the wolf "by going into another *world*" (emphasis added). The pigs discover that travel in, out of, and between worlds is possible by playing with the liminality and potentiality of the blank page and the possibility of exceeding frames and boundaries created by escalating the disjuncture between pictures and words, and thus creating new possibilities and a new story. They are thus able to transgress the limits of the original stories. The frames and sequential regularity of the story-panels, along with the familiarity of the narratives, help to establish the clear bounds of each of the largely incompatible and self-contained stories that the pigs are able to break through. What I might describe as a multiversal form of travel is enabled by activating the stories—or by the picture book pages that comprise them—to act as portals both into and between many otherwise wholly separate worlds.

Stories as Portals

While the three pigs disrupt old stories to create their own new story and, in this way, activate the stories-as-portals accessing many story-worlds, *The Red Book* and *A Child of Books*, render themselves as portals to other worlds, and, mind-contortingly, each acts a portal to its own self-contained world.

In *The Red Book*, by Barbara Lehman, a child, dressed for winter, walks through a gray, snowy city largely comprised of rectangular buildings with rectangular windows, rectangular vehicles, and rectangular streets. *The Red Book* is a wordless picture book. It conveys its story entirely through pictures contained within a series of squares framed in thick black line. This motif of squares is repeated in the many squares and rectangles of the city and the grid-pattern of a map featured in the book. Although the book is wordless, a narrative can still be "read" as the pictures, like a comic book, convey a sequence of actions and reactions that follow a legible, spatial logic and temporal order—even if that logic and temporality are surprising. As she's walking, the child discovers a red book in a snowbank on the sidewalk. She takes the book to school and, there, looks inside to see a map overlaid with a grid. As she looks closer and closer, she sees on the map, an island and, around the perimeter of the island, a beach, and on the beach, a boy walking. The boy discovers a red book on the beach. Opening it, he sees a gray, snowy city. As he looks closer and closer at the rectangular buildings with their rectangular windows, he is able to peer through the grid-like window of one building into the classroom in which the girl is sitting at her desk, looking at the boy on the beach in her book, looking at the girl in his book. The girl in his red book, the boy in her red book, and the girl in *The Red Book* all look out, as if directly at the reader, surprised, then happy. As her class is let out, the girl walks in the street, comes across a balloon vendor, and buys all the balloons. She is taken up into the sky, up out of the city. Her book slips out from under her arm and lands open on the pavement below. In its pages, we can see the boy, sitting on the beach, looking in his book seeing her fly away from the city until she can no longer be seen in the picture in his book. He sits with his head in his knees forlorn, book closed in front of him. But, behind him, the girl comes floating down with her many balloons. The boy and girl are happily united and, as the wind closes the girl's book, still laying on the city pavement, we see the boy's book on the beach being pulled into the sea by the rising tide. The girl's book, closed, is discovered by one of the girl's classmates, who cycles away with the book tucked under his arm, looking back, as if directly at the reader.

As in the other books considered later in this chapter, there is an equivalence suggested in *The Red Book*, between windows and books with both of these offering an entry into other worlds. Such ordinary objects as portals, accessible in the everyday world, offer young readers a way to imaginatively enter into another world and can also be seen in the stories for older children such as the drawing-room mirror through which Alice passes to access an inverted, looking-glass world in *Through the Looking-Glass, and What Alice Found There*; or, in the case of stories like *The Phantom Tollbooth, The Lion, The Witch, and the Wardrobe*, or *The Wonderful Wizard of Oz*, elements of daily life that are ordinary yet potentially mysterious or unknowable to children become portals to fantastical alternate dimensions. However, different than these other portals that are depicted as taking the characters in the story to another place contained within the narrative, *The Red Book* is a book about a red book and about itself *as* the red book, and thus creates for its reader an entry into the book itself. This book about itself acts both as a portal to a world and *is* the world into which the reader is transported. By repeated series of ever more zoomed-in scenes contained in the book's various grids, the reader is both absorbed figuratively by the captivating narrative and simultaneously absorbed or pulled into the logic of the book that contains access to another world, which itself, by interpellating the reader (or, breaking the fourth wall), contains access to our world. This book about itself acts both as a portal to a world and is the world into which the reader is transported. Thus, the reader is imaginatively transported into the world of the book and becomes part of the book.

Stories as Worlds

Like *The Red Book* with its red book cover, *A Child of Books* is another red-covered book—this one with a golden lock illustrated in the center of the red book on its cover—that is both about books in general while also about itself specifically. Whereas *The Red Book* produces multiversality through pictures and illustrates the possibility of traveling and creating realities through imagination, *A Child of Books* emphasizes the potency of words for creating and traveling to worlds. The textual and pictorial elements are, in some ways, inverted: the verbal narrative is conveyed by hand-drawn letters while the visual narrative is frequently conveyed by images made of typography. This illustrative text is comprised of text from classic children's stories arranged to form pictures. Narrated by the eponymous child of books to a "you" who is both the reader as well as the boy illustrated in the pictures, the child literally travels on and through the words. The child comes "from a world of stories" and "floats" upon her imagination, illustrated as a girl riding a sailed-raft across a sea made of typographically arranged sentences from sea-faring stories. She appears on a tsunami of words before an isolated, sad-looking boy, who appears quiet, timid, and lonely, standing in front of a single small house in an otherwise blank space. Together, the child of books and boy create/go on adventures, ultimately claiming for themselves the world they have made/discovered through their imaginative travel through words and stories. As the two children stand euphorically atop a desk globe, the child declares that "this is our world" and "we're made from stories," suggesting that stories create and are the world, and stories create people. In the final few pages of the book, this world that is stories is shown as a house that is a home. As the child narrates, "Our house is a home of invention." This world takes the form of a house with a red front door and prominent golden lock centered on the door, which becomes a red book with a golden lock—the front cover of *A Child of Books*. The world made of stories by the child of stories and the boy is the book the reader holds in their hands. The narrative is rendered both through the words of the story and also through the text of other stories assembled together as images that are parts of larger pictures. These pictures convey a story that links *A Child of Stories*, the book itself, to the book in the story that is a house wherein the house is the world.

Portals as Stories

The notion of a house as a whole world, stories or books as whole worlds, and books as houses, or vice versa, can also be seen in *Windows*, written by Julia Denos and illustrated by E. B. Goodale, which opens with a child and small dog peering out a picture window from inside their home. Along with them within the frame of the window, some elements of the interior of the room are visible, including a shelf of books and a framed picture of houses in moonlight. On the next page, as the story begins, the child prepares to go for a walk with their dog. Looking out the picture window into the street, they observe the windows of the houses and apartments lighting up as dusk falls. The child narrates the story and addresses a "you" that can be understood to be the reader. As the child and dog walk through their inner-city neighborhood in this "almost-night," they pass by urban animals and people engaged in various activities throughout the neighborhood. And, in the illuminated windows, the child sees a panoply of activities and diversity of people, animals, and things. The child then passes a house with security bars across the front door and windows. The house appears to be abandoned. The child, looking at the darkened windows, notes that some windows "are empty and leave you to fill them up with stories." Upon returning home, a loving adult waves them in: "you look at your window from the outside. Someone you love is waving at you, and you can't wait to go in. So you do." The child and dog go in and the child curls up with the adult on a chair in front of the window, reading a book together, while the dog curls up on a mat to sleep.

For the child, the windows of the abandoned house leave an emptiness to be filled by the observer's stories. The other windows, in contrast, are already filled with all the various activities and people and thus are filled with their own stories. The child, in wandering through the neighborhood, encounters story after story, in the glowing windows of each home. The windows of stories, often stacked on top of each other in apartment buildings is like a library bookcase, suggesting a literal version of the "Rows upon rows of Word Windows" described in *A Dragon in a Wagon*. The window of the child's own home is a story of their own that they can enter. This parallel between stories (or books) and windows (or homes) is enforced when the child, waved into their own story through the window of their home, enters into the home then directly enters into a book, reading together with their loving adult, and creating a story tableau in their window for others who might pass by.

In her biography at the end of the book, author Julia Denos notes that she would take walks at dusk, when she observed "Windows were glowing worlds beckoning me in while reminding me of home." In walking at the liminal time of dusk, the child is able to explore and story these many worlds and imaginatively enter into them. While Beekle inhabits a land of imaginary beings far away and must physically (and ontologically) travel to the familiar world as an epic, heroic journey of yearning and discovery, *Windows* explores imagination, discovery, and travel to other worlds through a more ordinary route—a nightly dog walk and the exercise of observation.

Multiversality

In *The Red Book*, *The Three Pigs*, and *Windows*, the framed elements—windows and panels—act as visual gateways, while the grids and repeated squares and rectangles create an equivalence among the various elements—including the actual book itself. Through this application of equivalencies, and therefore transitive properties, these framed elements activate a magical transferential property enabling transposition. While both *Windows* and *A Child of Books* directly interpellate the reader by framing the narrative as an address to a "you," in *The Red Book* and *The Three Pigs*, a similar interpellation is achieved when characters in both books seem to look directly at the reader, drawing the reader into the unfolding of the narrative. This has the effect of suggesting that the reader may also inhabit yet another world, in addition to the ones already being explicitly explored in each book.

In these works, the imagined clear distinction between "reality" and the realm of imagination is re-cast. That is, the fantastical events of *Beekle*, *The Three Pigs*, *The Red Book*, or *A Child of Books* is no less real in their unfolding or plausible in their logics than the ordinary events of *Windows*. Magical equivalences in all of these books—among imagination, being, stories, windows or houses, and whole worlds or universes—is not discounted as make-believe but gain an important immediacy. The imaginative is given greater reality-status over ordinary rules of how things are and ought to be. Indeed, the improbable is not explained away or re-framed according to these rules of ordinary life. Rather, taken together, there is no singular reality that makes proper sense of the make-believe or whimsical. The absurd and fantastical simply are viable worlds and do not need to be explained. Characters in these stories often observe other worlds (as in *Windows*) or cross through to other worlds through a portal. In doing so, they move out from inside a limited world—one that is marked by low affect, such as in *A Child of Books*, that is literally delimited and bounded, as in *The Red Book* or *The Three Pigs*, or that is insufficiently actualized, as in *Beekle*. And, exceeding the bounds of their world, they move out to a more creative, participatory, and complex reality.

The realist-minded adult reader might be tempted to interpret these works as suggesting that one can imagine stories and windows *as if* they are many worlds and that, therefore, the lesson to be learned is that children can travel in their minds with enough imagination. Yet, in each of these books, travel from the familiar world to a universe or dimension that is unfamiliar is *not* owing to a change in individual attitude or interpretation but to a shift in *access* to one or more actually-existing alternative realms. That is, these books do not provide easy resolution of other worlds accessed through stories or windows into the ordinary world. The girl in *The Red Book* does travel to another land which is somehow also in her book. Beekle is a real, imaginary, unimaginary friend who finds the friend who imagines him. The pigs escape their story and help rescue a dragon from another story who rescue them from the wolf of their own story. The child does travel on seas of words and enter into a house that is the book in which she travels on seas of words. In these stories, while the various worlds of imagination, make-believe, the fantastical, or the fictional is distinct in a number of notable ways from the everyday familiar world, it is a distinction that can be surpassed and is shown to be less clear and firm than presumed and, rather, can be understood to be multiversal—simultaneous and equally real, although in different ways, with the beings, logics, and spatialities of the many worlds not reduced down to the version of reality of any one world. Characters in these stories cross over thresholds and move out from inside a singular version of reality out to a more active, complex, and potentially interactive multiversality that invites thinking anew about the boundaries and creative possibilities of our ordinary world. These storied places, including each of our own storied places as readers, is a multiverse if we shift the boundaries of what it means to be multiversal. Taking on board the seamlessness between the many imagined and storied worlds in these children's books opens up productive possibilities for seeing lived social realities differently. The significance of this creative multiversal orientation is that it also makes a reimagining of our shared social world possible. These books foreground that our everyday realities are re/produced through inescapably creative practices.

Acknowledgment

Thanks to Kaia Nitchie whose research assistance and conversations helped inform this chapter, particularly in the background work on scientific theories of the multiverse and who also drew my attention to *A Dragon in a Wagon*. Thanks also to Heidi Bickis, Bonar Buffam, Tim Paulson, and Barret Weber, who each helped me think about versions of these ideas. Support for this research was provided by the Hampton Research Endowment Fund through the University of British Columbia.

References

Baum, L. Frank. 1900. *The Wonderful Wizard of Oz*, illustrated by W. W. Denslow. Chicago & NY: George M Hill Co.

Belk Moncure, Jane, and Linda Hohag. 1987. *A Dragon in a Wagon*. Mancato, MN: The Child's World, Inc.

bo Odar, Baran and Jantje Friese. 2017–2020. *Dark*. Wiedemann & Berg Television.

Carroll, Lewis. 1871. *Through the Looking-Glass, and What Alice Found There*, illustrated by John Tenniel. London: Macmillan.

Denos, Julia, and E.B. Goodale. 2017. *Windows*. Somerville, MA: Candlewick Press.

Harmon, Dan and Justin Roiland. 2013–2023. *Rick and Morty*. Williams Street; Harmonious Claptrap; Starburns Industries; Justin Roiland's Solo Vanity Card Productions; Rick and Morty, LLC; Green Portal Productions.

Jeffers, Oliver, and Sam Winston. 2016. *A Child of Books*. Somerville, MA: Candlewick Press.

Johnston, Ingrid, and Jyota Mangat. 2003. "Cultural Encounters in the Liminal Spaces of Canadian Picture Books." *Changing English* 10, no. 2: 199–204.

Juster, Norton. 1961. *The Phantom Tollbooth*, illustrated by Jules Feiffer. NY: Random House.

Lehman, Barbara. 2004. *The Red Book*. Boston: Houghton Mifflin.

Lewis, C. S. 1950. *The Lion, the Witch and the Wardrobe: A Story for Children*, illustrated by Pauline Baynes. London: Geoffrey Bles.

Park, Ondine. 2011. "Illustrating Desires: The Idea and the Promise of the Suburb in Two Children's Books." In *Ecologies of Affect: Placing Nostalgia, Desire, and Hope*, ed. T. K. Davidson, O. Park, and R. Shields, 169–193. Waterloo, ON: Wilfrid Laurier University Press.

Persichetti, Bob, Peter Ramsey, and Rodney Rothman. 2018. *Spider-man: Into the Spider-Verse*. Columbia Pictures Industries Inc.

Santat, Dan. 2014. *The Adventures of Beekle: The Unimaginary Friend*. NY & Boston: Little, Brown and Company.

Shulevitz, Uri 1996. "What is a Picture Book?" In *Only Connect: Readings on Children's Literature*, 3rd ed., ed. S. Egoff, G. Stubbs, R. Ashley, and W. Sutton, 238–41. Toronto; New York: Oxford University Press.

Ward, Pendelton. 2007–2018. *Adventure Time*. Frederator Studios; Cartoon Network Studios.

Wiesner, David. 2001. *The Three Pigs*. New York: Clarion Books.

22

THE ATLAS, THE RESURRECTIONS, THE RISING, AND THE TIME WAR

Multiverse(ions) of Love from Four Fictional Universes

Rebecca Gibson

Whether platonic or romantic, intangible concepts of love populate our fictional worlds. Making an excellent plot propellant, a dastardly confounding factor, or the strong foundation upon which the story rests, love threads its way through most of the stories we tell ourselves. This chapter will look at four fictional universes using the theoretical concept of agency/ agentive behavior, and show how the multiverses in each story rely on multiversions of love. Agency, or the ability to make choices within a fixed system, requires only a knowledge of those choices—using chopsticks vs. using a fork is an agentive choice. Love can provide a character with the motivation to choose more difficult choices, for example, a comfortable life vs. a painful death. Multiverses are often composed of a character attempting to right a wrong on behalf of the concept of love. If the person they love dies, or if they lose them, or if they love a thing, a concept, a version of the world that they have only imagined, their existence within a multiverse will allow them a do-over, so that they might fix whatever has broken.

In *Cloud Atlas*, these chances and choices are mostly outside the control of the main characters—people are reincarnated into others, are resurrected by the prescients, live, and die at the whim of other characters, which makes love all the more powerful—it is an act of faith in a world out of their control. In the *Matrix* franchise, similarly, the characters are at the whim of their machine overlords or the Agents which serve them. They

DOI: 10.4324/9781003480846-26

love despite knowing that such control is ever-present, and that they might lose their lives at any moment. In another universe, mostly outside of the characters' control, Johnny and Nick are each other's best friends. This friendship replaces most other love in Nick's life, as they work together to fix a dimensional anomaly that Johnny accidentally caused…until Nick finds out that Johnny has been manipulating him—she does not truly love him after all. Yet, when given the chance to reset their friendship, they both accept and try again. Contrasting the others is the tireless, timeless, deathless love of Blue and Red in *This is How You Lose the Time War*, both of whom know exactly who controls their lives (and potential deaths), and who fight to preserve their eternal love, weaving it into the fabric of existence, and defying hundreds of universes to prevail.

What compels our enjoyment of the multiverse concept? The chance to meet again; the chance for another chance; the chance to do things over and maybe, this time, get them right. When we know there is another universe, we have hope—we can envision another way for things to turn out and potentially have agency over the changes. Whether sequential, as in *Cloud Atlas*; inevitable, as in the *Matrix* franchise; invasive, as in *Beneath the Rising*; or parallel, as in *This is How You Lose the Time War*, these chances are driven by the characters' various loves.

Each multi-version allows us to ask the questions: what would we do for love? How far would we go—how many times would we break and reset time, space, the universe—to know that one person again? Is it a leap of faith? The healing of a rift? The braiding of a strand? Love and the agentive behavior it engenders, unites these plots, and the multiverses within them.

Themes of choice and agency thread throughout all multiverse concepts, from the Spiderverse and the Marvel Cinematic Universe, to the Doctor Who and Torchwood contiguous universe. In fact, it would be difficult to conceptualize a multiverse without those concepts. However, in many conceptualizations, the fracturing occurs from greed, hatred, violence, or acts of desperation, and while these concepts can be adjacent to love, then can also be exclusive of it.

Part One: The Atlas

Thus closes the final letter between the characters of Robert Frobisher (Ben Wishaw) and Rufus Sixsmith (James D'Arcy), in the 2012 *Cloud Atlas* (Wachowski, Wachowski, and Tykwer). With six overlapping, interwoven storylines, *Cloud Atlas* follows the lives of the main characters from each. Signified with a comet/shooting star birthmark, and characterized by the various ways their words persist through time, each main character connects not only backward to the one before and forward to the one after, but multi-directionally to other characters, time-lines, and versions of the storyline.

> I believe there is another world waiting for us, Sixsmith. A better world. And I'll be waiting for you there. I believe we do not stay dead long. Find me beneath the Corsican stars where we first kissed. Yours, eternally, RF.

In keeping with the tagline of the film ("everything connects"), the movie is set up as a series of interlocking vignettes, each one carrying with it a piece of the one it follows, and leading into the next one, via the use of actors playing multiple characters across timelines. Each actor's main character has a throughline/character arc, and those arcs—though sequential, are recursively linked, so that each decision regardless of where/when it is made in time, refracts throughout the multiverse, effecting every other character. Thus, while there are six distinct storylines, which are at discrete locations and times, the connections between them are often splintered and recombined.

Wishaw plays a cabin boy in the first timeline, Robert Frobisher in the second, a music store clerk in the third, a woman named Georgette in the fourth, no one in the fifth, and a tribesman in the sixth. The character of Frobisher has a starring role in the second timeline (his comet is shown streaking across his lower back as he lies in bed with Sixsmith), his creation of the Cloud Atlas sextet and symphony influences the rest of the multiverse, showing up as restaurant theme music in the fifth timeline, being purchased from the record store clerk by the main character of the third, and unifying the score of the whole movie.

However, his other characters are also impactful. The cabin boy is abused, demonstrating a ship captain's cruelty to another main character (Adam Ewing, played by Jim Sturgess) just when such a demonstration was needed to effect a change in allegiances. Georgette's affair with a separate main character (Timothy Cavendish, played by Jim Broadbent) sparks that character's brother to "save" him in a way that saves his life but takes his liberty. The store clerk sells the Cloud Atlas sextet recording to a third main character (Luisa Rey, played by Halle Berry), so that the world continues to have that music in it. And so on.

The importance of understanding the structure of Cloud Atlas is that the choices made by previous characters impact the trajectory of not only their subsequent incarnations, but the eventual trajectories and incarnations of those around them, too. When, for example, Broadbent's character Vyvyan Ayrs rejects Frobisher's advances so cruelly, he goes on to play a fiddle player in a later storyline who we see playing the Cloud Atlas sextet, while Wishaw plays the store clerk, both seeing Frobisher's music refracting through time.

Two concepts are firmly woven into all of the story's timelines: choice and love. Choice is the relatively easy one—cause and effect, belief that one's choices have lasting impact, and that character arcs drive change all are both easily understood by the viewer as movie tropes, and also easy to track through the film's (multiple) narrative(s). The narratives are set in the following times and places: 1849, an island in the South Pacific—main character Adam Ewing, a lawyer; 1936, Cambridge, England, and Edinburgh, Scotland—main character Robert Frobisher, a composer; 1973, San Francisco—main character Luisa Rey, a reporter; 2012, London, England, and Edinburgh, Scotland—main character Timothy Cavendish, a publisher; 2144, Neo Seoul, in what used to be South Korea—main character Sonmi-451 (Bae Doona), a genomically engineered slave called a fabricant; and "106 winters after the fall," an island in the South Pacific—main character Zachry Bailey (Tom Hanks), a goat herder.

Throughout these times, places, and characters, the theme of choice emphasizes the concept of agency, or the ability to exercise personal free will within the physical confines of the system one exists in. That is not to say that free will is infinite—far from it. A world's delineated choices are deterministic, that is the characters cannot break the bonds of physics or space/time/spacetime within their own worlds. However, within that predetermined framework, agency recognizes that free will exists as the ability to choose within those choices that are physically possible, and the concurrent ability to step outside of one's own social conditioning to envision choices that are proscribed by one's culture. Thus in 1849, a lawyer can help a runaway slave free himself—an act that breaks many social and legal conventions, but is the right thing to do. And in 2144 a slave can be freed, learn truths about herself and about freedom, and make the choice to be recaptured and die for her cause, because to die is always among the choices one can exercise.

The characters emphasize their choices and the struggles they go through to make them through communication between time periods. Adam Ewing writes in a diary, where he puzzles out that his doctor is trying to kill him while also contemplating the moral implications of slavery. That diary is then published, and read by Robert Frobisher, who asks his lover Rufus Sixsmith, via letters, to find him a clean copy: "the other half has gone missing. A half-finished book is, after all, a half-finished love affair." Those letters are cherished by Sixsmith—he is reading them when he is shot by an assassin. His body is found by Luisa Rey, who takes the letters to help her figure out who is behind planning a nuclear power plant explosion. Her story is written into a mystery manuscript by her friend Javier, which is then sent to Timothy Cavendish for publication—he reads it as he attempts to flee from angry mobsters. Cavendish writes his memoirs after escaping, which are then turned into a movie, viewed by Sonmi-451, which spurs her to fight for her own freedom, introducing her to the concept of illegal incarceration. Her deposition, taken by an Archivist who is visibly moved by her plight, is made into scripture, written in the brightly colored hair shorn from freed fabricant slaves, bound in the slave-collars removed from their throats. This scripture inspires the Valleysmen to be a kind and sharing society, and inspires Zachry to complete his transformation from coward to hero.

Within these overlapping textual journeys, the characters often speak of interconnectedness, of the effects of their choices, of making those choices because of love for or from others. Many ideas from Sonmi-451 show this, including "our lives are not our own; from womb to tomb we are bound to others, past and present, and by each crime and every kindness, we birth our future," and When asked if she loved her savior, a commander by the name of Hae-Joo Chang, she states she does and always will. The archivist picks up on her phrasing and asks if she believes "...in an afterlife? In a heaven or a hell?" She replies "I believe death is only a door. When it closes, another opens. If I cared to imagine heaven, I would imagine a door opening, and behind it I would find him there, waiting for me."

> to be is to be perceived, and so to know thyself is only possible through the eyes of the other. The nature of our immortal lives is in the consequences of our words and deeds that go on apportioning themselves throughout all time.

Many story beats are driven by choices made from love. Sixsmith finds Frobisher's body after the latter has died by suicide, his last letter confessing that the only things he loved in his "short, bright life" were Sixsmith, and music. When he could not have either anymore, there was nothing to live for, so he ended it. It is heavily implied that Sixsmith never found anyone else, but according to his niece "he considered love a kind of natural phenomenon. He believed that love could outlive death." Similarly, a secondary character in the 1973 storyline, Isaac Sachs, says "Fear, belief, love; phenomena that determined the course of our lives. These forces begin long before we are born and continue after we perish" while confessing his love for Luisa Rey to himself. While evil is prevalent throughout as well—generally personified by Hugo Weaving in various characters—we are reminded that choices based on love overcome it. That love carries on through our actions, our words, our impact on those around us rippling through eternity.

Part Two: The Resurrections

2021's culminating entry into the *Matrix* franchise, *The Matrix Resurrections* has Keanu Reeves and Carrie-Anne Moss reprising their roles as Neo and Trinity. In the original movie, 1999's *The Matrix*, we are introduced not only to the pair of lovers and the world of the Matrix simulation, but also to the concept of Agents—programs used by the machine overlords to control the actions of people plugged into the simulation. Anyone still plugged in can be taken over by an Agent program, turning them from innocuous citizens to single-minded killing program that targets those who have managed to unplug. This inevitability of overlapping realities for the characters is what creates this as a multiverse, particularly as we learn that the older Neo/Thomas A. Anderson has been resurrected multiple times, and set to work at Deus Machina creating his new Matrix games. He is in a treadmill simulation, where the same thing occurs day after day, and even when he attempts to die by suicide, they bring him back.

The directors of the first movie, Lilly and Lana Wachowski, collaborated with David Mitchell on *Cloud Atlas* and in multiple episodes of the Netflix Original series *Sense8* (2015–2018). While never officially confirmed, viewers can observe many in-universe referrants to their other properties, for example, The Agents, the prescients, and the sensates all sharing that body-hopping ability, and the inclusion of Tom Tykwer's musical score and directing on both *Sense8* and *The Matrix Resurrections*.

Let us turn our attention to how agency and love are shown to be themes in *The Matrix Resurrections*. The push and pull of choice and fate are prevalent in the whole series. Humans are plugged into pods, and controlled by a projected simulation that keeps them happy and stable while they are being used by machines to provide biothermal power. Within the simulation, they have no choice, only fate. But the unplugged can choose to fight against the machines once their minds are freed, and they recognize the simulation as nothing more than a lovely dream designed to keep them subjugated. Here we once again see agency—people can be freed and fight, people can self-free and fight, (and both of these are often a choice to die—the machines are ruthless and brutal), or people can choose to return to the dream-world of the Matrix simulation.

Every choice comes with consequences, within the simulation or without. For our protagonists, once Neo is freed he needs to choose to follow his destiny (convoluted though that sounds) to become "The One," the person who can bring down the reign of the Machines. He eventually does believe in himself and makes the choice, but only because Trinity believes it for him first—she is told it is her destiny to fall in love with a man, and that man would become the savior of the human world. When she realizes she loves Neo, she confesses her feelings to him along with the prophecy that foretold their romance, and he steps into his true role. While this synopsis is needfully brief, an important thing to know is that both Neo and Trinity are dead at the end of the third movie—2003's *The Matrix Revolutions*—and everyone thought that would be the end of the franchise. However, in 2019, Warner Bros. announced the fourth movie, *The Matrix Resurrections*.

In the intervening years, many changes occurred that influenced the tone of the movie; the Wachowskis both came out as transgender, Moss and Reeves embodied their characters more naturally, and while during the first three movies we were *told* that Neo and Trinity were in love, just as we were told that Neo is the One, in *Resurrections* we could feel it. Thus, the resumption of the multiverse mentioned above was based on that love. Neo's choices in *Resurrections* are based on a mature, yearning, sensitive love, and Trinity's choices reflect the lack of the same in her "real world" relationship.

They meet for the first time in *Resurrections* as their in-simulation alter egos, Thomas A. Anderson—a game developer who writes the Matrix games for the company Deus Machina, and Tiffany, a wife and mother of three, who builds motorcycles in her spare time. Both long for a life that is far different from the one they are living—Thomas wants to be his creation, Neo, and Tiffany...well, she is nothing like the Trinity we know from the first trilogy. Disaffected in her marriage, dominated by her husband and the ways in which being a wife and mother often take over a woman's life, in Thomas she sees a person who respects her and treats her kindly, perhaps for the first time in a while. Neither of them think they can leave their life, but in talking to each other over coffee, they realize how different things might have been, had they only made different choices. After all, choice makes us the people that we are; different choices make us different people.

Here again we see the theme of agency—they are making the choices available to them within the physical confines of their lives. Anderson cannot be Neo until he is re-awakened from the Matrix simulation by the already-awakened humans. Tiffany feels that pull of awakening less strongly at first, and resists becoming Trinity until it is almost too late. Even as the real-world crew prepares her consciousness to be transferred out of the simulation and onto a ship in the real world, she tells Neo she cannot be Trinity. He believes in her, but he does not force, cajole, or coerce her—the choice is hers. She does eventually make it, though. Her choice is motivated by love. Not just love of Neo, however, but also a deep love of herself. Love and respect go hand in hand, allowing her to finally make the choice to become her true self.

Part Three: The Rising

Premee Mohamed's *Beneath the Rising* trilogy (*Beneath the Rising*, 2020; *A Broken Darkness*, 2021; *The Void Ascendant*, 2022) involves a similar transformation to Trinity's, with self-love and the power of friendship as the catalyst, though through Mohamed's immense skill as a storyteller, the multiverse takes the entire three novels to get there. Following childhood best friends Joanna "Johnny" Chambers—a teenage super-genius autodidact whose inventions regularly make the world a better place—and Nick Prasad—a good-hearted, hardworking, but rather ordinary person, the books reveal that Johnny is paying a high price for the ability to make world-changing progress in multiple types of science, as is the world around her. Johnny has made a deal with extra-dimensional beings from another plane of existence—the de facto multiverse of the series. This multiverse is invasive, and yet partially created by Johnny's own actions oft seen only at the edges of consciousness, these beings of the void, composed mostly of tentacles, eyes, and teeth—take time off of her life for each scientific miracle she performs, just as each miracle strengthens their ability to break through from one part of the multiverse to another. She is killing herself to save the world because they offered her the power to do so, and she knew she could handle it.

But, of course, she cannot handle it. At least, from her perspective, not alone. And for the most part, she does not have to; Nick is always there to calm her, make sure she eats, play straight-man to her genius, and be the one person who she can show her true self to. They are best friends. Or are they? As we move further into the trilogy, we, along with Nick, discover that part of the bargain Johnny made with the beings was to ensure Nick's love, affection, and friendship remained hers, and hers alone. She has staked an extra-dimensional claim on him. However, extra-dimensional beings often lie—that is, they have a different concept of "truth," and also no incentive to play fair with mere mortals. The fact that they can see all versions of the current universe, and that they are invading the one containing this Johnny and this Nick in order to bring about the one they want, defines their truth; though there is a multiverse, the beings have a distinct bias toward one where Johnny flips to the dark side, and drags Nick along with her. The beings start to approach Nick, to work him against Johnny, though to be fair, he is already becoming quite disaffected with her selective affection, as she constantly holds him at arm's length but always expects him to be there when she needs it.

All of this comes to a head within a few pages of the beginning of the first book when Johnny's newest invention—a pollution free power plant that would solve fossil fuel dependence—draws its power from the alternate dimension itself, causing a rift between the worlds. As temporally and spatially shifted beings start descending through the rift, ordinary people get caught up in their attempt to find Johnny. Across the first two books, Nick and Johnny race both the beings and the organization to try to defeat extra-dimensional evil and close the rift before it can obliterate earth as they know it. They manage to do so at the end of the first book, but it reopens—thanks to both Johnny's power plant and the machinations of the society, and the second book ends with them just…failing. They do not close the rift, and Nick is shunted sideways to another part of the multiverse, where he is given the dubious honor of becoming his new world's prophet, based on his knowledge that there is more than one dimension and understanding of otherworldly things.

She, on the other hand, is determined to get back to their world, so she can complete the closing of the rift, even though in so doing she may obliterate herself. Even if self-obliteration is not the result, it would certainly result in the revocation of her powers—she would live out the rest of her life at a normal pace, as a normal person, without any super-science abilities, something she reluctantly accepts. Despite multiple ensuing confounding events, this is what eventually occurs, leaving our protagonists with a dilemma: how to move forward in their relationship. What trust there had been was shattered, what love there had been was built on a lie as well as being unidirectional, because Johnny could not love—the interdimensional beings took that ability away from her.

Although Johnny cannot love, however, Nick can. And he wants to, he would like to attempt a normal friendship with her. In the final scene, we see that Johnny retains the capacity to at least try, and they reset their friendship to an earlier version, introducing themselves as though they are strangers. This is what distinguishes Mohamed's work from others in the tentacled-alien-invasion genre is that the invasive multiverse full of interdimensional beings, while scary (to the characters), is not the main conflict. It is the catalyst for the conflict, but the conflict itself is interpersonal—something that the beings of the multiverse can never understand. The conflict is between Nick's desires and expectations of a normal human relationship, and Johnny's complete inability to ever want the same thing, let alone give it to him. While this inability is dictated by the extra-dimensional beings, we come back to the concept of agency: she eventually makes the choice to abandon those powers—to do the right thing—and to re-start their friendship over so that they can grow to love each other organically. Love and friendship, both his offer of it, and her acceptance of it, triumph, due to agency.

Part Four: The Time War

Contrast this offer of friendship with the enemies-to-lovers time-traveling, world-hopping Red and Blue from *This Is How You Lose the Time War*, by Amal El-Mohtar and Max Gladstone (2019). Blue from The Garden, and Red from The Agency meet each other over and over on the field of battle spanning millennia. Worlds twine in parallel past each other in the great braid of the multiverse, as Blue's faction attempts to bring the world closer to a glorious wilderness and Red's faction attempts to instill technological-based order. Put together in the form of a chapter—describing a mission and its failure or success—paired with a letter—first from Blue to Red, then alternating back and forth throughout the book—this book is a modified epistolary style that builds as the women take turns first taunting, then tempting, then tending to each other.

The exquisitely slow build of their relationship through pages and timelines, the way they dance together, each finding creative new ways to covertly contact the other so that no one from Agency or Garden finds out, provides us with an extreme contrast to the rest of our multiversions of love and agency. Who is good, here? Who is evil? Both kill for their causes, so that is no designating feature. We, the reader, can see the benefit of both Garden and Agency, and both organizations are ruthless, sending their operatives through time to weed, prune, raze, and electrocute. We are shown no purpose for the war, except to win, and herein lie the first threads of the destruction of the war itself, as Red and Blue share enough of each other to build a new loyalty—to each other. Their letters remain challenges to each other, but although they start out antagonistic, they do not remain so long, as each operative begins to tailor the missives to the other's taste, finding rare paper, intense scents and flavors, sealing wax (and whacked seals) to make a letter more memorable.

Throughout all of this, a Seeker trails the two through time and place, gathering up remnants from their correspondence. The Seeker knows what we do not—the only way to win the time war is for both Red and Blue to lose: Blue, killed by Red's final letter which was designed by Agency to tear Blue apart; Red, soon to die at the hands of her compatriots for colluding with Blue. But Blue resurrects herself, due to Red's collusion, and helps Red escape. Red turned herself into the Seeker, and went back in time and space to find Blue before she was Blue, to contaminate her, and keep her alive.

This parallel multiverse makes use of repetitions, reiterations, and the ability to work toward a common future from the splinterings of the past. As each agent, Red or Blue, works to secure their own future, they tweak various strands of their world to ensure that they have enhanced or deemphasized actions that would also shore up their enemy's past. An example of this is the destruction of Atlantis, which both lament must occur again and again for the future of both Agency and Garden, yet both agree that though they *loathe* Atlantis, the civilization produces things useful to the overall multiverse—mathematics, mystical thinking, poetry, even the wheel.

And here we see agency again: the choices both make, from their first letter to their final escape, are set up by their various organizations, but driven by free will and individual choice. Both make the choice to correspond, to lie to their commanding officers, to conceal their entanglement from their factions, and, eventually, to sacrifice themselves for the love that has grown between them, letter by letter. Actions driven by love create the driving force in their multiverse.

The story is, to put it succinctly, elegant; a dance of time, space, and love between two women on opposing sides of a war they no longer support, who decide to fight for each other instead of against. Even the Seeker, who interferes with Red directly twice, *is* Red, after she has figured out how she might win Blue's resurrection. There is no clear-cut line between good and evil, but there is a clarity of purpose for both characters: to be together. And they succeed. We, the readers, do not need to ask what happens next. We know that *everything* happens next.

How Love Strengthens the Multiverse Concept

Love is a true universal. Present as a concept in all human cultures, though varying in form between time and place, we know what it is to love, to want love, to feel it inside ourselves. Humans innately care for other humans—as social animals, we are cooperative, collective, and collaborative (Fuentes 2019). Platonic, romantic, filial, sexual, we are a species that comes together around this concept of love. We may not hate, we may not experience violence on the giving or the receiving end, but we know of love.

Presented as examples of this, we see four differing types of multiverse: sequential, inevitable, invasive, and parallel. In each of the four, love is the motivating factor for the characters to act as they do, and in each their actions alter and shape their world. The multiverse of each world exists independent of them, and would persist without them, but the stories being told to us would not. While all characters have agency, love—for ourselves, for others, for our ancestors or descendants—drives us beyond the mere act of making day-to-day choices. Love gives our agency a goal, and a meaning.

Thus when *Cloud Atlas* shows us love that can outlive death, or *The Matrix Resurrections* shows two people recognizing each other despite wearing different bodies, or *Beneath the Rising* allows two friends to start over, their deep love surviving an almost world-ending betrayal, or *This Is How You Lose the Time War* kills one lover at another's hand, only to have that be their salvation, we understand.

References

Derrickson, Scott. 2016. *Doctor Strange*. Marvel Studios.

El-Mohtar, Amal, and Gladstone, Max. 2019. *This Is How You Lose the Time War*. New York, NY: Saga Press.

Mitchell, David. 2004. *Cloud Atlas*. New York, NY: Random House.

Mohamed, Premee. 2020. *Beneath the Rising*. Oxford, England: Solaris Books.

Mohamed, Premee 2021. *A Broken Darkness*. Oxford, England: Solaris Books.

Mohamed, Premee 2022. *The Void Ascendant*. Oxford, England: Solaris Books.

Wachowski, Lana. 2021. *The Matrix Resurrections*. Village Roadshow Pictures.

Wachowski, Lana; Wachowski, Lilly. 1999. *The Matrix*. Warner Bros.

Wachowski, Lana; Wachowski, Lilly. 2003a. *The Matrix Reloaded*. Warner Bros.

Wachowski, Lana; Wachowski, Lilly. 2003b. *The Matrix Revolutions*. Warner Bros.

Wachowski, Lana; Wachowski Lilly; and Tykwer, Tom. 2012. *Cloud Atlas*. Cloud Atlas Productions.

Wachowski, Lana; Wachowski Lilly; and Tykwer, Tom 2015–2018. *Sense8*. Anarchos Productions.

23

FORGET-ME-NOT

Written by Rebecca Johns

INT. CHILD'S BEDROOM - DAY

The inside of a typical child's bedroom--toys and dirty clothes on the floor, logos of sports teams on the wall. A clock next to the bed reads "3:13 pm, Sunday, June 5th."

In the corner sits a cardboard box designed to look like a spaceship. It reads TIME MACHINE in childish scrawl.

APRIL FLETT (early 40s) comes to stand in the doorway dressed all in black. She bends over and starts picking up clothes, toys, putting away the mess.

She takes a pile of toy Matchbox cars off the floor and puts them in a neat row in a drawer, then breaks down. She lies down on the bed, clutching the pillow to her face.

Her husband TODD (early 40s) finds her there. He is still wearing his mourning clothes, including a distinctively ugly green tie. He too sits on the bed next to her, reaching out to touch her back, then pulling his hand back before he does.

DOI: 10.4324/9781003480846-27

                    TODD
         You want to come to bed? You might feel better
         if you slept.

April turns away.

                    TODD (CONT'D)
         Don't shut me out.

April stands up and walks out.

Todd pauses in the doorway, then goes across the hall
to their room. In a minute, he turns out the light.

INT. LIVING ROOM - LATER THAT NIGHT

    We see April sitting alone with a glass of wine while
    from deeper in the house, we hear Todd snoring.

                                        CUT TO:

INT. LIVING ROOM - DAY

April is sitting in the same place on the sofa. There's
now an empty wine bottle on the table in front of her.

Todd goes out to work while April remains on the sofa
in her robe.

                    TODD
         I'll be home by six. Call if you need me, okay?

April doesn't respond.

INT. LIVING / DINING ROOM - NIGHT

Todd comes home from work with Chinese food.

                    TODD
         Want some?

She won't look at him. He sits at the table and eats
alone.

                    TODD (CONT'D)
         The least we can do is talk to each other.

                    APRIL
         If I wanted to talk, I would.

April gets up and gets herself another bottle of wine, pours herself a glass, and returns to the sofa.

> TODD
> I'm suffering too, you know.

> APRIL
> Doesn't look like it.

> TODD
> What's that supposed to mean?

> APRIL
> Whatever. It means whatever. Enjoy your dinner.

Todd picks up his plate, goes to the sink, and throws it in. The sound of china breaking makes April jump.

> TODD
> If you're determined to push me away, it just might work.

He goes to the bedroom and slams the door.

INT. CHILD'S BEDROOM - CONTINUOUS

April gets up from the sofa and comes to the empty child's bedroom. She crawls into the cardboard box labeled TIME MACHINE. She closes her eyes and falls asleep.

INT. CHILD'S BEDROOM - MORNING

She rubs her face and crawls out of the box, and looks at the bed…

…to see her son JOSH (9) asleep under the covers.

Outside, there are bustling noises of the shower turning on, the clink of silverware. The floor of the child's bedroom is messy once more, covered with clothes and dirty towels. The row of Matchbox cars that April had put away are scattered once again across the floor.

> TODD (O.S.)
> Josh! Breakfast!

Hearing footsteps coming down the hall, she peeks outside the cardboard box to see Todd come through the
door, sit down on the edge of the bed, and start bouncing on the mattress.

> TODD (CONT'D)
> Rise and shine, sleepyhead!

> JOSH
> I'm up!

> TODD
> You're still under the covers.

Josh jumps up.

> TODD (CONT'D)
> Better. Now what do you have to say to me?

> JOSH
> Happy birthday, Dad!

> TODD
> Have you seen Mommy?

> JOSH
> Not yet.

As Josh scurries off, Todd catches sight of her peeking
out from inside the cardboard box. He bends down to
stick his face inside the "door."

> TODD
> There you are! Did you sleep in here all night?

> APRIL
> I…guess I did.

April sees the clock next to Josh's bed. It now reads:
"7:15 a.m., Wednesday, May 15." She turns and looks at
the TIME MACHINE.

INT. DINING ROOM - CONTINUOUS

Todd puts out a bowl with cereal and milk and starts
to tie his tie, a distinctively ugly green tie.

> APRIL
> I thought you said you were never going to wear
> that tie again.

> TODD
> When did I say that?

> APRIL
> Never mind.

> TODD
> It's my birthday tie. I'm wearing it on my
> birthday.

Todd finishes putting on his tie and looks at his watch.

> TODD (CONT'D)
> Gotta go, babe. I'm late. Josh! See you after
> school, bud!

He kisses her and dashes out the door.

When he's gone April sits down shakily at the kitchen
table, listening to the sounds of Josh brushing his
teeth, watching him come to the table, sit down, and
pour milk in his cereal.

He eats slowly, reading the back of the cereal box.
When he senses his mother staring at him, he looks up.

> JOSH
> What's wrong, Mom? Are you crying?

Slowly it begins to dawn on her what has happened.

> APRIL
> What do you say we skip school today? Go do
> something fun instead?

> CUT TO:

INT. - CAR

April and Josh eat ice cream and cheeseburgers in the
front seat.

> JOSH
> Are you sure Dad won't mind? He always gets mad
> when we eat junk food.

> APRIL
> It's our secret, okay?

> JOSH
> Okay.

INT. LIVING ROOM - NIGHT

As soon as they walk in the door, Todd pounces.

> TODD
> Where have you been? I've been calling you all day!

> APRIL
> Sorry, my phone must have run out of juice. We went out for the day.

> TODD
> The school called me. They wanted to know where Josh was, and *I couldn't tell them!* Didn't it occur to you to call him in as absent?

> APRIL
> You're right, I wasn't thinking. It won't happen again.

> TODD
> I thought we'd go out for my birthday. Instead I come home and no one's here.

> APRIL
> I'm sorry. Let's do something now. We can grab dinner…

> TODD
> It's too late now. It's bedtime. I mean, did you think about me even for a second?

April, chastened, gestures for Josh to have a bath and get ready for bed, but she's fearful of letting the day end too soon.

INT. CHILD'S BEDROOM - LATER THAT NIGHT

Josh is wearing his pajamas.

> JOSH
> I'm sorry Dad got mad, but I had the best day.

> APRIL
> Me too, buddy.

Todd walks past the bedroom door.

                    TODD
        Goodnight, mister. (to April)
        You coming to bed?

April glances at the TIME MACHINE.

                    APRIL
        I think I might stay here a bit longer.

                    TODD
        Goodnight.

He flicks out the light. April crawls back inside the
TIME MACHINE and beckons Josh to come inside with her.
He lies down next to her and pulls a blanket over them
both.

                    APRIL
        Where should we go tonight? Or should I say *when?*

                    JOSH
        How about to Egypt? To see the pyramids being
        built?

INT. CHILD'S BEDROOM - MORNING

April wakes up the same as she had the day before. Everything is the same: the noises outside the door, the
mess in the bedroom, the Matchbox cars on the floor. The
date on the clock once again reads 7:15 a.m. Wednesday, May 15.

                    TODD (O.S.)
        Josh! Breakfast!

Todd comes through the door, sits down on the edge of
the bed, and starts bouncing on the mattress exactly
the same as he'd done the day before.

                    TODD (CONT'D)
        What do you have to say to me?

                    JOSH
        Happy birthday, Dad!

                    CUT TO:

INT. LIVING ROOM - MORNING

Todd puts out a bowl with cereal and milk and starts to tie his tie, a distinctively ugly green tie.

> APRIL
> I like your tie.

> TODD
> You always said you hated my birthday tie.

> APRIL
> I changed my mind. You should wear it every day.

Todd finishes putting on his tie and looks at his watch.

> TODD
> Josh! See you after school, bud!

He kisses her and dashes out the door.

When he's gone April sits down at the kitchen table, watching Josh come to the table, sit down, and pour milk in his cereal.

> APRIL
> What do you say we skip school today? Go do something fun instead.

> JOSH
> Sure!

She picks up her phone and starts dialing.

> APRIL
> Hold on, I'll call the school and tell them you're sick.

> CUT TO:

EXT.- BASKETBALL COURT

April and Josh play a game of pickup. Josh makes a three- pointer and April gives him a high five. Her phone, on top of her gym bag, buzzes, and "TODD" pops up on the screen. It goes to voicemail.

EXT. HOUSE- NIGHT

Once again, Todd is fuming when they come in.

                    TODD
Where have you been? I've been calling you for
hours!

                    APRIL
Sorry, my phone must have run out of juice.

                    TODD
I came home and no one's here. Do you know how
frantic I've been?

As Todd continues ranting about how irresponsible she
is, April and Josh look at each other and smile.

                    TODD (CONT'D)
Is this funny? I come home thinking we're going
to spend my birthday together as a family.

                    APRIL
I'll make it up to you, I promise.

INT. HOUSE- CHILD'S BEDROOM

Todd walks past the bedroom door.

                    TODD
Goodnight, mister. (to April)
You coming to bed?

                    APRIL
I think I might stay here a bit longer.

Todd flicks out the light. April crawls back inside the
TIME MACHINE and beckons Josh to come inside with her.
He lies down next to her and pulls a blanket over them
both.

                    APRIL (CONT'D)
Where should we go tonight? Or should I say
when?
                    JOSH
How about the future? Maybe the day I become an
NBA basketball player?

                    APRIL
Not the future. Let's stick with the past.

                    JOSH
          Maybe...to see Michael Jordan in the playoffs?

EXT. - BEACH

The next day, April and Josh chase each other along the
sand and skip rocks along the waves.

INT. - LIVING ROOM - NIGHT

Todd looks up from a plate of Chinese food as they come
home. He is irritated.

                    TODD
          Have fun?

                    APRIL
          What are you upset about now?

                    TODD
          Now? When was I upset earlier?

                    APRIL
          Never mind.

Todd is chewing aggressively.

                    APRIL (CONT'D)
          You are angry about something. I called the
          school to tell them I was keeping Josh home.
          I called you so you'd know where we were going
          to be.

                    TODD
          Nothing. It's nothing.
          I thought maybe you'd include me. I would have
          liked to have come with. Did you forget it was
          my birthday?

                    APRIL
          It was just a spontaneous thing. I didn't think
          you'd mind.

                    TODD
          That's what I mean. I don't want to be an
          afterthought. Not to you.

She stands up and goes to the kitchen for Chinese food, pauses, and kisses him.

INT. - CHILD'S BEDROOM - LATER THAT NIGHT

Josh climbs into bed. Todd walks past the bedroom door.

> TODD
> Goodnight, mister. (to April)
> You coming to bed?

April glances at the TIME MACHINE.

> APRIL
> Sure. Just for a minute, anyway.

April stands in the doorway a minute longer, then flicks out the light.

> APRIL (CONT'D)
> Goodnight, buddy. I love you.

> JOSH
> I love you too, Mom.

INT. PARENTS' BEDROOM - NIGHT

Todd lies down on the bed, April next to him. They embrace, lying like that together for a long time until, inadvertently, April falls asleep.

INT. PARENTS' BEDROOM - MORNING

She wakes in her own bed, but something is different. There are no sheets on the bed, no pillows. The closet is empty.

INT. LIVING ROOM - CONTINUOUS

The furniture is all in place, but it's dusty, silent. The cupboards in the kitchen are open, empty.

INT. CHILD'S BEDROOM - CONTINUOUS

Again the room is empty—no clothes, no toys. The bed is a bare mattress. There's no clock on the bedside table.

April panics when she sees the spot where the TIME MA-
CHINE used to sit. It's gone, too.

INT. LIVING ROOM - CONTINUOUS

She searches the house, but there's no sign of the TIME
MACHINE.

EXT. FRONT YARD - CONTINUOUS

There's a "FOR SALE" sign stuck in the front lawn. No
sign of Todd. He's clearly been gone for some time.

EXT. BACKYARD - CONTINUOUS

She finds the remains of the cardboard TIME MACHINE cut
into pieces and stuffed in the recycling. She gathers
up the pieces and weeps.

24

FINDING YOURSELF

Multiversal Identity Crisis in Ted Chiang's 'Anxiety Is the Dizziness of Freedom'

Amy Coles

Current popular multiverse narratives depict parallel realities as exciting and outlandish: perhaps a crossover event featuring a number of favorite versions of the same character all on one screen, or maybe worlds where hot dogs replace fingers and a pig and T-Rex can be Spider-Man. These star-studded and wacky imaginings present the multiverse as a place where every reality is possible. But what if the majority of parallel universes look almost *exactly* like yours filled with people *exactly* like yourself? The significance of the multiverse is less in the boundless possibilities out there but rather in how these inform our understanding of who we are. A multiverse filled with an infinite number of you, all alike and yet living separate lives, suggests that free will and determinism actually stand in cohesion; our lives have a distinct number of possible paths, but our choices determine which path we will take. The idea of both limitless and yet limited futures sparks a philosophical question, which I am calling the "multiversal identity crisis," that is at the core of Ted Chiang's short story "Anxiety is the Dizziness of Freedom" (2019, published in the collection *Exhalations* in 2020).

Like many writers of science fiction, Chiang uses his stories to express "concerns about technoscience's implications for human life" (Ahn 2020, 82), but this focus is largely on technology's impact on a *psychological* level. While Chiang explains he "want[s] there to be a depth of human feeling in [his] work, ... that's not [his] primary goal as a writer"; instead,

DOI: 10.4324/9781003480846-28

it "has to do with engaging in philosophical questions and thought experiments, trying to work out the consequences of certain ideas" (Rothman 2017). One of the primary inspirations on Chiang's writing is the idea of entropy put forward by Roger Penrose, a mathematical physicist: "the physical property associated with disorder, randomness, or uncertainty" (Tremblay 2023, 24). This randomness and uncertainty is at the very heart of "Anxiety is the Dizziness of Freedom," in which the existence of endless versions of a person causes characters to confront and compare their own identities to answer the greater question of what their choices mean within the deterministic landscape of the multiverse.

The title of Chiang's story is borrowed from Soren Kierkegaard's 1844 essay on anxiety, in which he describes it as being the "dizziness of freedom" (61): "anxiety is freedom's actuality, the possibility of possibility" (42). Within Chiang's multiverse, the endless myriad of parallel versions of the character causes this same anxiety. Though the story explores several characters' experiences with the multiverse, it primarily follows recovering drug addict Nat, who works at a store selling prisms—a type of multiversal communication device. Alongside her supervisor Morrow, the pair use the devices to scam vulnerable customers.

The prism, or "Plaga interworld signaling mechanism" (Chiang 2020, 273), works by connecting a device from one world with another device in a corresponding branch. However, unlike typical multiverse narratives these parallel universes can only be communicated with, not visited:

> a prism wasn't like a radio connecting the two branches; activating one didn't power up a transmitter whose frequency you could keep tuning into. It was more like a notepad that the two branches shared and each time a message was sent, a strip of paper was torn off the top sheet. Once the notepad was exhausted, no more information could be exchanged and the two branches went on their separate ways, incommunicado forever after.
>
> *(274)*

Moreover, communication between prisms is limited only between realities whose branches have not yet diverged. For example, "no prism would ever allow communication to a branch that had split off prior to its moment of activation, so there'd be no reports from branches where Kennedy hadn't been assassinated" (294). The purpose of the prism, then, becomes almost entirely self-involved. However, this self-involvement takes many different forms. In one case a prism is used to create "a supremely realistic military simulation" (295). This runs into issues when the army general's alternate version, or "paraself" (271) as the text terms it, "intended to use him in exactly the same way"—the flaw being that "every branch was of paramount importance to its inhabitants, no one was willing to act as a guinea pig for anyone else" (295). Multiversal media also gains popularity in Chiang's story, as "hard-core sports fans collected information from multiple branches and argued about which team had the best overall performance", and "authors faced competition from pirated copies of books they might have written" (296–7).

However, the most successful use of the prism is for those suffering bereavement (302). This idea is demonstrated in the plot through the characters of Scott and Roderick, a married pop star and actor who are involved in a car accident. In the story's universe, Scott has survived, however, a prism is found which "connect[s] to a branch where there was no parallel Scott, only a grieving Roderick" (322–3). This is one of the scams Nat becomes involved in; knowing that Scott will pay any rate to be virtually reunited with his loved one, Nat and Morrow view this as an opportunity to become obscenely rich. Thus, the use of the prism as a device for grief encompasses the view within the story that these devices are "promoting unhealthy behavior in their customers" (302), both by encouraging a lack of closure and an immorality of those who prey on the vulnerable. However, while prisms perpetuate unhealthy coping mechanisms, it is their insight into parallel—and potentially superior—lives which poses a greater threat to the psyche by both unveiling the different paths each character's life could have taken while simultaneously reinforcing the fact that those doors are now closed.

The Multiversal Identity Crisis

While the major plot of the story revolves around Nat, it is through Dana's narrative that Chiang offers a broader range of repercussions of the multiverse upon individuals. Dana is a therapist and counselor to a prism addicts group, which Nat also attends; through this role, Dana meets multiple people suffering from multiversal "identity crises", who "[feel] that their sense of self was undermined by the countless parallel versions of themselves" (Chiang 2020, 311). Collectively, they represent the damaging effect that prisms can have on self-perception. While previously the prisms had been used for multiversal media consumerism or as a mechanism to cope with (or ignore) grief, "as prisms with larger pads became available, data brokers began offering personal research services for people who wanted to learn about the other paths their lives might have taken" (300–1). Yet, it is noted in the story that this does not provoke a universal identity crisis:

predictably, some individuals became depressed after learning that their parallel selves had enjoyed successes that they themselves hadn't… However, most people decided that they liked more things about their life than they did about their parallel selves' lives, and so concluded that they had made the right decisions.

(301)

The device, then, becomes something that does not *cause* identity crises directly, but exacerbates something that is already there; those who are predisposed to the anxiety of selfhood are the ones who fall victim to it. In Chiang's story, the multiverse becomes an engine through which concepts of identity and purpose, as well as the determinism of the universe, can be directly interrogated through a comparison with the characters' many variants.

However, as prisms become more affordable the desired audience changes, "target[ing] new parents, […] to buy one now, activate it, and store it until their child was an adult, at which point the child could see how her life might have gone differently" (310). Targeting a new generation who would be raised with this multiversal technology is a futuristic and extreme update to those growing up with social media. The prism, then, becomes analogous to new technology, hyperbolizing these issues in a fictional future to impress upon readers the danger of similar technology in the present. Chiang presents this idea to an extreme degree by having children raised with the technology to judge their future success (or failure) by the alternate lives of their paraselves. While older generations may escape these fatal comparisons by having experienced life without such technologies, there are still those who will succumb to the device (as is true of social media) and suffer the same crisis of identity.

Whereas social media encourages comparison of both materialism and appearance, the exact likeness of the paraselves means that it is specifically the different choices made and paths which each variant's life had led them down which contribute to the multiversal identity crisis. The first of Dana's patients introduced is a woman called Teresa, who struggles with "maintaining a long-term romantic relationship" as she is "prone to seeking better alternatives" (275); immediately, then, she is presented as a woman who is dissatisfied with what she has and a desire to always have more. After encountering an ex-boyfriend whose proposal she had refused five years earlier, she learns he is now happily married and begins to question how her life would have been different had she accepted the marriage proposal. With this in mind, she turns to a prism store to seek answers from the multiverse. That Teresa is already predisposed to "seeking better alternatives" strengthens the idea that this "technological innovation is an outlet for preexisting tendencies rather than an agent of psychological change" (Hughes and Eisikovits 2022, 11); her life has already been hindered by conscious attempts to be the happiest and best version of herself, the prism is merely a way to explore these anxieties. The multiverse provides endless possibilities of what Teresa's life could be, overwhelming her to the point that she cannot make any decision for fear of it being the wrong one and causing her unhappiness. Paradoxically, by believing that the wrong choice will bring her unhappiness, Teresa can never be happy. In viewing the multiverse as a world of lost possibilities rather than one of limitless potential, Teresa traps herself within a loop of unhappiness. Ironically, then, Teresa is too concerned with free will; although Chiang writes of its synergetic existence with determinism, the belief of free will alone further enforces the multiversal identity crisis.

Jorge is another victim of identity crisis; following an incident at work where he "punctured all four tires of his manager's car" (Chiang 2020, 286), Jorge is drawn to prisms to understand if this aggressive behavior is part of his inherent nature or a freak accident. Unlike Teresa, who is concerned with being the *happiest* version of herself, Jorge's fear is about being the *worst* version of himself. Like Teresa, it is not the prism that initiates his identity crisis, but it does become a way of restoring his sense of self. This is, however, a potentially damaging use of the prism. By securing his identity through an assessment of his paraselves' identities, Jorge rejects responsibility for his actions as a mere "freak accident" (333). However, Dana encourages him to assume responsibility, explaining that while "it was obviously out of character for [him]…it was still something [he] did" (333). Although the discovery that none of his paraselves participated in this behavior relieves Jorge's anxiety about whether he is inherently a violent person, it is only by accepting that despite his good nature he *did* commit a violent act that he can prevent himself from repeating this behavior in the future. In understanding that he *could* be the most violent version of himself, Jorge *chooses* not to be; Chiang demonstrates, through Jorge's character, free will's ability to change our futures and strengthen our sense of identity.

Although Dana's clients begin therapy for reasons existing beyond the multiverse (but often lead to use of prisms), the members of the prism addicts group that she leads require help directly *because* of prisms. For those in the help group, the anxiety comes from already knowing that there is a better paraself in another universe, and the jealousy stemming from that. As Nat explains, the identity crises arise because "with a prism, it's not other people, it's you. So how can you not feel like you deserve what they have?" (283). These characters become addicted to contacting their superior paraselves; not unlike virtual reality or the metaverse, the prism allows them to vicariously live the life they feel they should have. One member, Lyle, becomes jealous of his paraself after activating a prism to decide whether or not to accept a new job offer. He accepts the offer while his paraself stays in the previous job, however the new job becomes boring while his paraself receives a promotion.

This jealousy increases when his paraself begins a relationship with a woman named Becca. When Lyle attempts to date his universe's Becca, his conversation fixates on his rivalry with his paraself, effectively ruining his chances at the relationship. By convincing himself that he can only be happy by living the exact same life as his seemingly happier paraself, Lyle traps himself in a loop of unhappiness not dissimilar to Teresa. Simply put, by believing he is a worse version of himself, he *becomes* a worse version of himself. However, acknowledging that "keeping [his] prism would just keep [him] in that mind-set, wanting to prove something" (313), Lyle gives up the device, recognizing that his reliance upon his prism only encourages jealousy. Like Jorge, Lyle understands that trying to unify himself with his paraselves leads to disappointment and confusion of selfhood; by freeing himself from the prism, he *chooses* to become happier.

These multiversal identity crises narratives told through Chiang's story "comment on the existential burden of freedom in the modern world" (Hughes and Eisikovits 2022, 10), and the overwhelming weight of our actions. The struggle to find happiness seen in both Teresa and Lyle comes from the knowledge that one decision can hold so much weight on the outcome of our lives, as "the more possibilities we construct, the more counterfactual scenarios there are to negotiate, the more difficult it becomes to process them in our working memory" (Kukkonen 2010, 47). Yet, while these limitless scenarios cause a confusion of identity, they also provoke another form of identity anxiety, one which questions our purpose and the meaning of our decisions within a seemingly deterministic universe.

Finding Meaning in the Multiverse

The debate between free will and determinism has long been a recurring feature in Chiang's work. He explains in an interview that he "believe[s] that the universe is deterministic, but that the most meaningful definition of free will is compatible with determinism" (Rothman 2017). Conceivably then, while the universe itself has preset outcomes that will occur no matter the choices we make, we can also consider that there are a number of preset outcomes as opposed to one singular possibility; consequentially the important choices we make –that is, our *moral* choices – determine the ultimate outcome. The question of free will within the story is pertinent to overcoming the multiversal identity crisis; if a person's decisions matter and they have the freedom to choose, then they are in control of creating the person they want to be. While the story tells us that many major events are impacted by minor changes in weather (Chiang 2020, 285), Chiang also uses the multiverse as "a proving ground for one's character. The story implies that a person's character is not defined or determined by any one event but by the cumulative sum of his or her actions" (Shephard 2022, 2). Although in Chiang's world, there are deterministic outcomes that will occur no matter the choices a person makes, these choices do impact the person they can become. The issue of free will sparks its own kind of identity crisis, raising the question of our individual purpose within the universe - or, on a greater scale, the multiverse.

Dana explains the paradoxical comfort and weighty responsibility that comes with free will during a prism addicts meeting:

> we like the idea that there's always someone responsible for any given event, because that helps us make sense of the world…But not everything is under our control, or even anyone's control.
>
> *(Chiang 2020, 299)*

To believe our actions have a great impact upon the universe gives us a semblance of control, however, it also creates unnecessary anxiety toward each decision that we make. This anxiety is amplified in Chiang's world where both knowledge of and contact with the multiverse is commonplace; for these characters, the weight of their decisions does not only impact them, but possibly every other branch as well. Moreover, if all decisions are being made across the multiverse—including the worst possible decision—how much impact does the choice made in the current universe have? The story addresses this anxiety by explaining that choices *do* matter, representing Chiang's belief of a harmonic world containing both free will *and* determinism:

> many worried that their choices were rendered meaningless because every action they took was counterbalanced by a branch in which they had made the opposite choice. Experts tried to explain that human decision-making was a classical rather than quantum phenomenon…it was quantum phenomena that generated new branches, and your choices in those branches were as meaningful as they ever were. Despite such efforts, many people became convinced that prisms nullified the moral weight of their actions.
>
> *(311)*

In Chiang's story, the multiverse's impediment of any meaning in decision-making can prompt a descent into nihilism.

Glenn Oehlsen, the son of an elderly woman who was scammed by Nat and her supervisor Morrow, is the best representation of this nihilism within Chiang's story. After Morrow convinces Mrs. Oehlson to leave her money to her paraself in her will (an impossibility which sees Morrow and Nat receive the money instead), Glenn enters the prism store armed and demanding his mother's money be rightfully returned to him. Morrow calls the man's bluff, believing that he will want to avoid imprisonment; however, as Glenn comments, "there's some timeline where [he] shoot[s Morrow] right now", and "if it's going to happy anyway, why shouldn't [he] be the one to do it?" (319). Though at first he begins to leave, he turns at the door to ask "what difference does it make?" before shooting "Morrow in the face" (319). This destructive behavior is the product of nihilism; Glenn believes that if *any* paraself is going to shoot Morrow then he has more reason to do it because it is going to happen anyway. He exemplifies the Edgar Allan Poe phrase referenced in the story, "the imp of the perverse," used "to describe the temptation to do the wrong thing simply because you could" (311). Thus, Glenn shooting Morrow demonstrates the very truth of free will in Chiang's novel; when faced with the opportunity to be a better person, he actively chooses to be the worst.

Dana believes she has made a similarly bad decision which has impacted the life of someone close to her. This decision is made as a teenager when she and her friend Vinessa smuggle Vicodin on a school trip to Washington, D.C. (305). A teacher finds them with the pills during a surprise room check, and Dana claims they are Vinessa's: "there was a moment when Dana could have taken back what she said, when she could have confessed the truth, but she didn't" (306). For Vinessa, who was already a free-spirited troublemaker—although also a very promising student—this decision seems to have a negative impact upon her life:

> Vinessa was suspended [...]. It was as if, before that night, Vinessa has been balanced on a knife's edge; she could have become either what society considered a good girl or a bad girl. Dana's lie had pushed her off the edge onto the side of being bad.
>
> *(306)*

Dana spends her life feeling guilty for the choice she made, believing that she is the reason for Vinessa's lack of success in life. In her guilt she gives Vinessa money for education and business ventures - though these go nowhere.

However, at the end of the story, Dana receives a tablet from an anonymous sender containing multiple videos from her paraselves, all explaining the different decisions they had made yet the same outcome of Vinessa's fate. In a universe where Dana takes the blame, Vinessa starts to ignore her because of her own feelings of guilt and joins a new friendship group who encourage the same bad habits she becomes involved in the current Dana's universe. In another instance where Dana confesses the entire truth so that both she and Vinessa will take the blame, Vinessa is furious with her and begins acting out, ultimately getting suspended. After watching these videos, Dana realizes that *"if the same thing happens in branches where you acted differently, then you aren't the cause"* (339; italics in original). Ironically, then, determinism becomes a comfort; knowing that Vinessa would have had the same outcome no matter what decision she made gives Dana a more secure sense of her own identity and removes her guilt. Once more we see Chiang's idea of a combined free will and determinism: while Dana's choices could not change the course of Vinessa's life, Vinessa's choices *could*. She is no longer to blame, and the decisions she makes cannot entirely affect the life of someone else because *they* must take responsibility for *their* actions.

Chiang's story reflects the importance of taking responsibility for one's own actions; although this is represented through many of the characters, it is Nat who undergoes the greatest personal development. As a recovering drug addict, Nat already understands the importance of "tak[ing] responsibility for the things she did" (310). However, though she is making comparatively better choices, her engagement with Morrow's scams also exhibit immoral behavior. Yet, Morrow's murder sends her into an existential crisis, questioning "whether [her] decisions matter" (327). Haunted by the words of Glenn, Nat reflects during a prism addicts meeting. Dana informs her that that "if you act compassionately in this branch, that's still meaningful, because it has an effect on the branches that will split off in the future" (328). The multiversal opportunities to make the worst possible decision give all the more reason to make the right one; rather than believing that if there has to be a worst version of yourself it may as well be you, you can acknowledge that this worst version exists, but it will *not* be you. Instead, you can be the best you, and *create* a better you for all your futures.

The multiverse thus becomes a sort of mirror, but instead of displaying a single reflection it shows all possible selves; it is up to us to choose who we want to be in every universe. Moreover, in Chiang's universe "each time you do something generous, you're shaping yourself into someone who's more likely to be generous next time":

it's not just your behavior in this branch that you're changing: you're inoculating all the versions of you that split off in the future. By becoming a better person, you're ensuring that more and more of the branches that split off from this point forward are populated by better versions of you.

(329)

Through this, not only is free will evident in Chiang's universe as every action directly impacts parallel branches, but is imperative in the creation of identity. While there may be wider largescale events that are predetermined, the choices we make as individuals directly impact who we are as people. This revelation changes the course of Nat's life, encouraging her to "[imagine] what a better person might do, and [do] that instead" (336), consequently funding the tablet that Dana receives. Using the money she has made from selling the prism to Scott which will connect him with his deceased husband, Nat pays for parallel Danas to be contacted to relieve her guilt.

Through this multiverse narrative, Chiang presents free will as something that is both comforting and anxiety-inducing. For Dana, the belief that her choices have ruined her friend's life causes an identity crisis which causes her to question her morality; for Nat, the possibility that her choices have no meaning almost drives her to nihilism. Yet, as in *The Butterfly Effect* (2004), each choice made "spans ever-new branches" (Ryan 2006, 659), proving that decisions *do* have meaning. Yet, more important than the impact these choices have on the wider world is the impact they have on the self. Through this story, Chiang explains that the importance of choice is how it shapes us as individuals. Thus, the multiversal technology that Chiang presents becomes an opportunity for positive change; as James Hughes examines, "rather than the loss of responsibility we assume the technology implies, Chiang gives us its amplification" (Hughes and Eisikovits 2022, 10).

The Multiversal Guide to Self-improvement

On first glance, Chiang's story seems to present the multiverse and its surrounding technologies as deeply problematic tools; however, these devices merely reflect preexisting problems within society exacerbated through new technologies, they "[do] not fundamentally change human nature" (Hughes and Eisikovits 2022, 8). Thus, Chiang's engagement with the multiverse becomes a thought-provoking questioning of free will and identity, in which the technology that may encourage a dismantling of these ideas and initiate the multiversal identity crisis ultimately provides the means to overcome it. As he comments in his accompanying notes to the story, even if the many-worlds interpretation is correct, it doesn't mean that all of our decisions are canceled out. If we say that an individual's character is revealed by the choices they make over time, then, in a similar fashion, an individual's character would also be revealed by the choices they make across many worlds (Chiang 2020, 349–50). While Teresa's anxieties and Glenn's nihilism are not resolved by the end of the story, Chiang provides us with a far more optimistic narrative in which characters like Dana, Jorge, and Lyle overcome their crises *through* the multiverse, using the prism as a mirror which ultimately secures their own sense of identity. By highlighting Nat's narrative arc specifically, Chiang showcases a character who is more at risk of nihilism, yet finds solace in the multiverse by learning that her positive actions in her own universe will have a domino effect in other universes as well.

Chiang tells us that we must take responsibility for our actions; by placing this message into a multiverse narrative, he amplifies the importance of a person's decisions upon not only themselves, but all of their paraselves. Although nihilism initially seems inevitable, the story ends with a message of hope. Moreover, Chiang informs us that the worst version of yourself is not necessarily the least successful version, but the most immoral. True goodness—and possibly even happiness—comes from knowing that a determinist world means the possibility of doing bad but *choosing* not to, thus giving these choices purpose. In actuality, determinism is less about a single preset destiny and more of a choose your own adventure; there may only be a handful of options, but it is up to you to choose the right one.

Ultimately, Chiang's story uses speculative scientific technology as a mirror, forcing us to confront ourselves as we always have been. The technology does not create a dystopic world, but merely enhances it, "point[ing] towards a synthesis of utopian technological optimism with dystopian pessimism…to synthesize and present a narrative tension that does not make technology either hero or villain" (Hughes and Eisikovits 2022, 6). However, it also offers us a way out. Chiang explains that "while some dystopian stories suggest that doom is unavoidable, other ones are intended as cautionary tales" (Marcus 2020). The multiverse need not create unnecessary identity spirals unless we choose to spend our lives miserable, believing that in another life we may be happier. Knowing that we have limitless possibilities can cause a dizziness of identity, but it can also offer up a world of opportunities in which we can be anything. It is up to us to choose who we want to be.

References

Ahn, Sunyoung. 2020. "The Everyday Life of Artificial Intelligence: The Humanism of Ted Chiang's *The Lifecycle of Software Objects*." *Science Fiction Studies* 47, no. 1: 73–92.

Chiang, Ted. 2020. *Exhalations*. London: Picador.

Hughes, James J., and Nir Eisikovits. 2022. "The Post-Dystopian Technorealism of Ted Chiang." *Journal of Ethics and Emerging Technologies* 32, no. 1: 1–14.

Kukkonen, Karin. 2010. "Navigating Infinite Earths: Readers, Mental Models, and the Multiverse of Superhero Comics." *Storyworlds: A Journal of Narrative Studies* 2: 39–58.

Marcus, Halimah. 2020. "Interview: Ted Chiang Explains the Disaster Novel We All Suddenly Live In." *Electric Lit*. Mar 31. https://electricliterature.com/ted-chiang-explains-the-disaster-novel-we-all-suddenly-live-in/.

Rothman, Joshua. 2017. "Interview: Ted Chiang's Soulful Science Fiction." *The New Yorker*. Jan 5. https://www.newyorker.com/culture/persons-of-interest/ted-chiangs-soulful-science-fiction.

Ryan, Marie-Laure. 2006. "From Parallel Universes to Possible Worlds: Ontological Pluralism in Physics, Narratology, and Narrative." *Poetics Today* 27, no. 4: 633–674.

Shephard, W. Andrew. 2022. "Chiang, Ted." In *The Encyclopedia of Contemporary American Fiction 1980–2020*, edited by Patrick O'Donnell, Stephen J. Burn, and Lesley Larkin, 1–6. London: Wiley.

Tremblay, Jean-Thomas. 2023. "Homeostasis and Extinction: Ted Chiang's 'Exhalation'." *SubStance* 52, no. 1: 22–29.

25

ON WORMHOLES, LINK CABLES, AND THE LIMITATIONS OF TRANSMEDIA

Fan Interpretations of Trans-Spatial Discourses in *Pokémon* Video Games

Ross Garner

This chapter offers a provocation against current intellectual orthodoxies relating to the study of commercial media storyworlds by postulating how concepts derived from multiversal discourses may be useful in understanding the textual structures and audience negotiations of a high-profile media franchise. The franchise selected for examination is *Poké-mon*, the Japanese-originating intellectual property (IP), which spans myriad mediations including computer games, trading cards, anime, manga, toys and plush, apparel, and experiences. Specifically, the chapter focuses on the *Pokémon* video games' deployment of multiversal discourses as a function of how *Pokémon*'s storyworld develops. This is not to disavow that *Pokémon* narratives in alternative media have explored multiversal tropes (see Switzer 2021 and Valdez 2021 on the *Pokémon* anime), but rather that video games advance audience knowledge about what this chapter names "the *Poké-verse.*" The appropriation of multiversal concepts by commercial popular culture properties like *Pokémon* "expands the reach of scientific and philosophical concepts" (Katerynych 2024, 152). If, as video games journalist Ben Sledge (2022) argues, "multiverses are all the rage this year," what insights might be gained by using these ideas to interrogate the form and reception of an IP of vast textuality? This chapter argues that fan negotiations of *scientific* multiversal concepts linked to trans-spatial travel and wormholes reintroduce less prominent academic debates linked to *textual* interpretation such as pleasure (Eco 1994) and affect (Grossberg 1992).

DOI: 10.4324/9781003480846-29

Marie-Laure Ryan and Jan-Noël Thon (2014) argue that "storyworlds can always sprout branches to their core plots that further immerse people, thereby providing new pleasures" (19). Given a storyworld's potential for attracting and sustaining audience interest, it is perhaps unsurprising that Elizabeth Evans (2011) has noted that "the presence and importance of a fictional world offers the space for transmedia storytelling to be explored" (13). Evans makes two important points here: first, that storyworlds have taken on new significance and commercial value in the twenty-first century. This is because, as Ryan and Thon (2014) argue, storyworlds allow for characters with established popularity "to migrate from medium to medium in any imaginable order," therefore enabling multiple narrative installments (2–3). Secondly, Evans indicates how perspectives indebted to what Henry Jenkins (2006) theorized as convergence culture have become the aforementioned intellectual orthodoxy for contemporaneous academic analyses of commercial storyworlds. Much work has subsequently been undertaken examining any combination of the following issues: industry configurations that enable or constrain how an IP becomes disseminated across media forms in specific national contexts (Phillips 2012; Steinberg 2012; Jenkins 2014); the narrative strategies and media affordances that underpin those disseminations (Harvey 2015; Mittell 2015, 292–318); and the extent to which audiences debate, rework, promote, and/or share information about such properties through digital platforms (Stein 2015; Booth 2016; Williams 2020). These explorations of the contours and configurations of transmedia storyworlds typically approach their subject matter "in anthropological terms" by conceptualizing intellectual properties like *Star Wars* or DC Comics superheroes as "something you *do*, not just something you read or watch or 'consume'" (Buckingham and Sefton-Green 2004, 12). In contrast, as this chapter demonstrates, approaching the *Poké-verse* in terms of how audiences have negotiated its construction of multiversal discourse shifts the focus back to debates concerning interpretation, demonstrating how "a storyworld is …a dynamic model of evolving situations, and its representation in the recipient's mind is a simulation of the changes that are caused by the events of the plot" (Ryan 2014, 33). Such a change in focus allows for positions relating to textual meaning, pleasure, and affect to take on renewed resonance in the analysis of contemporary cross-media properties.

Contextualizing the Poké-verse: Genre, Depth, and Commerce

From a genre perspective, individual *Pokémon* video games utilize conventions from adventure stories (Buckingham and Sefton-Green 2004, 19–21). These roots are ascribed to *Pokémon*'s author-figure, Satoshi Tajiri, via romanticized narratives of the franchise's origins: "As a boy, his favourite pastime had been insect and crayfish collecting, an activity involving interactions both with nature (exploration, adventure, observation, gathering) and society (in exchanges and information sharing with other children)" (Allison 2004, 41). Players of *Pokémon* video games mirror these activities by journeying between and exploring a set of pre-defined spaces while accomplishing a range of tasks. These include game-wide challenges like completing the Pokédex (an in-game data repository listing every species native to the fictional setting) or vanquishing the plans of nefarious scientific organizations, to smaller cumulative undertakings such as completing puzzles or defeating other trainers *en route* to becoming a Pokémon Master. Inquisitiveness and meticulousness are frequently rewarded throughout a game's narrative, as both crucial items and rare Pokémon are usually only found by thoroughly exploring each game's towns and surrounding wild areas. Individual *Pokémon* video games subsequently exemplify what Alain Boillat (2022) names "world-centred" fiction where "users exploit the interactivity of the medium by discovering – along a more or less predetermined path – various environments" as a structuring part of the gameplay.

Additionally, *Pokémon*'s storyworld generically incorporates science fiction codings by featuring "cognitive innovation[s]" (Suvin 1977), like the presence of Pokémon species (68). These sentient-yet-fantastical creatures which co-exist alongside human characters differentiate the storyworld "from the author's and implied reader's norm of reality" (68). Individual *Pokémon* video games also contain other innovations within their storyworld(s) which encourage their classification as science fiction texts. These include extrapolations from scientific principles connected to multiversal discourse such as the narrative devices that enable trans-spatial travel discussed below.

However, while *Pokémon* video games appropriate scientific discourses ranging from multiversal concepts to "evolution, geography, astronomy, history, and physiology" (Sefton-Green 2004, 142), this does not mean that these engage these ideas in a detailed or developed manner. This is largely because, in terms of target audience, *Pokémon* video games are intended to be "challenging but doable even by children as young as four" (Allison 2004, 42). Consequently, in-depth knowledge of scientific concepts is not required for fear of alienating the youngest players. Rather than requiring players to utilize detailed intertextual knowledge drawn from educational and/or cultural experiences, *Pokémon* video games instead emphasize players acquiring and mobilizing what Ryan (2014) names "*intradiegetic elements*, which exist within the storyworld" (37). This includes information concerning Pokémon species types, move sets, and item locations. Thus, although "participating in the culture is, to some extent, a question of learning, of becoming proficient, and of using knowledge to succeed" when completing the game, the games emphasize *intradiegetic* over *intertextual* knowledge (Sefton-Green 2004, 142). This observation extends to other areas of the games, such as characterization, as almost all characters encountered while exploring the game's storyworld are types, often with limited dialogue, rather than complex, developed, or conflicted characters. This is also the case with non-playable characters who you encounter more than once, such as your in-game rivals and antagonists.

Capitalist industrial factors also structure the form and development of *Pokémon*'s storyworld—both across and within specific video game titles. Cumulatively, the storyworld constructed across *Pokémon* video games exemplifies what Mark J. P. Wolf (2012) names an 'open' structure. That is, the *Poké-verse* operates as "a world in which canonical material is still being added; such a world is still growing and developing, as it accrues more information, detail, and narrative" (Wolf 2012, 270). Central to how the *Poké-verse* expands is the concept of Generations. Initially a fan-generated term, the linear progression from one Generation to another is commercially motivated as it refreshes the property by releasing new "core" video game titles for a Nintendo console. Each new Generation reveals hitherto unseen regions, Pokémon species, and human characters, thus growing the *Poké-verse*. For example, Generation I covers *Pokémon Red* and *Pokémon Blue*, the first computer games that were released for Nintendo's Game Boy, and so the territories, inhabitants, and events associated with the Kanto region. Currently, the release of *Pokémon Scarlet* and *Pokémon Violet* in November 2022 for the Nintendo Switch marked the start of Generation IX and the *Poké-verse*'s expansion to include the inhabitants and topographies of the Paldea region (Table 25.1 summarizes the different *Pokémon* Generations). Through continually adding new regions, characters, narratives, and titles to the *Poké-verse*, revenues can be boosted and market share can be retained.

Individual Generations also include what fans refer to as "upper version" games. These can either be an additional title that offers variation on the initial release, or remakes of previous titles that feature updated and expanded gameplay features and/or storylines. Remakes like *Pokémon Omega Ruby* and *Pokémon Alpha Sapphire* (ORAS) thus support the *Poké-verse*'s open status, by adding new storylines and diegetic areas to what has previously been established about the Region depicted. At the same time, these second types of "upper version" games represent what

TABLE 25.1 List of *Pokémon* Generations, their corresponding video game titles, and popular Pokémon species introduced in each

Generation	Release Year	Region Name	Core Video Games	'Upper' Video Games	Popular Pokémon Species Introduced
I	1996	Kanto	*Pokémon Red* and *Pokémon Blue*	*Pokémon Yellow*[a]	Pikachu, Bulbasaur, Charmander, Squirtle, Jigglypuff, Eevee, Mew, Mewtwo, Articuno, Zapdos, Moltres
II	1999	Johto	*Pokémon Gold and Pokémon Silver*	*Pokémon Crystal*[a]	Chikorita, Cyndaquil, Totodile, Espeon, Umbreon, Lugia, Ho-Oh, Entei, Raikou, Suicune
III	2002	Hoenn	*Pokémon Ruby and Pokémon Sapphire*	*Pokémon Emerald*[a], *Pokémon FireRed*[b] and *Pokémon LeafGreen*[b]	Treecko, Torchic, Mudkip, Gardevoir, Milotic, Groudon, Kyogre, Rayquaza, Latias, Latios
IV	2006	Sinnoh	*Pokémon Diamond and Pokémon Pearl*	*Pokémon Platinum*[a], *Pokémon HeartGold*[b] and *Pokémon SoulSilver*[b]	Turtwig, Chimchar, Piplup, Bidoof, Garchomp, Dialga, Palkia, Arceus, Darkrai, Giratina
V	2010	Unova	*Pokémon Black and Pokémon White*	*Pokémon Black II*[a] and *Pokémon White II*[a]	Snivy, Tepig, Oshawott, Zorua, Chandelure, Reshiram, Zekrom, Kyurem, Genesect
VI	2013	Kalos	*Pokémon X and Pokémon Y*	*Pokémon Omega Red*[b] and *Pokémon Alpha Sapphire*[b]	Chespin, Fennekin, Greninja, Sylveon, Dedenne, Xerneas, Yvatal, Zygarde, Hoopa

(*Continued*)

TABLE 25.1 (Continued)

Generation	Release Year	Region Name	Core Video Games	'Upper' Video Games	Popular Pokémon Species Introduced
VII	2016	Alola	Pokémon Sun and Pokémon Moon	Pokémon Ultra Sun[a] and Pokémon Ultra Moon[a], Pokémon: Let's Go! Pikachu[b] and Pokémon: Let's Go! Eevee[b]	Rowlet, Litten, Popplio, Lycanroc, Mimikyu, Solgaleo, Lunala, Necrozma, Tapu Bulu, Tapu Koko, Tapu Fini
VIII	2019	Galar	Pokémon Sword and Pokémon Shield	Pokémon Brilliant Diamond[b] and Pokémon Shining Pearl[b], Pokémon Legends: Arceus[c]	Growkey, Scorbunny, Sobble, Wooloo, Greedent, Eternatus, Zacian, Zamazenta, Urshifu, Calyrex
IX	2022	Paldea	Pokémon Scarlet and Pokémon Violet	Pokémon Legends: Z-A[c]	Sprigatito, Fuecoco, Quaxly, Lechonk, Pawmi, Koraidon, Miraidon

Sources: https://bulbapedia.bulbagarden.net/wiki/Generation [Accessed 24/04/24] and https://Pokémon.fandom.com/wiki/Generation [Accessed 24/04/24].

Key:

[a] Initial game with updated storylines and features
[b] Updated version of a previous Generation title for a new platform
[c] New storyline set in a previously seen location

Matt Hills (2015) calls "*periodic* re-commoditization" as titles with established recognition and popularity are repackaged for a pre-established market and reimagined for the newest pieces of Nintendo hardware (24).

Hoenn via Kalos: "The Delta Incident," Inferential Walks and Affect

The first allusion to multiversal discourse within the *Poké-verse* took place in Generation VI's *ORAS* titles. During "The Delta Episode," an additional "post-game" storyline, the Hoenn region becomes threatened with destruction by a meteor. One proposed solution—which is almost immediately rejected—is to open a portal and send the space rock to an alternate dimension, "a world where Kalos' ancient superweapon was never fired, and so Mega Evolution does not exist" (Wilcox 2023, para. 4). Mega Evolution is a plot device added to Generation VI's core titles, *Pokémon X* and *Pokémon Y*. It is thus associated with the Kalos region and involves certain species of Pokémon evolving beyond their previously assumed final form to become even more powerful for a limited period of time. Mentioning the "ancient superweapon" that triggers Mega Evolution in *ORAS* exemplifies how remaking "upper" titles within specific Generations can employ what Wolf (2012) names "retroactive linkages," or "connections between two worlds which were conceived and made separately" and often occur "in the work of authors who have created two or more imaginary worlds and wish to bring them together into one larger creation, so they can be considered as a form of world-building" (216). In this instance, Game Freak and Nintendo are using narrative devices introduced in newer Generations to revise previous titles by constructing trans-spatial multiversal discourses. Retroactive linkages can "be done for commercial reasons, such as …hoping to increase …sales" of previous titles by making these align with franchise developments that post-date the production of the original, in this instance Generation III, titles (219). However, "The Delta Episode" has taken on extra significance amongst fans of the *Pokémon* video games, functioning as a locus for affect and pleasure.

Writing on fan attachments to commercial popular culture, Lawrence Grossberg (1992) argues that:

> affect …operates within and, at the same time, produces maps which direct our investments in and into the world; these maps tell us where and how we can become absorbed – not into the self but into the world – as potential locations for our self-identifications, and with what intensities.

This 'absorption' or investment constructs the places and events which are, or can become, significant to us. They are the places at which we can construct our own identity as something to be invested in, as something that matters.

(57)

For Grossberg, affect is an embodied state which is difficult to communicate with others, but not wholly subjective. Instead, affect is guided by the structures of contemporary consumer culture toward the intellectual properties—and, going further, parts of those properties—that resonate with particular subject positionings. It is for this reason that, Grossberg (57–8) argues, the parts of popular culture that become meaningful to fan groups "are like investment portfolios …They tell us …how to live within emotional and ideological histories." Despite being a momentary line within 'The Delta Episode'—the kind of dialogue that might exemplify fans investment in trivia (see Jenkins 1992, 10; Eco 2007 [1986], 68)—the suggestion that the *Poké-verse* is constructed as a multiverse has become a locus for affective attachments to *Pokémon*. Sledge (2022) indicates this, writing that 'The Delta Episode' "confirms that every remake exists in a parallel universe to its original game. … if every Pokémon game exists in the same multiverse, what's to stop them from colliding again in the future?" (para. 5) The possibility of trans-spatial movement, and thus the storyworld of games titles existing parallel to each other, alters fan understanding by assigning canonical status to until-then fan speculation concerning the ontology of the *Poké-verse*.

The significance afforded to the momentary intrusion of trans-spatial discourse in *ORAS* can be accounted for in two ways. First, it has become part of a shared mattering map amongst *Pokémon* fans that provides pleasure by inviting fans to undertake what Umberto Eco (1994) discusses as "inferential walks" within the *Poké-verse*. Writing on textual interpretation, Eco employs a metaphor that equates reading fiction to walking in a forest, whereby inferential walks constitute either "imaginary walks outside the wood" or, alternatively, acts of "lingering" where the reader "stops to ponder" on a particular implication of the story (50). Inferential walks are pleasure-inducing moments for readers (or players of *Pokémon* video games) as they allow for deeper connection to the storyworld by reflecting on underdeveloped narrative threads such as the previously unconfirmed interconnections between the different regions and titles that make up the *Poké-verse*. Such forms of discourse arguably heighten the *Poké-verse*'s cult appeal by allowing it to be considered as a "*hyperdiegesis* …a vast and detailed narrative space …which nevertheless appears to operate according to principles of internal logic and extension" (Hills 2002, 137). Rather than an "open" storyworld that expands in relation to the commercial demands of Nintendo and Game Freak, the allusion to trans-spatial travel made in "The Delta Episode" has allowed *Pokémon* video game fans to re-negotiate their understanding of the *Poké-verse* and confirm their affective investments in (until then) subculturally circulating interpretations of the property. Of course, the validation bestowed on *Pokémon* video game fans remains structured by commercial forces as a result of *ORAS* being retrospectively linked to plot devices introduced elsewhere in Generation VI. Nevertheless, this incident demonstrates how constructing multiversal discourse between video game titles has become important to fan investments in and understandings of the *Poké-verse*.

Second, "The Delta Episode" provided pleasure by inviting fans to take inferential walks which speculated about potential future developments within the games' storyworld. "The Delta Episode" altered the ontological status of the textual *Poké-verse* as, rather than understanding this as a storyworld that unfolds in a linear and serialized way, the possibility of inter-dimensional travel implies more complex spatio-temporal underpinnings. As Wolf (2012) argues, "retroactive linkages can alter the context and canonicity of a work, and change how an audience sees a particular world and the overarching narratives taking place within it" (219) Following canonical confirmation that the *Poké-verse* is constructed as a multiverse where trans-spatial movement can take place, fans could subsequently speculate on possible ways in which these plot devices could be used within future Generations. It is perhaps unsurprising, then, that Generation VII made trans-spatial movement a core component of its storyworld.

From Hoenn to Alola: Ultra Wormholes and Fallers

Generation VII's *Pokémon Sun* and *Moon* is where trans-spatial multiversal discourses "begin to have their impact truly felt" (Wilcox 2023, para. 6) on the *Pokémon* video games by being incorporated into what Ryan (2014) names the "*Physical laws* …[or] principles that determine what kind of events can and cannot happen in a given story" (35) Generation VII's core titles adjust the physical laws regarding how a Region is typically constructed in a *Pokémon* video game, by introducing Ultra Wormholes, gateways to other words where unusual Pokémon, Ultra Beasts, exist (Grosso 2022, para. 4). In *Pokémon Sun* and *Moon*, Ultra Wormholes served commercial purposes of product differentiation by introducing a hitherto unseen way of encountering new Pokémon species. Rather than searching open areas of the gameworld and/or completing puzzles, the introduction of trans-spatial movement allowed for specimens like Guzzlord and Celesteela to be encountered in a novel way. Yet, while assisting with differentiating Generation VII's core titles from their predecessors, Ultra Wormholes, Ultra Beasts, and the full integration of multiversal discourses into the *Poké-verse*, were aligned with rare and hard-to-catch specimens, elevating them from the established playmodes of traveling around the depicted region. Mirroring fan responses to "The Delta Episode," the inclusion of trans-spatial multiversal discourse in this Generation elevated the status of these existents and their storyline, positioning them as loci for fans' affective and imaginative pleasures.

The association between constructions of trans-spatial multiversal discourses as pleasurable and affect-sustaining events within *Pokémon* video games has endured since Generation VII. For example, "The Crown Tundra," part of the additional paid-for downloadable content for Generation VIII's *Sword* and *Shield*, employs Ultra Wormholes as a method through which players can encounter powerful and rare Mythical and Legendary Pokémon from previous Generations. What's more, appearances of Ultra Wormholes and Ultra Beasts within the augmented reality game *Pokémon Go* were introduced to mark 2022's *Go Fest* celebrations. Since then, their appearances have remained rare, time-sensitive occurrences. By doing this, the *Poké-verse*'s construction of trans-spatial multiversal discourse has consistently been coded as something atypical and so used to either attract high traffic to the title or, as in the case of "The Crown Tundra," encourage players to pay a premium to participate with these.

Sun and *Moon*'s addition of Ultra Wormholes functions more than offering product differentiation by innovating on the well-established play modes of *Pokémon*. This is because Generation VII's "upper" titles (*Ultra Sun* and *Ultra Moon*) included a storyline featuring "transnarrative character[s]" in the form of antagonists Team Rainbow Rocket (Wolf 2012, 66). Transnarrative characters are those that "imply a geographical linkage between worlds, which enables characters to cross from one to the other," and this type of coding is one of the "most common ways to link worlds together" within fiction (216). Team Rainbow Rocket are an amalgamation of chief antagonists from previous *Pokémon* Generations who have traveled by wormholes to Alola and are led by Giovanni, the original leader of Team Rocket from Generation I. The possibility of battling this team of previous über-villains is thus only possible because of how Generation VII adds trans-spatial travel to the physical laws of the storyworld. What's more, the inclusion of Ultra Wormholes is indicative of how including multiversal discourses in media franchises can "invit[e] viewers to contemplate existential questions, and foster... critical thinking." (Katerynych 2024, 152) This potential is reflected in Wilcox's (2023) reflection on the Team Rainbow Rocket storyline:

it is hard to call defeating them a true success because the various victorious team leaders Giovanni recruited are merely transported back to their own worlds, and Giovanni himself clearly aims to renew his schemes in another reality at the end of the post-game story.

(para. 9)

The ambiguous conclusion of the Team Rainbow Rocket storyline exemplifies how multiversal discourses can encourage players to reflect on philosophical issues. In this instance, the games raise ideas linked to the futility of conflict and whether absolute victory can ever be achieved. These opportunities for pause and reflection exist, as suggested above, due to the cross-generational appeal intended for each title. However, as statements made about Team Rainbow Rocket and "The Delta Episode" imply, plot points like these become loci for fan pleasure and affect by encouraging players to partake in inferential walks where deeper exploration of the game's themes can occur.

From the Intradiegetic to the Extradiegetic

Fan negotiation of trans-spatial multiverse discourses has also taken place beyond the intradiegetic elements encoded into *Pokémon* game narratives. An alternative discourse, adopting what Ryan (2014) names an *"extradiegetic"* focus, referring to aspects that "are not literally part of the storyworld but play a crucial role in its presentation," is also identifiable (37). This discourse sees fans combining reflections on hardware with a philosophical stance to address the technologies that have historically supported gameplay. Prominent amongst these discussions have been Link Cables. Link Cables are cords that pre-dated Generation I, but which enabled two separate Game Boy units to connect with each other for the purpose of swapping Pokémon and completing the in-game Pokédex. Link Cables have, on the one hand, been important because using these to exchange Pokémon between players has been the only way to get certain species to evolve. This technique dates to Generation I where, for example, the ghost-type Pokémon Haunter could only evolve to Gengar, its final and most powerful form, by exchanging it across game units via the Link Cable.

On the other hand, Link Cables have been important to the game's design because of the duality that characterizes the release of each Generation's core games, where certain Pokémon species are only catchable in one or the other title. For example, in the most recent *Pokémon Scarlet* and *Pokémon Violet*, the ghost-type Pokémon Drifloon can only be found in *Scarlet*, whereas Misdreavous, a fellow ghost-type first seen in Generation II, is only obtainable in *Violet*. Individual *Pokémon* games (whether core or upper) thus consistently and deliberately demonstrate divergences in terms of the existents (see Ryan 2014) assigned to the diegesis of individual titles (34–35). The consequences of these intra-game divergences could cynically be read from a commercial perspective. From this perspective, Game Freak and Nintendo could be accused of co-opting fan "commodity-completist practices" by spurring players to purchase two copies of new *Pokémon* games (Hills 2002, 4). However, this explanation sits alongside an alternative, authorially endorsed interpretation, one that positions *Pokémon* video games as encouraging prosocial values between players:

> Disturbed by this current tendency toward atomism, both in gaming and in society at large, Tajiri aimed to design his game to promote social interaction. …To acquire all 151 Pokémon on one's Game Boy, one needs to make exchanges with other children.
>
> *(Allison 2004, 42)*

Commercial and pro-social priorities thus intertwine around this element of how the *Pokémon* video games have been designed. Nevertheless, the pro-sociality embedded into the *Pokémon* video games has also led to fan speculation about whether in-game trading equates to trans-spatial interactions:

> Not only are the mainstream titles all set in different universes, but also that each copy of each game is its own universe as well. The Link Cable connects these different universes through rifts in time and space. In fact, a scientist at the Devon Corporation casually says that they use this technology all the time in everyday devices, such as warp panels.
>
> *(Holt 2018, para. 9)*

Other fan commentators have similarly positioned the trading practices enabled by Link Cables as "the first argument towards a Pokémon multi-verse" (Friend 2023, para. 3). These constructions of extradiegetic trans-spatial discourse are another example of players of *Pokémon* video games deriving pleasure from the franchise by partaking in inferential walks that allow them to explore the games on a deeper, philosophical level.

However, developments in both console and consumer technology have resulted in diminished opportunities for this inflection of extradiegetic multiversal discourse. The release of *Pokémon: Let's Go! Pikachu* and *Pokémon: Let's Go! Eevee* in November 2018 marked both the end of Generation VII and an institutional change of the primary platform for the *Pokémon* franchise from the Game Boy to Nintendo's hybrid Switch console. This change has resulted in pro-social aspects that give way to extradiegetic multiversal discourse, such as in-game trading, being retained. However, *Scarlet* and *Violet* demonstrate moving toward constructing the in-game world not as finite iterations contained within individual user accounts but instead as a shared sandbox. Players of *Scarlet* and *Violet* can now use their player avatars to meet up with fellow players and engage in in-game forms of sociality such as partaking in raids for rare Pokémon or staging picnics. These changes demonstrate tensions emerging between previous constructions of extradiegetic and intradiegetic trans-spatial discourse as, while the latter remains in place as each player of *Scarlet* or *Violet* must have their own copy of the game and in-game account, the technologically enforced barriers between are muddied through the ability for players to sync-up their game with their peers. Consequently, the shift in how pro-social values are coded into *Scarlet* and *Violet* represents a diminishing of extradiegetic trans-spatial thinking and its associated pleasures for *Pokémon* fans.

Conclusion

Multiversal discourses are a form of multi-coded, trans-spatial movement. On the one hand, they can be aligned with accounts of the *Pokémon* video games as commercial forms. From this perspective, the introduction of multiversal trans-spatial movement assists industrial practices of periodic re-commoditization and product differentiation by building on the property's science-fiction underpinnings to introduce novel ways for constructing storylines, encountering new Pokémon species, and expanding the *Poké-verse*. On the other hand, from an audience perspective, constructions of trans-spatial multiversal discourse have functioned as loci for fans' affective and imaginative engagement with *Pokémon* video games. These constructions have been read as supporting fan theories about the structure of the *Poké-verse* at both intra- and extradiegetic levels and enabling deeper engagement with a series of games whose status as a cross-generational property do not encourage this level of reading. More than purely assisting with disseminating and popularizing complex scientific concepts (see Katerynych 2024), what underpins each of the discussions outlined in this chapter are issues of cultural value. That is, *Pokémon* fans use trans-spatial multiversal discourses to accrue meaning and status to the property by linking new Generations to what has come before, building coherence to the *Poké-verse*, and thus downplaying *Pokémon*'s commodity status.

A multiversal approach to analyzing transmedia properties also allows for a shift of focus within academic inquiry away from the anthropological tendencies that discourses of media convergence and participatory cultures have advanced, instead moving back toward issues of subcultural interpretation. For example, a multiversal perspective involves asking alternative questions to those related to an approach informed by discourses of media mix. Marc Steinberg (2012) defines media mix as a concept associated with Japanese-originating properties like *Pokémon* which relates to "the cross-media serialization and circulation of entertainment franchises" (vii) For Steinberg (2012), the center of individual media mixes "is the dynamically immobile character image" (6) and how this becomes reproduced across multiple media forms ranging from digital platforms like video games through to multiple material commodity forms. Analyses of media mix thus foreground how characters like Pikachu, Mewtwo, and Fuecoco become "embedded in specific material environments (the living room and the TV set for one...)" and become a ubiquitous part of social and cultural space (80). In contrast, the multiversal mode of inquiry taken by this chapter switches its focus back to situated audience readings of textual material and how these open up spaces for speculation relating to mechanics and ontology of the fictional world and the technology that sustains and enables contact with this.

Alternatively, a multiversal-informed approach to *Pokémon* video games encourages academic movement away from studies of the affordances and narratives that different media iterations of a transmedia property offer (see, e.g., Harvey 2015). Instead, questions concerning how and why audience subcultures afford significance to and affectively invest in how and where interconnections between different installments occur can be addressed. This involves focusing less on questions regarding the characteristics of individual media to instead looking at the nodes constructed for connecting texts together, how audiences understand and negotiate these, and the pleasures that audiences derive from how multiple texts are made to interconnect.

What might a future set of inquiries into popular culture properties rooted in mutiversal discourses look like? To some extent, these ideas are already being taken up in research methodologies linked to digital culture, where examining large data sets through the lens of "a 'multiverse' of reasonable datasets and analytical decisions" has been seen to increase the validity of data interpretations (Pipal, Song, and Boomgaarden 2023, 256). This chapter has argued for a different approach to multiversal inquiry, treating these as discursive concepts that are constructed and discussed by (fan) audiences rather than as an underpinning philosophy for research. By adopting the chapter's mode of inquiry, both to other corners of the *Poké-verse* and/or alternative media franchises, the opportunity for a better understanding of how fan philosophizing appropriates academic concepts, in what contexts this type of discourse occurs, and what purposes such philosophizing serves can be better understood.

References

Allison, Anne. 2004. "Cuteness as Japan's Millennial Product." In *Pikachu's Global Adventure: The Rise and Fall of Pokémon*, edited by Joseph Tobin, 34–49. Durham: Duke University Press.

Boillat, Alain. 2022 *Cinema as a Worldbuilding Machine in the Digital Era: Essays on Multiverse Films and TV Series*. New Barnet: John Libbey Publishing Ltd.

Booth, Paul. 2016. *Digital Fandom 2.0: New Media Studies*. New York: Peter Lang.

Buckingham, David and Julian Sefton-Green. 2004. "Structure, Agency, and Pedagogy in Children's Media Culture." In *Pikachu's Global Adventure: The Rise and Fall of Pokémon*, edited by Joseph Tobin, 12–33. Durham: Duke University Press.

Eco, Umberto. 1994. *Six Walks in the Fictional Woods*. Cambridge, MA: Harvard University Press.

Eco, Umberto. 2007 [1986]. "*Casablanca*: Cult Movies and Intertextual Collage." In *The Cult Film Reader*, edited by Ernest Mathijs and Xavier Mendik 67–74. Maidenhead: McGraw-Hill Education.

Evans, Elizabeth. 2011. *Transmedia Television: Audiences, New Media, and Daily Life*. London: Routledge.

Friend, Devin. 2023. "*Pokémon* is One of the Earliest Multiverses." *GameRant*, October 11. https://gamerant.com/pokemon-multiverses-version-exclusives-remakes-gen-7/.

Grossberg, Lawrence. 1992. "Is There a Fan in the House?: The Affective Sensibility of Fandom." In *The Adoring Audience: Fan Culture and Popular Media*, edited by Lisa A. Lewis, 50–65. London: Routledge.

Grosso, Robert. 2022. "*Pokémon Legends: Arceus* Proves the *Pokémon* Multiverse Exists." *TechRaptor*, February 9, 2022. https://techraptor.net/gaming/features/pokemon-legends-arceus-multiverse-exists.

Harvey, Colin B. 2015. *Fantastic Transmedia: Narrative, Play, and Memory across Science Fiction and Fantasy Storyworlds*. Basingstoke: Palgrave MacMillan.

Hills, Matt. 2002. *Fan Cultures*. London: Routledge.

Hills, Matt. 2015. "From "Multiverse" to "Abramsverse": *Blade Runner, Star Trek*, Multiplicity, and the Authorizing of Cult/SF Worlds." In *Science Fiction Double Feature: The Science Fiction Film as Cult Text*, edited by J.P. Telotte and Gerald Duchovnay, 21–37. Liverpool: Liverpool University Press.

Holt, Valerie. 2018. "The *Pokémon* Multiverse Will Blow Your Mind." *Gamers*, May 29. https://vocal.media/gamers/the-pokemon-multiverse-will-blow-your-mind.

Jenkins, Henry. 1992. *Textual Poachers: Television Fans and Participatory Culture*. London: Routledge.

Jenkins, Henry. 2006. *Convergence Culture: Where Old and New Media Collide*. New York: New York University Press.

Jenkins, Henry. 2014. "The Reign of the "Mothership": Transmedia's Past, Present, and Possible Futures." In *Wired TV: Laboring over an Interactive Future*, edited by Denise Mann, 244–268. New Brunswick: Rutgers University Press.

Katerynych, Petro. 2024. "Unlocking the Marvel Multiverse: The Cosmic Nexus of Science, Philosophy, and Fiction through the Infinity Stones." *Philosophy and Cosmology* 32: 141–154.

Mittell, Jason. 2015. *Complex TV: The Poetics of Contemporary Television Story-telling*. New York: New York University Press.

Phillips, Andrea. 2012. *A Creator's Guide to Transmedia Storytelling: How to Captivate and Engage Audiences Across Multiple Platforms*. New York: McGraw-Hill.

Pipal, Christian, Hyunjin Song, and Hajo G. Boomgaarden. 2023. "If You Have Choices, Why Not Choose (and Share) All of Them? A Multiverse Approach to Understanding News Engagement on Social Media." *Digital Journalism*, 11, no. 2: 255–275, DOI: 10.1080/21670811.2022.2036623

Ryan, Marie-Laure. 2014 "Story/Worlds/Media: Tuning the Instruments of a Media-Conscious Narratology." In *Storyworlds Across Media: Toward a Media-Conscious Narratology*, edited by Marie-Laure Ryan and Jan-Noël Thon, 25–49. Lincoln: University of Nebraska Press.

Ryan, Marie-Laure, and Thon, Jan-Noël. 2014 "Storyworlds across Media: Intro-duction." In *Storyworlds Across Media: Toward a Media-Conscious Narratol-ogy*, edited by Marie-Laure Ryan and Jan-Noël Thon 1–21. Lincoln: University of Nebraska Press.

Sefton-Green, Julian. 2004. "Initiation Rites: A Small Boy in a Poké-World." In *Pikachu's Global Adventure: The Rise and Fall of Pokémon*, edited by Joseph Tobin, 141–164. Durham: Duke University Press.

Sledge, Ben. 2022. "Before the MCU, *Pokémon* had its own Multiverse." *The Gamer*, July 9, 2022. https://www.thegamer.com/pokemon-multiverse/.

Stein, Louisa Ellen. 2015. *Millennial Fandom: Television Audiences in the Trans-media Age*. Iowa City: University of Iowa Press.

Steinberg, Marc. 2012. *Anime's Media Mix: Franchising Toys and Characters in Japan*. Minneapolis: Minnesota University Press.

Suvin, Darko. 1977 [2010]. "Science Fiction and the Novum." *Defined by a Hol-low: Essays on Utopia, Science Fiction and Political Epistemology*, 67–91. New York: Peter Lang.

Switzer, Eric. 2021. "The *Pokémon* Multiverse is Happening, but I'm Already Dis-appointed." *The Gamer*, December 19, 2021. https://www.thegamer.com/pokemon-anime-multiverse-ash-movie-crossover/.

Valdez, Nick. 2021. "*Pokémon*'s Anime Introduces its Own Multiverse." *Comic-book*, December 5. https://comicbook.com/anime/news/pokemon-journeys-multiverse-anime/.

Wilcox, Matthew. 2023. "Is There a *Pokémon* Multiverse? How All *Pokémon* Games May Be Connected." *ScreenRant*, May 21. https://screenrant.com/pokemon-multiverse-story-dual-releases-remakes-ultra-space/.

Williams, Rebecca. 2020. *Theme Park Fandom: Spatial Transmedia, Materiality and Participatory Cultures*. Amsterdam: Amsterdam University Press.

Wolf, Mark J. P. 2012. *Building Imaginary Worlds: The Theory and History of Subcreation*. London: Routledge.

26

PLAYING ACROSS PLANES WITH THE D&D MULTIVERSE

Maria K. Alberto

Initially released in 1974, the tabletop role-playing game (TTRPG) *Dungeons & Dragons* (D&D) has recently undergone a cultural renaissance. D&D has featured in big-ticket pop culture texts (think Netflix's *Stranger Things*) and is often tied to the increasingly popular "actual play" format, wherein players record and release their gameplay sessions as new media texts (think *Critical Role* and *Dimension 20*, among others). Simultaneously, current D&D parent company Wizards of the Coast has been leaning hard into transmedia strategies, working to position and market D&D as an accessible hobby or a lifestyle product. Some of this transmedial approach takes the form of mainstream hits: think immensely popular videogame *Baldur's Gate III*, a party-based RPG; or the 2023 movie *Honor Among Thieves*, a fantasy action/heist movie; or even the 2024 shoe line launched in partnership with Converse. However, as I argue in this chapter, another key element of how D&D is positioned and marketed occurs via its multiverse.

Formally, as mentioned in official texts like game manuals, the D&D multiverse is "a vast cosmos" uniting "an endless variety of worlds... connected in strange and mysterious ways to one another and to other planes of existence" (5e *Player's Handbook* [PHB] 5); this multiverse is also described as "a vast array of planes and worlds where adventures happen" (5e *Dungeon Master's Guide* [DMG] 4). But practically, I argue, the D&D multiverse also lets Wizards of the Coast connect different intellectual properties (IPs) and genres—and thus, the multiverse functions as both a

DOI: 10.4324/9781003480846-30

spatial schema within the fiction of D&D as well as an organizing schema for D&D as a commercial product. The concept of a multiverse enables Wizards of the Coast to maintain, for example, that sci-fi settings with spaceships, aliens, and interstellar travel exist in the same imaginative realm as concepts more recognizable from fantasy, such as magic, gods, and elves. This co-existence of distinct genre tropes, the explanation goes, is made possible because all of these places and potentials are located within the same expansive multiverse—even if contact between different locations is rare and difficult to achieve, and most don't even realize that the others exist. But additionally this connection, however tenuous and circumstantial, also creates the impression of planned, diegetic cohesion within D&D itself. This way, new settings can be introduced without interrupting existing ones; meanwhile, inspirations from disparate genres, classic fantasy texts, mytho-historical origins, and more, can all be said to coexist within a single commercial product.

In this chapter, I delve deeper into what the D&D multiverse is, where players encounter it, what it is meant to achieve, and why that matters in a cultural moment when multiverses—both conceptually and linguistically—have come to be the fictional mode du jour.

What In The World(s) Is The D&D Multiverse?

While the term "multiverse" has become more visible in popular culture recently, the D&D multiverse is not a new phenomenon. In fact, it has been a part of D&D for nearly as long as the game system itself has existed.

The underlying concept first appears in 1977, when D&D co-creator Gary Gygax laid out a plan for planes of existence in an article for *Dragon*, an official D&D magazine. In his "Planes: The Concepts of Spatial, Temporal and Physical Relationships in D&D," Gygax puts a new idea to the readers of *Dragon* for "your careful consideration and thorough experimentation" (28). For "game purposes," he states, "the DM is to assume the existence of an infinite number of co-existing planes" (4), which include the "Prime Material Plane" where most "human-type life forms" live, in addition to astral and elemental planes, heavens, hells, limbo, and more (4, 28). At this early stage of the concept, Gygax is most interested in explanations for magical weapons. While parts of this 1977 article do discuss how a multiverse comprised of *planes of existence* will allow player characters to adventure across worlds, Gygax specifically explores how this schema enables an explanation for why some creatures can only be harmed with magical weapons: for example, demons and certain types of undead might exist on simultaneous planes and only certain materials, like silver, can reach into or across these dimensions of existence, and thus, actually harm such an entity. Yet even so, Gygax is already realizing that having multiple planes of existence in play for any D&D game "will cause a careful rethinking" of the events that take place within it, and thus planes "vastly expand the potential of all campaigns which adopt the system" (28). But because its potential is so far-reaching, adopting the schema of a multiverse of planes will also mean "that I must revise the whole of D&D to conform to this new notion" (28).

And indeed, when the actual term "multiverse" made its D&D debut the next year in the first-edition *Player's Handbook* (1978), this text's invocation of "the fantastic 'multiverse' of ADVANCED DUNGEONS & DRAGONS" (120) is explained in terms of how "There exist an infinite number of parallel universes and planes of existence" (120). Since this text is geared toward an audience of D&D players, though, the AD&D 1e *Player's Handbook* is also quick to remind readers that "how 'real' each [plane] is depends entirely upon the development of each" by their DM (120). But the *Dungeon Master's Guide* of the same edition tells quite a different story—although, interestingly, this rulebook does not use the term "multiverse" at all, unlike its player-facing counterpart. Where the planes are covered only briefly in an appendix of the first-edition *Player's Handbook*, they receive an entire section in the AD&D 1e *Dungeon Master's Guide*, whose readers are likewise informed that:

The term "the planes" encompasses all the alternate levels of reality that may be encountered in the ADVANCED DUNGEONS & DRAGONS game. The planes are more than a different part of a standard campaign, or a different planet to adventure on. The many known planes have very different physical and magical laws than most adventurers are used to. These planes each have unique rewards as well as unique dangers.

(235)

These early iterations from Gygax's *Dragon* article and the first codified game rulebooks establish several important things about the D&D multiverse, which would then be reiterated in successive texts up to the present. These include, first, that the D&D multiverse is dependent on the notion of *planes*: those different levels or dimensions of existence, which may coexist, overlap, and/or be linked with one another. Second, there is also the way that the D&D multiverse is deeply concerned with categorizing and facilitating certain spatial and material relationships that help make a D&D game interesting: these include travel, exploration, and combat. Third, the D&D multiverse brings together some wildly disparate inspirations, and does so by making them coequal and coeval: for example, early lists of the D&D planes include Elysium, Olympus, Arcadia, Hades, and

Tartarus from Greek mythology alongside Gladsheim from Norse mythology, Nirvana from Buddhist tradition, Gehenna from Judeo-Christian tradition, and a version of Dante's nine circles of hell, plus new locations created specifically for D&D (AD&D 1e DMG 237).[1] Then finally, the D&D multiverse has a complex relationship with the abstruse notion of "being real"—a concern that, elsewhere, I have also argued becomes evident with the notion of D&D's "canon" (Alberto "So It's Canon Now"). That is, as the AD&D 1e *Player's Handbook* puts it: "how 'real' each [plane] is depends entirely upon the development of each" by the DM (120). Or, in other words, whether a DM decides to use the planes, and which ones, is what makes these levels of reality exist or not in a specific D&D game.

As of the fifth edition (5e), official D&D materials typically present the multiverse as a testing ground for higher-level player characters or a background for higher-level adventures. For instance, the *5e Dungeon Master's Guide* explains how more powerful adventurers' "path extends to other dimensions of reality: the planes of existence that form the multiverse" (43), which are "not simply other worlds, but [instead] dimensions formed and governed by spiritual and elemental principles" (43). Building from the versions introduced in previous editions, fifth edition D&D offers five categories of plane(s), which can be summarized from the *Dungeon Master's Guide* "Chapter 2: Creating a Multiverse" as follows:

1 The **Material Plane** (aka, the "real world" or primary world where most people [characters] live) and its two reflections or **"Echoes"**: the Feywild, a place of fey wonder and heightened emotion, and the Shadowfell, a place of gloom and numbing rhythm
2 Two **Transitive Planes** that let characters, creatures, and spells travel between other planes
3 Five **Inner Planes** based on classical Elements plus elemental Chaos

---

1 While the planes have since been revised to focus more on fantasy and fictional names, the use of real-world mythological and cultural references remains in place today. For example, an official resource on D&D Beyond (as of this writing in 2023) directs readers interested in creating a cleric player character to the appendix "Deities of the Multiverse," which opens with "Each world in the D&D multiverse has its own pantheons of deities" (par. 4) and then presents a "classic" D&D pantheon as well as Celtic, Egyptian, Greek, and Norse counterparts. The latter, this online resource suggests, are "fantasy interpretations of historical religions from our world's ancient times" and prioritize "deities that are most appropriate for use in a D&D game, divorced from their historical context in the real world and united into pantheons that serve the needs of the game" (par. 8). Simply by their re-location into the D&D multiverse, then, we are meant to understand familiar names as semi-fictionalized signifiers that can not only coexist with one another, but also, be mixed and matched as players desire.

4 Sixteen **Outer Planes** based to varying degrees on moral alignments
5 The **Positive and Negative Planes**, which essentially correspond to larger forces of life and death (43)

Following its rundown of that basic and "assumed" cosmology enumerated above, the fifth edition *Dungeon Master's Guide* also explains that any campaign requires "at minimum" three types: planes where fiendish creatures, celestial creatures, elemental creatures, and gods originate; planes where souls go after death; and planes among which creatures and spells can travel (43). These three types are needed because each provides a rationale for some part of a D&D game. For instance, the first type explains where certain beings originate, as when gods may be found upon good or evil planes, and subsequently, grant their followers either good or evil magic. Then the second type of plane, in the form of an afterlife, provides a place where fallen characters end up: story reasons for this might include remembering or rescuing loved ones from such a place, while game mechanic reasons can include summoning a spirit to provide information (such as the spell Speak with Dead) or resurrecting characters (such as spells like Resurrection and True Resurrection). Finally, the third type, which concerns travel between planes, enables powerful magical spells like Etherealness, Plane Shift, and Astral Projection as well as further plot hooks and story beats.

As a whole, then, "The construction of 'other worlds' and their connection by means of travel lie at the very heart of D&D" (Perlini-Pfister 284), just as much as more easily recognizable features like dice rolls, fantastical characters, adventures, locales, or magic. This, I argue, is because when magic and gods are already part of the mundane world for D&D characters, then something that is even more rare or wondrous than magic and gods is needed to strike these adventurers as extraordinary. So, even beyond offering a rationale for the way magic weapons work or where certain enemies like demons originate from, the planes can provide this heightened notion of wonder, by offering a way to access other parts of existence utterly unlike most D&D characters' own already-magical world.

What Takes Place In The D&D Multiverse?

As we have just seen, players can get a sneak peek into and an initial taste of the D&D multiverse through basic introductions in core rulebooks like the *Player's Handbook* and *Dungeon Master's Guide* or also via paratexts like articles in *Dragon*. In these cases, though, the D&D multiverse is typically being presented more like a backdrop, and here the more typical or foregrounded adventure takes place on the prime material plane where most fantasy peoples live. In such instances, players and their characters may not ever learn much about other planes, let alone interact with them beyond basic spells. Conversely, though, there are also D&D texts that foreground the multiverse, either as an important feature *in* a specific adventure or else as its own setting *for* adventure. Moreover, a closer look at these depictions reveals the most critical stakes of the multiverse in D&D, which we've already seen a little of with the explanation of silver weapons and the co-existence of gods from various primary-world religions. Namely: the multiverse provides a way for D&D as a franchise to have its cake and eat it too, combining all desired genres, inspirations, and even changes in the game system itself into one "place" that contains and explains all of them together.

To demonstrate, let's consider a few of these examples that go beyond a mere introduction to the planes of existence that comprise the D&D multiverse.

For starters, many D&D adventures hinge on entities or even regions from planes beyond the prime material one that most adventurers hail from. Early examples include the adventure modules (i.e., prewritten adventures) *Vault of the Drow* (1978) and *Queen of the Demonweb Pits* (1980), later combined with several others into *Queen of the Spiders* (1986). In these modules, the player characters can locate and enter a portal into the realm of Lolth, a demonic spider goddess who rules an abyssal plane. Another popular adventure module making use of planes began as *Ravenloft* (1983), a Gothic-inspired and vampire-haunted adventure whose setting has been adapted multiple times, including most recently as the polemical *Curse of Strahd* (2016). Ravenloft itself is a setting described as a "demiplane of dread and desire, a world whose misty fingers can reach into any other campaign setting and draw unsuspecting heroes into its midst" (Appelcline par. 2); in its original TSR incarnation, the Gothic-inspired castle that "spawned" this setting "lies lost within the Ethereal Plane, in the Demiplane of Dread" (Nesmith, Hayday, and Connors 12). In most iterations of this adventure, including *Curse of Strahd*, the story begins with player characters being caught up by magical mists that transport them from their home plane into this one. Elsewhere again, *Tales of the Outer Planes* (1988) collates eleven prewritten mini-adventures, each taking place on a different plane and most starting from a hub called the World Serpent Inn, an interdimensional tavern featuring a godlike bartender and "a number of 'convenient' exists to other planes" (Christian et al. 2).

Beyond adventures, one of the earliest D&D texts to focus specifically on the planes themselves is *Manual of the Planes* (1987). This *Manual* was described by its primary author Jeff Grubb as a look into the proverbial 'overstuffed closet' of the D&D game system, after years of treating "the known planes of existence as holding bins for every idea and adventure that did not quite belong in the Prime Material Plane" (3). This *Manual* is also the first codification of the planes that make up the D&D multiverse, providing an overview of each one as well as explanations of how to travel there, how to survive, how time works, what entities might be encountered there, and how combat and magic operate (Grubb, *Manual*, 5–7). Here Grubb also proposes a "Grand Unified Theory of Spells" that enumerates the planar aspects of much common magic (*Manual*, 3) and sets in motion an organizing schema that would eventually be called the Great Wheel, which in fifth edition is considered the "default cosmological arrangement" of D&D (Crawford et al. 44), though other possible schemas include the World Tree based on the Norse Yggdrasil or the World Axis of fourth edition, and DMs may implement whichever they like for their own campaigns (Crawford et al. 43-44).

Grubb's *Manual* also inspired a foray into campaign settings—that is, pre-existing storyworlds or "the shared setting of... a series of interconnected adventures" (Barton 20)—that foreground the planes of the D&D multiverse. First and foremost, Grubb himself took the helm on *Spelljammer: AD&D Adventures in Space*, which introduced magically driven spaceships—the nominal spelljammers—and "fantasy physics" that enabled characters to travel among fantasy storyworlds (3): the result is a fantasy-sci-fi mashup that Curtis Carbonell has called a "D&D-meets-Jules Verne, meets-the-Era-of-High-Sails-in-Space" setting (97). Then just months later, David "Zeb" Cook also cooked up the campaign setting of Planescape, which takes a multiversal route inspired more by weird fiction than sci-fi. As laid out in *Planescape Campaign Setting: Sigil and Beyond* (1994), this new setting introduces a center to the D&D multiverse—the city of Sigil, which is ruled by the enigmatic and god-like Lady of Pain. Cook's initial *Planescape* manual was also more high-concept than previous D&D offerings, even using in-universe slang to introduce concepts and rules to the reader: "Here's the real chant, and pay attention, berk" (8).

The most interesting factor uniting these various adventure modules, accessories, and campaign settings, though, is that they all concern "harmonization" (Carbonell 98). That is, all of these texts seek to organize the most far-flung and widely separated concepts included within D&D's IP into one consistent, coherent, and compatible end result. While each takes a different route to get there, the end goal is the same: the objective of presenting "everything" in D&D as being interconnected and available to play with, in whatever fashion or combination that players desire.

This sort of organizational project is visible right upfront in most of the texts we have just considered. For instance, Grubb's foreword to *Manual of the Planes* specifically reports that "One of the basic assumptions of this tome is that what has been written about in the past is true, and our job is to explain it" (3), since, apparently, "While the [D&D] game can absorb any amount of new material, casting off pre-existing material often damages the system" (3). Similarly, Grubb recollects that the original *Spelljammer* was tasked to "Maintain the spirit and play of the AD&D game *and* tie it in with years of previous work without invalidating anything" (3). Meanwhile, Planescape is all about "the meaning of the multiverse" and how to tie this into players' experiences during gameplay (7).

Thus, whether through codification, science-fantasy spaceships, or weird fiction, the D&D multiverse has always been meant to bring different inspirations and hypotexts together into one overarching schema. Moreover, this interest continues into newer texts like *Spelljammer: Adventures in Space* (2022) focusing on the Astral Plane, where "With the help of magic, spelljammers can cross the oceans of Wildspace, ply the silvery void known as the Astral Sea, and hop between worlds of the D&D multiverse" (Wizards, "Spelljammer," par. 2), as well as *Planescape: Adventures in the Multiverse* (2023), which promises "both adventure locations and springboards to adventures across the multiverse" (Wizards, "Planescape," par. 4). Entirely new versions are also available, as seen with *Journeys through the Radiant Citadel* (2022). Conceptualized and offered as D&D's first book entirely by authors of color, *Radiant Citadel* introduces a stronghold housed in the Ethereal Plane, from which player characters can travel to new settings that are not strictly set in any specific campaign setting. These mini-adventures are all based on specific cultures and cultural references from our own primary world, ranging from Iran to Central America to New Orleans USA, and *Radiant Citadel* thus provides authors with opportunities to showcase their own cultures and lived experience, offsetting the more typical—and often, presumed default—whiteness of D&D designers, settings, and even players.

Who Really Needs A D&D Multiverse?

As other chapters in this collection have shown, the multiverse is a popular construct that springs up in a wide range of contexts across fiction and popular culture, especially in the wake of increasingly transmedia storytelling. Additionally, the multiverse has particular valences for speculative fiction, most notably science fiction and fantasy. For instance, what is arguably the first-ever science fiction text, Margaret Cavendish's 1666 *The Blazing World* (see Venter's chapter), depicts a voyage to another world only accessible via the North Pole of ours; genre successors from *Gulliver's Travels* to *20,000 Leagues Under the Sea* to *2001: A Space Odyssey* propound their own fantastic voyages, though not always to new worlds in the cosmological sense. Meanwhile, back over in fantasy, Ryan Vu identifies Poul Anderson's novel *Three Hearts and Three Lions*—a text cited in Gygax's famous list of influences for D&D, Appendix N—as one of the inspirations that brought this idea to D&D specifically (285).

What most, if not all, of these classic genre texts have in common is an interest in other worlds—whether characters access these through portals or spaceships, or whether we as readers are simply observing worlds very different from our own—and in world-building, or the process of constructing that fictional setting. D&D certainly shares in these larger interests from its non-game counterparts, but with the new addition that in tabletop role-playing, worlds and world-building are collaborative and participatory. For example, whether players are using an official campaign setting or one of their own devising, playing D&D inherently entails what Mark J.P. Wolf has called a "participatory world"—one that invites audiences to participate in creating it, "[making] permanent changes that result in canonical additions to the world" (281). Specifically, Neal Baker has argued that official game-texts simultaneously "supply organizational world infrastructures (timelines, etc., packaged for ready-made campaign settings)" and also "offer generative world-building material" for players to use themselves (84). Moreover, D&D campaign settings are unquestionably "shared worlds," which are those developed by multiple authors, "accessed" through multiple mediums of text, and that grow and change over time "as if it were a character itself" (Mackay 29).

We might sum up thus: D&D is a mode of fantasy that, unlike its print or filmic or even videogaming counterparts, anticipates—and even depends on—audiences to *intervene into* the worlds of its multiverse. For instance, DMs are encouraged to modify the cosmology provided by official game-texts in order to suit the needs of their own campaigns; likewise, players are informed that whatever they learn of the multiverse in official D&D texts, those "realities" are not actually true in their game unless their DM chooses to implement or depart from them. In turn, though, this inherently participatory nature also means that the possibility space of the D&D multiverse differs substantially from its counterparts in other SFF texts.

For example, Christopher Bagger argues that in American superhero comics, queer superheroes—or those who get queered—are often written into alternate universes so that "these characters are excluded from the 'actual' or 'canonical' of a given text" (171). A bisexual son of Superman who doesn't exist on our own Earth, just an alternate reality version, isn't as actual or even "real" as characters who exist in that main world. This approach, Bagger maintains, enables cultural producers like DC to claim they are advancing desirable social justice issues like queer representation, while also not alienating long-standing audiences who don't want to see such major changes to classic characters, timelines, or storylines (171–2).

D&D, though, has a very different relationship to its own "'actual' or canonical'" (Bagger 171). With a work like DC's superhero comics, both the main world(s) and the alternate one(s) of that given multiverse are entirely in the hands of its producer, who decides what worlds exist, what relationship(s) they have with each other, who populates them, and so on. With D&D, however, the game's creators can produce all the material they hope to sell, but players are quite literally meant to tell their own stories using its game system—thus impacting, intervening into, or even ignoring what we would normally call the canon of the franchise. Also, unlike even videogames, D&D does not come with a computational black box that would compel its players to use game pieces, rules, or metaplots in particular ways; because of this, D&D makes use of what I have called *procedural metanarratives* to persuade players that the rules are needed to make a fantasy storyworld work (Alberto 40).

From the beginning, then, D&D game designers and parent company have understood that the D&D multiverse is very much in players' hands— and the potential that a D&D multiverse represents comes down more to selling various mechanisms of possibility, not possibility itself.

References

Alberto, Maria K. 2024. "So It's Canon Now: Texts and/as Truths in Transmedia Franchise Dungeons & Dragons." PhD diss., University of Utah.

Appelcline, Shannon. 2016. "Ravenloft: Realm of Terror (2e) [entry]." *DriveThruRPG. net*, December 27. https://www.drivethrurpg.com/en/product/17474/Ravenloft-Realm-of-Terror-2e.

Bagger, Christopher. 2019. "Multiversal Queerbaiting: Alan Scott, Alternate Universes, and Gay Characters in Superhero Comics." In *Queerbaiting and Fandom: Teasing Fans through Homoerotic Possibilities*, edited by Joseph Brennan, 171–176. Iowa City: University of Iowa Press.

Baker, Neal. 2021. "Secondary World Infrastructures and Tabletop Fantasy Role-Playing Games." In *Revisiting Imaginary Worlds: A Subcreation Studies Anthology*, edited by Mark J.P. Wolf, 83–95. Abingdon: Taylor & Francis Group.

Barton, Matt. 2008. *Dungeons and Desktops: The History of Computer Roleplaying Games*. Boca Raton: AK Peters/CRC Press.

Carbonell, Curtis D. 2022. *Dread Trident: Tabletop Role-playing Games and the Modern Fantastic*. Liverpool: Liverpool University Press.

Christian, Deborah A., Vince Garcia, et al. 1988. *Tales of the Outer Planes: Official Game Adventure*. Lake Geneva: TSR, Inc.

Cook, David "Zeb." 1994. *Planescape Campaign Setting: Sigil and Beyond*. Lake Geneva: TSR, Inc.

Crawford, Jeremy, Christopher Perkins, James Wyatt, and Mike Mearls. 2014. *Dungeon Master's Guide [5th Edition]*. Renton: Wizards of the Coast.

Gygax, Gary. 1978. "Planes: The Concepts of Spatial, Temporal and Physical Relationships in D&D." *Dragon* 1, no. 8: 428.

Gygax, Gary. 1978. *Player's Handbook* [Advanced D&D 1st edition]. Lake Geneva: TSR, Inc.

Gygax, Gary. 1979. *Dungeon Master's Guide* [Advanced D&D 1st edition]. Lake Geneva: TSR, Inc.

Grubb, Jeff. 1987. *Manual of the Planes*. Lake Geneva: TSR, Inc.

Grubb, Jeff. 1989. *Spelljammer: AD&D Adventures in Space*. Lake Geneva: TSR, Inc.

Mackay, Daniel. 2001. *The Fantasy Role-Playing Game: A New Performing Art*. Jefferson: McFarland.

Mearls, Mike, Jeremy Crawford, James Wyatt, Robert J. Schwalb, and Bruce R. Cordell. 2014. *Player's Handbook* [5th Edition]. Renton: Wizards of the Coast.

Nesmith, Bruce, Andria Hayday, and William W. Connors. 1994. *Ravenloft Campaign Setting: Realm of Terror [AD&D 2e]*. Lake Geneva: TSR, Inc.

Perlini-Pfister, Fabian. 2012. "Philosophers with Clubs: Negotiating Cosmology and worldviews in DUNGEONS & DRAGONS." In *Religions in Play: Games, Rituals, and Virtual Worlds*, edited by Philippe Bornet and Maya Burger, 275–294. Zurich: Theologischer Verlag Zürich.

Vu, Ryan. 2017. "Fantasy After Representation: D&D, Game of Thrones, and Postmodern World-Building." *Extrapolation* 58, no. 2–3: 273–301. https://doi.org/10.3828/extr.2017.14.

Wizards of the Coast. n.d.-a "Planescape: Adventures in the Multiverse Digital + Physical Bundle." *DnDStore.wizards.com*. https://dndstore.wizards.com/us/en/product/820944/planescape-adventures-in-the-multiverse-digital-plus-physical-bundle.

Wizards of the Coast. n.d.-b "Spelljammer: Adventures in Space." *DnD Wizards.com*. https://dnd.wizards.com/products/spelljammer.

Wolf, Mark J.P. 2014. *Building Imaginary Worlds: The Theory and History of Subcreation*. Abingdon: Taylor & Francis.

27

CARDS AGAINST MONSTROSITY

The *Arkham Horror* Multiverse

David Scott Diffrient

Although the much-discussed literary output of H. P. Lovecraft has inspired countless artists, dramatists, filmmakers, musicians, and writers across different cultural contexts and media industries since his death in 1937, relatively little scholarly attention has been given to the centrality of his published work to the growth of tabletop gaming over the past four decades. At first glance this is surprising, given the sizable number of board games and card games based on the controversial American author's signature brand of cosmic horror. That ever-growing list of titles encompasses several recent crossover hits and cult favorites such as *Elder Sign* (2011), *Mansions of Madness* (2011), *Eldritch Horror* (2013), *Cthulhu Realms* (2015), *Cthulhu: A Deck Building Game* (2016), *Mountains of Madness* (2017), *AuZtralia* (2018), *Arkham Horror: Final Hour* (2019), *Cthulhu: Death May Die* (2019), and *Unfathomable* (2021) as well as earlier precursors from the 1980s, including the story-driven card game *Dark Cults* (1983), the single-player role-playing game (RPG) *The Thing in the Darkness* (1984), and the cooperative pulp-adventure game *Arkham Horror* (1987). The latter, created by the prolific designer Richard Launius and published by Chaosium (a company that had released the first edition of the *Call of Cthulhu* RPG in 1981 and has gone on to publish six more editions since then), was reimplemented by the Minnesota-based publisher Fantasy Flight Games (FFG) in 2005, with spiffier components, improved gameplay, and a larger map of the titular city to go along with its daunting set of updated rules. That map, set on a mounted board and featuring

DOI: 10.4324/9781003480846-31

creepy locations like the Black Cave, the Graveyard, and the Witch House (where gates to alternate dimensions might open and monsters are sure to spawn), can be regarded as a literally graphic illustration of the way that Lovecraftian horror had expanded globally and spread itself considerably across the tables of amateur investigators during those intervening years. Nevertheless, not until the 2010s would those developments make their way into scholarly accounts of non-digital gaming—a fact that underscores the longstanding disregard for analog amusements and supposedly "childish" pastimes like board gaming within the discipline of media studies.

Indeed, a more sustained look at this phenomenon reveals that the relative dearth of academic studies concerning Lovecraft-inspired tabletop games is not surprising at all, given the general unwillingness of researchers to consider any kind of playable media that is not a video game worthy of in-depth analysis. Notably, Paul Booth, one of the few media scholars to seriously take up the subject, devotes the first chapter of his pioneering book *Game Play: Paratextuality in Contemporary Board Games* to Launius's *Arkham Horror* (2015: 21–43). Booth uses *Arkham Horror*'s seemingly paradoxical mix of difficult-to-parse rules, sequentially organized players' turns, and ludic randomness (or potential for chaos) as a means of showcasing how the unknowable structure—or what he calls "unstructure"—of the dead author's living universe seeps into the game's mechanisms (Ibid.: 4). Those mechanisms, which entail moving characters into adjacent areas on the modular map, flipping cards to check for any potentially deadly encounters in their locations, and rolling dice to perform skill checks (including those for Fight, Lore, Luck, Sneak, Speed, and Will), are operationally important to the players' long-term strategic and short-term tactical goals. Yet they ultimately lead to game endings in which—even if the investigators are victorious (having sealed off the gates to the Other World)—the mysteries of the Cthulhu mythos remain intact and the questions surrounding the presence of otherworldly creatures called Ancient Ones go unanswered.

This chapter builds upon Booth's foundational work in tabletop gaming to consider the continued relevance of Lovecraftian horror within this largely overlooked area of media studies. Significantly, in the years immediately following the publication of his book (leading up to its follow-up, *Board Games as Media*, from 2021), several more Lovecraft-inspired games were published by FFG and other companies, including a few that further substantiate Booth's claims about the productive unknowability and generative unstructure of the Arkham universe. I argue that the word "universe" can now be switched out with the more accurate term "multiverse" given the cosmic underpinnings of the author's short stories and novels and the seemingly infinite array of scenarios, campaigns, and character-ability combos that comprise those games' practically cosmo-logical view of the world as something far too large and confounding to ever make sense of fully (or, in gamer-speak, "to grok"). Chief among those releases is a more streamlined but equally challenging third edition of *Arkham Horror* that was published by the same company in 2018, giving newcomers to the board gaming hobby as well as seasoned veterans familiar with its less-polished progenitors the opportunity to gather clues about unspeakably evil monsters while playing as a team of fearless investigators and immersing themselves in a deeply involving, emergent narrative more akin to the procedural rhetoric and algorithmic affordances of contemporary video games than to the simple mechanisms of traditional roll-and-move board games.

Around that same time, Fantasy Flight Games' *Arkham Horror: The Card Game* (2016; hereafter *AH:TCG*), designed by Nate French and M. J. Newman, was beginning to generate more than a few glowing reviews from critics who singled it out as possibly the best Lovecraft-inspired game ever made. This particular "Living Card Game" (LCG), more than any other addition to the company's ever-growing "Arkham Horror Files" line (a collection of games bound together by the Cthulhu mythos), has since gone on to spawn hundreds of fan-created websites, podcasts, Discord channels, and blogs detailing the minutia of playable content across its multiple sets of story expansions. Along with that ongoing series of narrative campaigns and the correspondingly vast collection of tutorial videos and buyer's guides seeking to ease the minds of anyone who might be intimidated by the complicated terminology or sheer scale of this LCG, at least two dozen recently published novels and novellas (from Rosemary Jones, Charlotte Llewelyn-Wells, Josh Reynolds, S A Sidor, and other contemporary authors paying tribute to Lovecraft's work) have further expanded the already vast diegetic worlds of *Arkham Horror* while inspiring fans of the franchise to create their own artwork and fiction.

That paratextual expansion of the primary text is not unlike the conjectural play with narrative possibilities that defines *AH:TCG*'s "pre-game" activity, and which carries through to the literally expansive spread of materials laid out on the table during setup. From a narrative standpoint, the mental act of forecasting what *might* happen during a scenario, in the moments leading up to the first draw from the shuffled Encounter Deck, puts parallel universes into play; just as playing any of the other entries in the "Arkham Horror Files" line (e.g., *Arkham Horror, Mansions of Madness, Elder Sign, Eldritch Horror*) featuring many of the same characters (e.g., parapsychologist Agatha Crane, student Amanda Sharpe, drifter "Ashcan" Pete, salesman Bob Jenkins) creates opportunities to imagine alternative histories of the people and places at the heart of this

Lovecraftian universe. The ludic worlds of Arkham—its aboveground and underground network of protoplasmic horrors (with slimy Shoggothian tentacles slithering into the farthest reaches of the world)—are further expanded through the aforementioned works of fiction. Far from being uninspired or derivative, the recent writings of Jones, Llewelyn-Wells, Reynolds, and Sidor add more than a little "flavor" to those recurring characters' backstories and help to remedy, through the same moves toward racial diversity and gender inclusivity that the game designers have been undertaking, the deeply problematic, ethnocentric treatment of "otherness" in the decades-old source material.

Focusing on the LCG and its many campaign expansions as my main case study, I seek to show how the cosmic sprawl of the *Arkham Horror* multiverse manifests both prior to and during actual gameplay. This occurs, for instance, when players construct their decks from a massive assortment of investigator, asset, event, skill, and weakness cards, or whenever they reach into a draw bag to pull out a chaos token that will partially determine if they fail or succeed at a challenge; including any that might result in "horror tokens"—the game's brain-shaped components indicating a person's gradual loss of sanity—being placed on character cards. Although actual "insanity"— a Lovecraftian trope that saturates this multiverse—is not likely to result from players' random encounters with cultists, flesh-eaters, goat spawns, and ravenous ghouls, a feeling of "madness" might flicker into consciousness whenever those players' *badness* at the game—their failure to play their proverbial cards right—interrupts the narrative flow and halts their progress through a particularly difficult campaign, or simply one that is made more complicated through the incorporation of new keywords (beyond those such as "Aloof," "Bonded," "Fast," "Hunter," "Massive," "Peril," "Retaliate," and "Surge," which can be found in the base game's three introductory scenarios). Using my own repeated failures—my frustrations of losing at *AH:TCG* numerous times due to poor hand-management and my occasional misunderstanding of terms—as a kind of autoethnographic anchor, this chapter highlights not only the instrumentality of user-created "support" texts (including online archives of "starter tips" and deck-building suggestions) to one's ability to derive pleasure from this decidedly maddening sort of play, but also their centrality to the endless world-building that comes from "ludifying Lovecraft" in the first place (Booth 2015: 21–43).

Toward a Transludic Multiverse

If the reader were to purchase a new copy of the *AH: TCG* base set from an online or brick-and-mortar retail store today, they would most likely be acquiring the 2021 revised edition of the 2016 game, one that ups the

maximum player count from two to four and which features a few cards •
not included in the original (for a total of 245 player cards and 111 scenario cards), along with a "chaos bag" from which to draw the tokens that
serve as the game's dice alternative. Additionally, the cards for the five investigators included in the revised core set—Roland Banks ("The Fed"),
Daisy Walker ("The Librarian"), "Skids" O'Toole ("The Ex-Con"), Agnes
Baker ("The Waitress"), and Wendy Adams ("The Urchin")—showcase
artwork that differs from the previous release's illustrations, indicating an
ever-evolving set of aesthetic, mechanical, and textual "moves" or fine-
tunings on the publisher's and designers' part. Rounding out the box are
40 heart-shaped damage tokens, 27 brain-shaped horror tokens, 57
double-side clue/doom tokens, 61 resource tokens, and 44 chaos tokens,
along with a 24-page "Learn-to-Play" rulebook, a 32-page reference guide,
and an 8-page, beautifully illustrated yet text-heavy campaign booklet. Inside of that booklet are the setup instructions and narrative framework for
three scenarios comprising the introductory "Night of the Zealot" campaign: "The Gathering," "The Midnight Masks," and "The Devourer Below." In providing multiple "Intros" and "Resolutions," which players are
asked to selectively read based on the narrative outcomes of their investigators' previous adventures, these scenarios establish a pattern for all the
campaign expansions that follow. They also hint at the ways in which this
Lovecraftian textual universe will be transformed into a *multiverse* through
play, through the piecemeal actualization of *ludic potentialities*.

That transformation is itself an expansion of sorts, insofar as the spread-
ing out of physical materials atop the table—itself a real-world analog to
the metaphysical otherworld of Lovecraftian horror—entails not only an
expenditure of resources (actual money paid for expensive expansion sets
as well as in-game money that continues to line the investigators' pockets
each round), but also an ongoing accrual of branching narratives, all of
which trace their origins to a single source. "Night of the Zealot," the
three-part campaign included in the core set of *AH:TCG*, epitomizes this
in-game expansiveness. "The Gathering" begins by telling players that their
investigators are inside a single room—the cramped study of the lead investigator's house—only to nudge them toward a larger set of options, spatially and strategically, as details of the missing townspeople slowly
accumulate over the course of the session. The front of the first double-
sided agenda card, fittingly titled "What's Going On?!," includes a brief
written description of the "strange chanting coming from the parlor down
the hall" as well as the sound of "dirt churning…beneath the floor." Like
most of the agenda cards included in subsequent campaigns, this printed
text hints at the unspeakable terrors that lie in wait. But it does so by directing our attention both beyond and below a physical setting that is already
starting to dematerialize before our investigators' eyes. The accompanying

act card, which sits to the right side of the agenda card (and together forms the pages of an open book), reminds us that the characters are initially trapped inside that study, unable to access its only door which, for reasons that are unclear, has suddenly vanished. This odd situation has left them with few other options besides arming themselves with weapons and other assets in preparation for what is to come and to search for clues amidst the clutter of the room. Once enough clue tokens have been placed atop that card (two per investigator), it can be flipped to its reverse side. That previously hidden side of the card informs players that the missing door has somehow traveled from the wall to the floor, and that by pulling back a mud-stained rug their investigators can now exit the study and move into an adjoining area.

First-time players might feel disoriented not only by the large number of game-state changes occurring within this opening passage, which necessitates the removal of the study from the game and the putting into play of those other areas of the house—the hallway, cellar, attic, and parlor—which, only minutes earlier, had been set aside during setup, but also by the confusing layout of the house itself. The agenda and act cards contribute to that uncertainty. They show, on the left side, large feral creatures crawling up through bathroom floor tiles and, on the right side, the exterior of the house, alongside text concerning the cell-like study. We are at once inside and outside a home that, we are told, belongs to us, but which is inhabited by unwanted, unknown beings. Or, rather, we are neither inside nor outside that place, but instead somewhere else, somewhere *between*, occupying a liminal zone in which the very structure of the building has begun to crumble, and, through its cracks, another world is beginning to peak through. Anyone who has played *Arkham Horror: Second Edition*, which predated this card game by eleven years, might experience a sense of déjà vu when they begin their first session of "Night of the Zealots," for the "unstructure" of the former—the "feeling of randomness generated by an unknowable structure" (which, ironically, is only compounded by a lengthy set of complex rules)—seems to have infused the inner and outer landscapes of this latest Lovecraftian paratext (Booth 2015: 31).

Importantly, both input randomness and output randomness are crucial to *AH:TCG*. Though the "bad luck" of a failed skill check can be mitigated through card play, output randomness—the unforeseen effect of a player's action (which is widely believed to rob a person of his or her agency at the table)—is part of its "unstructure." While similar to some cube-pushing, engine-building eurogames, insofar as players have control over a fairly robust action selection system and place their "workers" (small character cards rather than meeples) onto discrete locations arranged on the table, *AH:TCG* also embraces "chaos" by forcing players to randomly draw tokens from a bag in order to complete skill checks. That, in addition to the randomization of shuffled Encounter cards (as well as the players' own Investigator deck), means that, regardless of any luck-mitigation that comes from playing cards with specific skill icons, a significant amount of unpredictability offsets the "perfect information" that is so highly prized by hardcore eurogamers. Through its transludic appeals to fans who might conceive of the already loose concept of "playing" in a different, more elastic, way than do many critics and reviewers, *AH:TCG* begins well *before* the first investigators' phase takes place. A more expansive definition of "play" would account for the ways that games such as this involve players in a series of "What if?" questions in anticipation of *what might transpire* (with regard to the possible outcomes of their strategic choices) once those people at the table take turns playing cards that they deliberated upon, speculated about, and finally selected long before the first round kicked off.

What if I choose a Guardian character like Zoey Samaras ("The Chef") over a Seeker character like Rex Murphy ("The Reporter") in the lead-up to the *Arkham Horror* campaign "Edge of the Earth"? How might putting assets such as the .45 Automatic, the Machete, and the Flashlight into Zoey's deck impact her chances for success, relative to other equipment (for instance, the Knife, the Kukri, Zoey's Cross, and First Aid) which might otherwise take their place? Should I incorporate additional Basic Weakness cards into her roughly 30-card Investigator deck to cover the XP costs of selecting stronger, more effective cards, such as the Cyclopean Hammer (a relic melee weapon that deals additional damage to non-Elite enemies)? Knowing that "Edge of the Earth" is set in the uncharted reaches of Antarctica, I might ask myself which other gear would come in handy as Zoey, a resourceful character who grew up an orphan (after her parents died in a fire), learns to navigate that treacherous new environment—one that is clearly inspired by Lovecraft's 1931 novella *At the Mountains of Madness*—with or without the aid of allies and fellow investigators. Here, in posing questions about a given character's readiness for hypothetical challenges in the forthcoming scenarios and larger campaigns, I and other players of *AH:TCG* undertake a multiversal journey into the heart of the unknown; a universe of sprawling potentiality made up of *other universes* in which untold numbers of other Zoeys fight for their lives through synergized card combos. Thus, by building a deck of cards for her or any of the fifty-seven other investigators currently available and included in FFG's official releases, a player does not narrow the range of *Arkham Horror*'s playable synergies so much as open up an already expansive field of possibilities, in terms of in-game card play and the vast number of ways in which one character's assets, events, and skills might interact with those of other characters.

Writing four years prior to the release of *AH:TCG*, Stewart Woods points out that "board game systems are anything but closed" (2012: 6). Fittingly, his use of the term "board game" is as open as the systems themselves, for it can refer to any playable thing with physical pieces placed on a table, including card games lacking an actual "board" (Ibid.: 5). The fact that this particular game does *not* have a board, and that the play area itself is bounded only by the physical dimensions of the flat surface on which the cards are laid, means that its transludic appeals are related not only to time (the pre-game and in-game temporalities mentioned earlier) but also to space. In true Lovecraftian fashion, a person can "play with space," or be spatially playful, by putting location cards either close together, in relatively tight formation, or far apart, so as to exacerbate the great distance that must be traversed by investigators before arriving at places where clues might be discovered and enemies might spawn. Although each scenario comes with a "Suggested Location Placement" for the cards unique to that scenario, players are free to arrange those locations any way that would like, creating a fluid, amorphous area of positive and negative space that shifts over time (owing to hands accidentally brushing up against those cards or rearranging them to make room for more heroes and monsters). The physical distances between Path of Thorns, River Canyon, Rope Bridge, and Serpent's Haven (four of the many location cards in the campaign expansion "The Forgotten Age") might be signified by the placement of those cards, but they are never fixed. Instead, they are left to the whims of players whose game-table layouts will look different from those of others. While it is easy to accurately calculate the distance between real-world locations such as London's Big Ben, Kensington Gardens, Tower Bridge, Traitor's Gate, and Westminster Abbey, the arrangement of cards bearing these locations' names (in the campaign expansion "The Scarlet Keys") is up to the player(s), who can create their own map(s) of London based on how they imagine the city to be laid out within the fictional world of what truly feels like a *living* card game.

Unlike well-known Collectible Card Games (CCGs) such as Richard Garfield's *Magic: The Gathering* (1993) and Tsunekazu Ishihara, Kouichi Ooyama, and Takumi Akabane's *Pokémon Trading Card Game* (1996), which encourage consumers to "blindly" purchase booster packs containing random sets of trading cards (as part of their pre-game deck-building) in order to be competitive in duel-like battles at the table, the LCGs upon which *AH:TCG* is partially based (*Call of Cthulhu: The Card Game* [2008], *A Game of Thrones: The Card Game* [2008], and *The Lord of the Rings: The Card Game* [2011], which designer Nate French had worked on prior to 2016), rely upon a different distribution model, supplementing their core sets with expansion packs whose contents are the same, consumer to consumer (Arnaudo 2024: 45). For example, every box of the "Dunwich Legacy" expansion (either the original "deluxe" release and accompanying "Mythos Packs" from 2017 or the repackaged campaign and investigator sets from 2022) will have the same combination of 521 cards. As Marco Arnaudo points out, players of living card games "can have all of the intellectual engagement that comes from creating their own deck, without any of the disparity that may derive from unequal income and access to certain cards" (Ibid.). Moreover, like serialized stories told in the pages of monthly or bimonthly magazines, living card games grow or "live" over time through those regularly scheduled releases, but players know in advance that the expansions they will have purchased are just like any others bearing the same title. And yet, one person's "Dunwich Legacy" pack is decidedly *not* like any other person's pack of the same title, in terms of *how* the contents of those boxes will be put to use by individual consumers. Speaking from my own experience, and judging from what more advanced specialists of *AH:TCG* have posted in their online strategy guides (for instance, on the long-running YouTube channel PlayingBoardGames), players often customize their decks of 30-35 investigator cards by pulling from more than one expansion. They also tend to tailor their choices according to their own playstyles, predilections, and tastes—for particular characters and/or classes (Guardian, Mystic, Rogue, Seeker, or Survivor). For me, deck-building is not only a kind of world-building, but also a means of personal expression on an otherwise unassuming table where the Lovecraftian universe of nearly unpronounceable horrors (e.g., Ghatanothoa, Nyarlathotep, and Tsathoggua) is merely the pretext for the transludic textuality of my own unfolding.

Of course, *AH:TCG*'s primary sources of inspiration are the published poetry, short stories, and novellas of an American author whose contributions to the genre of horror fiction are as impossible to deny as is his well-documented racism. On the latter topic there has been much scholarly work over the past two decades, including that which attempts to leverage Lovecraft's demonizing of ethnic minorities and "swarthy'" foreigners ("freaks of alien blood") against his literary craft as a pioneer of "weird" fiction (Ellis 2010: 132). Lovecraft's anti-immigrant attitudes and loathsome statements about people of color, indicative of a "racist neurosis" that was not his alone during the forty-six years he was alive, infect his writing to such a degree that even his greatest admirers are hard-pressed to separate the art from the artist. Holding both perspectives at once—balancing the writer's antiquated beliefs about racial purity and impurity with a recognition of his significance as the creator of "the first 'open-source fictional universe'" (Nevins 2013)—seems as counterintuitive as trying to adapt the "myths, occultism, and secret histories" of that universe into a tabletop game. As Mark Jones argues in his study of "the Lovecraftian being in popular culture," an RPG like *Call of Cthulhu*, which has remained in print ever since its 1981 publication, might be "eclectically omnivorous in its absorption" of those weirdly defining yet indefinite elements, but it, like all rules-heavy games, imposes a "rational system onto its fantastical reality." As such, "playing *Call of Cthulhu* brings the cosmic horror of Lovecraft's mythos into continual collision with a rule-bound rationalist and pseudorealist world" (Jones 2013: 229). Nevertheless, owing to the "somewhat arcane and cabalistic nature of the role-playing experience," that ludic intervention into the equally mythic legacy of the author "becomes strangely symptomatic, both of Lovecraft's fictional world and his reputation in contemporary culture" (Ibid.).

James Perkins Mastromarino, an NPR correspondent, puts it more directly when he states that the games in FGG's "Arkham Horror Files" line "take the nihilism and racism out of Lovecraft" (2022). Quoting Katrina Ostrander, creative director of story and setting for FGG's parent company Asmodee, Mastromarino notes that game designers "have attempted to revise Lovecraft's trademark 'fear of the unknown' for a more inclusive 21st century," often by adopting a more sensitive perspective on playable heroes of various genders, ethnicities, nationalities, and abilities hailing from all corners of the world. In *AH:TCG*, players can choose to be characters who would have been unthinkable as investigators in Love-craft's fiction, such as Amina Zidane, a refugee from French Algeria who works the switchboard for the Miskatonic Valley Telephone Company; Minh Thi Phan, a Vietnamese secretary born in Korea under Japanese rule; Akachi Onyele, a Nigerian shaman; and Jim Culver, an African American jazz trumpeter. Ostrander admits that some of the line's most popular characters (for instance, Lily Chen, a disciplined martial artist who first appeared in the 2008 "Kingsport Horror" expansion of *Arkham Horror: Second Edition*), tread on familiar ground as potentially stereotypical rep-resentations. However, she and designer M. J. Newman have been careful to avoid any identity-based material that might be offensive to players and have relied on a "sensitivity panel and cultural consultants to avoid such pitfalls" (Ibid.). The fact that someone like Stella Clark, a trans person of color who, as the letter carrier of Arkham, is a fixture of that fictional com-munity, was singled out as the main character in Jess McGatha, Tyler Stu-art, and Seth Wolfson's immersive multimedia board game *Arkham Horror: The Road to Innsmouth* (and is a playable character in *AH:TCG*) says much about how far removed from their source material these com-paratively "optimistic," transludic texts of the twenty-first century are.

Other titles in the "Akrham Horror Files" series include a few of the ones mentioned above: designer Nikki Valens's *Mansions of Madness*, the second edition of which (published in 2016) requires the use of a digital app in order for players to send their characters "through the veiled streets of Innsmouth and the haunted corridors of Arkham's cursed mansions" in pursuit of answers to ancient mysteries; the comparatively lightweight, fast-paced dice-chucker *Elder Sign*, which, codesigned by Richard Launius and Kevin Wilson, places one to eight investigators inside the Miskatonic University Museum where, in a race against time, they must locate rare artifacts and arcane symbols with the goal of closing portals to an otherworldly beyond; and *Eldritch Horror*, a globetrotting variation on the original *Arkham Horror* board game that expands the geographical scope of its spiritual successor while narrowing the decision space for greater ease of play and fine-tuning several mechanisms that were deemed "fiddly" in the original. The paradoxically streamlined expansiveness of *Eldritch Horror*, which Valens codesigned with Corey Konieczka (of *Battlestar Galactica: The Board Game* [2008] fame), gives one to eight players the opportunity to venture far beyond the fictional Miskatonic Valley that first appeared in Lovecraft's 1921 short story "The Picture in the House" and imagine themselves as adventurers traveling to such far-places as the Antarctica, Buenos Aries, Rome, the Himalayas, and Tokyo. That vastness of scale anticipates some of the more recent campaign expansions for *AH:TCG*, including "The Scarlet Keys," which presents a non-linear series of *Choose-Your-Own-Adventure*-style assignments for the investigators, who are tasked with tracking down the titular paradimensional objects before they fall into the hands of the Red-Gloved Man. Given its structure, "The Scarlet Keys" is among the most replayable of all expansions, though all of the other campaigns lend themselves to multiple plays for anyone seeking to test out different investigators and card combinations as well as difficulty levels (more negative-numbered chaos tokens in the bag). Because no two games will ever be the same, owing to randomized elements (including card shuffles) and different input randomness on the players' part, each session contributes to a greater "multiversification" of this literally expanding transludic text, which spreads all of the possible "lives" of its growing cast of characters across parallel worlds—multiple Arkhams—that hover just above the table.

As suggested above, the *Arkham Horror* universe is vast, a sprawling assortment of creative enterprises and commercial goods that attest to its widespread appeal beyond the relatively small world of tabletop gaming. It encompasses a seemingly endless stream of publications and productions across several overlapping cultural spheres, including fiction (e.g., the novels and short stories of Lovecraft as well as countless professional and fan-written pieces inspired by his work), nonfiction (e.g., scholarly studies, reference guides, and encyclopedic collections of all things Cthulhu), toys (e.g., stuffed plushies, painted miniatures, and action figures), and audiovisual media (e.g., music, films, television series, and video games that, if not directly related to the board and card games, help to flesh out their narratives through sounds and images that might otherwise only exist in the imaginations of players). However, it is at the table, where the growing number of cards and tokens in *AH:TCG* are shuffled, drawn, and revealed, that its already large universe becomes truly multiverse-like in scale: a cosmically big decision space full of metaphysical concepts that can nevertheless be imagined through physical components laid out on a flat surface whose limited play area belies the game's unlimited array of deck-building options.

References

Arnaudo, Marco. 2024. *The Tabletop Revolution: Gaming Reimagined in the 21st Century*. Jefferson, NC: McFarland & Co., Inc.

Booth, Paul. 2015. *Game Play: Paratextuality in Contemporary Board Games*. New York, NY: Bloomsbury Academic.

Booth, Paul. 2021. *Board Games as Media*. New York, NY: Bloomsbury Academic.

Ellis, Phillip A. 2010. "The Construction of Race in the Early Poetry of H. P. Lovecraft." *Lovecraft Annual*, No. 4, 124–135.

Jones, Mark. 2013. "Tentacles and Teeth: The Lovecraftian Being in Popular Culture." In: *New Critical Essays on H.P. Lovecraft*, edited by D. Simmons. New York, NY: Palgrave Macmillan, 227–247.

Nevins, Jess. 2013. "To Understand the World Is To Be Destroyed By It: On H.P. Lovecraft." *Los Angeles Review of Books*. May 5: https://lareviewofbooks.org/article/to-understand-the-world-is-to-be-destroyed-by-it-on-h-p-lovecraft/

Woods, Stewart. 2012. *Eurogames: The Design, Culture and Play of Modern European Board Games*. Jefferson, NC: McFarland & Co., Inc.

PART IV

Multiversal Constructs Across the Self and Others

28

TEACHING THE AMERICAN MULTIVERSE

Power, Agency, and Identity in an Age of Consumer Choice

Dustin Abnet

These are the words that greet students in my research seminar, "The American Multiverse" that I first taught in the Department of American Studies at California State University, Fullerton in Spring 2023:

> Americans in the 21st century live in a multiverse built on an ideology of consumer choice. Our world is saturated with constructed and virtual worlds in which consumers can immerse themselves. Whether in fan experiences such as the (now defunct) Star Wars: Starcruiser Hotel, the interactive environments of gaming, the "filter bubbles" that shape our consumption of (mis/dis)information, the conspiracy theories that dominate our political discourse, or the "metaverses" promoted by tech titans, Americans have unprecedented options for living, working, and playing in radically divergent worlds where conventional rules of existence do not necessarily apply. Meanwhile, the popular culture Americans consume offers often-apocalyptic stories of colliding realities and universes and mirror-verse doppelgängers threatening our real selves. In 21st century, America, both the reality of the world and the individual self seem contested in unique and potentially liberating and/or terrifying ways.

Accepting the existence of the multiverse, the course reimagines the concept as a framing device through which students will examine both contemporary experiences and the stories people tell about their worlds. It treats the multiverse as an allegory for the fracturing of "American"

DOI: 10.4324/9781003480846-33

culture in our neoliberal, postmodern, and digital worlds (Rodgers 2011) and asks students not whether the multiverse is real but what cultural work does the concept perform and how can we use it to navigate through our divided and often violent society.

The multiverse, the course argues, has become a dominant and useful allegory for understanding the contested nature of American life and identity in the twenty-first century. In particular, it shows how Americans have used the concept to navigate the complex relationships between power, culture, identity, and agency in a country whose ideologies and rhetoric celebrate individual choice, but whose structures often undermine the ability of many of its people to feel like their choices have meaning. It is, I think, telling that the two top multiverse films in 2022, *Everything Everywhere All at Once* and *Doctor Strange in the Multiverse of Madness*, centered their universe-hopping on two non-White, immigrant women, Evelin Quan Wang and America Chavez, who use the agencies granted by their powers to navigate worlds not designed by or for them while forging new forms of identity and stronger connections to their communities. Numerous scholars have shown how the multiverse is an ancient concept rooted in many cultures (Rubenstein 2014; Siegfried 2019); however, it is also a concept that speaks to specific audiences in particular moments in time. In this course, we explore one of those as a means of better understanding the contested nature of contemporary American identity.

As a scientific theory, the multiverse idea enables physicists and cosmologists to reconcile divergent observations and mathematics (Greene 2011, 108–51). As a cultural allegory it serves much the same function by enabling people to make sense of the increasingly contested nature of reality, the divergent ways that we interpret and inhabit the world, the liminality of identity and experiences, and the fluctuating nature of power and agency in a hypercapitalist world shaped by colonialism, racism, misogyny, homophobia, and other forms of oppression. The multiverse is a story told to explain a culture with far more voices and perspectives than were possible to encounter in the middle of the twentieth century. It is a grand narrative for a culture skeptical of grand narratives (Jeffries 2021, 95–108), an explanatory device for a culture that celebrates the disruption of institutions but longs so much for the comforts of mid-twentieth century homogeneity that it threatens to revert to fascism (Rasmussen 2022, 4–7). It is this longing for unity amid diversity, truth in uncertainty, meaning amid the chaos of market choice, that gives the multiverse its conceptual power. This essay, like the course, explores that power with an eye to helping students navigate the conflicts and possibilities of American culture in the twenty-first century.

The Course

Teaching the American Multiverse offered me the opportunity to synthesize other courses I teach on popular, consumer, and digital cultures while giving students the chance to demonstrate their understandings of core American Studies concepts by designing, conducting, and writing research papers within the confines of the course theme. In a course on the multiverse, such freedom of choice poised a potential problem. Students could interpret the theme narrowly which I feared would yield a nightmare: 20+ papers on the MCU. Alternatively, students could embrace the possibilities of the concept and write about nearly anything; knowing myself, I probably would have allowed it.

To make the class feasible, I suggested that students focus on one of five topics: 1. media depictions of artificial/alternate worlds and realities, 2. the "magic circles" created by gaming spaces, 3. digital and virtual spaces including social media, 4. immersive fan experiences that turn fictional worlds into physical spaces, or 5. conspiracy theories that posit alternative views of reality. Each of these, I argued, provided opportunities to connect the knowledge students had gained in their degrees to people's multiversal experiences. In addition, final papers had to include three elements, each of which is central to American Studies methodologies: (1) a structural analysis of their chosen alternative space/world/reality; (2) an analysis of consumers' experiences within or reception of that space; and (3) an examination of the historical and contemporary contexts that shape the cultural work of their chosen topic. Students could vary the balance between these sections, but at least some effort had to be given to each component.

I further confined the course by framing the multiverse as an outgrowth of what the historian Daniel T. Rodgers (2011) has called "the age of fracture," a period at the end of the twentieth century in which mid-century notions of the limitations on human agency, "gave way to conceptions of human nature that stressed choice, agency, performance, and desire" (3). In economics and politics, this meant the rise of neoliberalism and the promotion of deregulation, privatization, and other efforts to remove the state from an ever-expanding idea of what "the market" should handle. In culture, it meant the questioning of dominant narratives, the breaking down of binaries, and the deconstruction of texts. For conceptions of the self, it meant an increasingly emphasis on the fluid, malleable, and performative nature of identity (10–12). Yet, this rhetoric of fluidity and liberation often came paired with more coercive state controls over particular populations. The age of deregulation has also been the age of the prison industrial complex and the war on drugs. The weakening of the welfare state came with greater surveillance of the poor (Kohler-Hausmann 2015; Eubanks 2018).

It is in this gap between rhetoric and experience; between freedom for some, policing for others; that the American multiverse thrives. We saw this perhaps most vividly in *Everything Everywhere All at Once* by noting how Evelin's ability to draw on the power of the multiverse develops amid her struggles against the distinctly non-neoliberal bureaucracy of the IRS. By exploring such texts, we learned that the multiverse was not simply an allegory for the fracturing of American life but a concept explicitly used to navigate and survive it.

Even so thematically and contextually confined, the multiverse remained vast. Conducting their research required students to draw on a variety of disciplinary perspectives and methodologies, including anthropology, history, sociology, geography, ethnic and gender studies, and literary, media, cultural, and game studies. To analyze a particular phenomenon, they might need to understand complex theories of economics, culture, identity, the cosmos, and reality itself and rapidly shift between theories of postmodernity, subjectification, neoliberalism, and performativity. To be successful in classroom conversations, they had to be just as comfortable discussing the Afro and Queerfuturisms of artists such N.K. Jemisin, Ytasha Womack, and Janelle Monae as they were Jean Baudrillard, QAnon, and Twitch streamers. Consequently, course materials and activities had to introduce students to a broad range of frameworks and potential sources. The frameworks and pedagogical strategies I offer here reflect my actual classroom practices as well as adjustments I intend to make to future iterations. Based on our class conversations, I suggest three interrelated but distinct themes of the American multiverse: (1) questioning the nature of reality, (2) imagining alternative ways of being, and (3) movements between worlds and identities. Teaching the class was challenging and appropriately chaotic, but it proved a deeply rewarding experience for both me and the students as we collectively developed the ideas necessary to navigate our fractured world.

Questioning the Nature of Reality

We began the semester with a light question: the nature of reality. So many American political debates in the Trump/Biden era—Do masks/vaccines work? Who won the 2020 Presidential election? Are Democrats running a pedophile ring out of a non-existent basement in a pizza parlor?—focus on these questions that students immediately grasped the importance of the theme. But traversing the multiverse requires familiarity with a more epistemological approach. For that reason, we turned to philosophy, pairing selections from David Chalmers's (2022) *Reality+* on virtual worlds with a screening of *The Matrix* to illustrate that questions about how we know what reality is long predate our contemporary moment but have been made particularly acute in our hyperreal world of mass and digital media. After classroom discussions, we analyzed how people had deployed the film's red pill/blue pill dichotomy as an example of how consumers use media as a language to comprehend and express their own senses of reality (McCulloch 2019, 237–64).

Our conversations quickly turned to a much more common conversation in AMST, how, in the dramatically shortened words of Foucault (1995), "power...produces reality" (194). We initially discussed the issue in relation to narratives that falsely pit science against faith. As philosopher Mary-Jane Rubenstein (2014) argues, the science of the multiverse "replaces God with what is perhaps an equally baffling article of faith: the actual existence of an infinite number of worlds, eternally generated yet forever inaccessible to us" (17) As she succinctly quotes the physicist Bernard Carr in a line that came up frequently throughout the rest of the semester: "If you don't want God, you'd better have a multiverse" (1). The existence or non-existence of the multiverse was not a question of *reality*, we agreed, but of *faith* and power.

Americans historically have been told to place their faith in institutions such as organized religion, government, media, and universities to determine the nature of reality. Yet, the combination of neoliberalism, postmodernism, and the rise of the internet around the turn of the twenty-first century undermined such faith. As communications scholar Alice E. Marwick (2013) has recounted, the Silicon Valley promotors of "Web 2.0" technologies sold social media as a tool for mass democratization, participatory journalism, and the empowerment of individuals (21–25). This ethos that merged punk and hacker anti-institutionalism to the market logics of neoliberalism became one of the ideological foundations of our current multiverse. As a very excited Lev Grossman (2006), future writer of the multiverse fantasy series *The Magicians*, infamously put it in his essay for Time's selection of "You" as the 2006 Person of the Year the new web is "a tool for bringing together the small contributions of millions of people and making them matter. Silicon Valley consultants call it Web 2.0, as if it were a new version of some old software. But it's really a revolution." In our era of mis/disinformation, deepfakes, and conspiracy theories, such optimism seems naïve, but such rhetoric undergirds the logic of the American multiverse: we are the arbiters of our own realities, the makers of our own worlds. We choose to take the red or blue pill.

Of course, questioning the nature of reality is by no means a new development. We explore this in class by pairing contemporary and historical examples of debates over the nature of reality. For instance, when discussing conspiracy theories, we read Richard Hofstadter's 1964 essay

"The Paranoid Style in American Politics" and a contemporary account of QAnon. When examining "authenticity" on social media we watched Matt Spicer's (2017) film *Ingrid Goes West* about an Instagram user's efforts to mold her life after that of an influencer alongside Elia Kazan's 1957 film *A Face in Crowd* about a mass media celebrity's power over his audience. In both cases, students were struck by how relevant the mid-century text was to the present. But we also noted differences. Here we found media scholar Henry Jenkins's analysis of transmedia storytelling as "the art of world making" particularly relevant. As Jenkins (2006) writes, contemporary consumers have to work together "chasing down bits of the story across media channels, comparing notes with each other via online discussion groups, and collaborating" (21). The interactivity of this "research" and sense of community is precisely what scholars of contemporary conspiracy theorists have pointed to a source of its resonance (Hon 2022, 163–84). It is also, we determined, what enables Ingrid to turn the parasocial media relationships explored in *A Face in the Crowd* into stalking and obsession. Alternative realities draw on a long history of conspiracy theories, hyperreal depictions, and parasocial relationships; that's part of what gives them their power. But it is the multi-platformed-research and connection-forming process required in our contemporary media landscape that turns them into interactive, alternative worlds for consumers.

Imagining Alternative Ways of Being

Our second major theme was how the multiverse enables people to imagine alternative ways of being, a point made vividly with *Everything Everywhere All at Once*'s hot dog universe. Here, we were drawn to forms of culture that have enabled people from marginalized social groups to imagine alternatives to colonialism, racism, misogyny, homophobia, and other forms of oppression. As Afrofuturist artist and scholar Ytasha Womack has noted, "Imagination is important because it is liberating....When you are in challenging environments, you're not socialized to imagine. And so to claim your imagination — to embrace it — can be a way of elevating your consciousness" (Miller 2024). Creating such imaginary realities has been particularly important for such groups because, per literary scholar Ebony Elizabeth Thomas (2019), "When people of color seek passageways into the fantastic, we have often discovered that the doors are barred. Even the very act of dreaming of worlds-that-never-were can be challenging when the known world does not provide many liberating experiences" (2). In the classroom, we explored this issue by examining forms of culture that reclaimed oppressive forms as tools of Black and Queer liberation. For instance, we examined how Janelle Monae's early albums and music videos reclaimed the vamp robot from both Fritz Lang's 1927 film *Metropolis* and a much longer history of Euro-Americans imagining black and female robotic slaves (Abnet 2020, 297–8).

We were similarly intrigued by how American comics have imagined alternative bodies. As literary scholar Ramzi Fawaz (2016) has shown, post-1950s comics experimented with alternative forms of bodies such as those of the Fantastic 4, the Hulk, and, especially, the X-Men that were, as he notes, "implicitly queer" because of the ways they positioned the body in flux, "constantly moving among different identities, embodiments, social allegiances, and psychic states." When interpreted by fans, such radical imaginings, he argues, "made fantasy a political resource for recognizing and taking pleasure in social identities and collective ways of life commonly denigrated as deviant or subversive within the political logics of cold war anticommunism and an emergent neoconservatism (11, 10, 4)." To examine the topic further, we discussed the kinetic and joyful animation found in the 2018 film *Spider-Man: Into the Spider-Verse*, as it explored the divergent bodies of the Black Latinx Miles Morales, original Spiderman, Peter Parker, the queer Gwen Stacey, and Spiderpig (see also Fitz-Whittemore chapter in this volume). Watched in conjunction with Fawaz's arguments, it suggested the liberatory power of multiverse stories to connect alternative bodies to alternative ways of being.

Numerous students designed projects that allowed them to explore the movement of alternative and queer bodies through alternate worlds. One student, for instance, examined the Disney Channel *Halloweentown* film series which is set in an alternative dimension filled with monsters who left the human world because of the prejudice and violence they faced. Centered on a family of human witches who can move between the worlds, the series uses the seemingly monstrous bodies of the titular town's residents to explore the liberatory potential of finding a space to develop your own sense of identity. Another student explored the alternative family arrangements in the animated, reality-breaking sitcom *The Amazing World of Gumball* about the adventures of an anthropomorphic blue cat and his adopted goldfish brother. Others explored queer fans' responses to the Netflix revival series *She-Ra, Princess of Power* and the animated series *Steven Universe*. In each case, students pointed to the way that fans used such works to explore their own senses of self amid larger cultural efforts to limit and regulate particular ways of being.

Students were particularly excited to examine these possibilities in games such as *Stardew Valley, The Legend of Zelda: Breath of the Wild*, and *The Sims*. Each student who analyzed a game relied on the concept of the "magic circle." Originally articulated by Dutch scholar of play Johan

Huizinga to refer to the unique fictional worlds created by gameplay, the term was adopted by early game studies scholars to describe the seemingly bounded worlds of video games. As critic of the concept Mia Consalvo (2009) describes, "Inside the magic circle, different rules apply, and it is a space where we can experience things not normally sanctioned or allowed in regular space or life" (Consalvo 409). Yet, as scholars have noted, the concept is flawed because no game can ever be isolated from context. Designers and players always bring their prejudices, ideologies, and conceptions of reality to the world of the game. The magic circle is never as isolated from the ordinary rules of reality as people might desire it to be.

Nevertheless, the concept of the magic circle remains potent because it captures a longing for spaces in which people can explore what philosopher C. Thi Nguyen (2020) has called "alternative agencies" (2) The act of playing games, Nguyen argues, enables us to "take on temporary agencies—temporary sets of abilities and constraints" that allow us to explore different powers and limitations and embody different ways of being (4). In the classroom, we saw this theme reflected in the introductory materials to the 2022 First Nations-themed RPG *Coyote & Crow*. There, the designers included a suggestion for indigenous players:

> the intention of this game is not to simply take the reality of our lived world and transpose it onto a future fictional world. This is a work of alternate history fiction. The last 700 years of our real world history never happened. We encourage you not to overlay your tribe's recent past onto this different future, but instead think in terms of what could be, of what might have been.
>
> *(Alexander 2021, 10)*

Games, such sources suggested, are not magic circles; they don't create their own worlds distinct from our own any more than alternative futurisms do. But they do make it possible to imagine and explore alternative ways of being.

Moving Between Worlds and Selves

The third theme we were drawn to is how people move between different worlds and identities. Multiversal literature has long dealt with liminal spaces such as C.S. Lewis's "Wood Between the Worlds" from *The Chronicles of Narnia* or Grossman's the "Neitherlands" from *The Magicians*, and tools such as the subtle knife in Philip Pullman's *His Dark Materials* series. Such spaces and technologies are critical to navigating the multiverse, but, we were also drawn to powers like those of America Chavez or

Evelin Wang that allow people to move between worlds on their own or draw on the abilities of their alternative selves because they suggested the importance of tensions over identity to our multiversal moment.

Chavez is a useful example. Though none of her struggles over identity were included in *Multiverse of Madness*, they are a particular theme in the comics written by queer Latinx writer Gabby Rivera. According to literary scholar Melissa Castilo Planas (2021), Rivera's Chavez "is at its heart a story about a young woman's search for her roots" as a brown-skinned, queer woman with many features that mark her as Latinidad, even though she is actually an alien. Across Rivera's series, Chavez uses her home at the multiversal Sotomayor University as a base from which to move through the multiverse while exploring what it means to be identified as an immigrant member of a diasporic community amid the oppressions of racism, colonialism, and capitalism. As Rivera noted in an interview in the *Washington Post*, the comic

> is definitely going to tackle America's ancestry and ethnicity. But it won't be as neat as some folks might want it to be. For me, being Latina is really damn complicated, especially when it comes to tracing my roots….She's going to explore what it means to be brown across the dimensions. And like many people who've had to leave home at a young age, she's dealing with that feeling of disconnect, the you're a foreigner here and out of place when you go 'home' type of feeling.
>
> *(Betancourt 2017)*

In the context of a country increasingly obsessed with a border wall, the ability to transcend even the boundaries of the universe is a powerful form of agency directly tied to a diasporic identity, a theme that resonated powerfully with many of my students.

Students immediately grasped a connection between such multiversal powers and the code-switching required of people from marginalized groups. Yet, drawing on the work of social media scholar dana boyd, they also noted that shifting performances of the self is a requirement for everyone in the age of digital media (boyd 2014). This was a key inspiration for Phil Lord, writer and producer of the Spider-Verse films. As he told CNN Reporter Thomas Page (2022):

> I think we're living multiple lives in parallel dimensions sort of all the time…We're living an online life — or lives. Then we're living a work life that's on a screen … Then there's a home life, and then one with your friends. Trying to resolve those things is going to be something we're all thinking about all the time.

In an age of digital media empowered by fast capitalism, having multiversal powers like Chavez or Wang almost seems to be a requirement for survival.Movement between worlds and selves is also a key feature of fandom. Jenkins's (2013) influential *Textual Poachers* examined how fans read "intertextually" by bringing different fictional universes together in fan creations (37). After examining some examples from video game modding communities, we used the tagging system of fanfiction depository Archive of Our Own to explore how fans connected the subjects of student's research papers to other universes. Fans also, however, also perform what the what media scholar Victoria Godwin (2018) calls, "liminal play," which she defined as a "a self-reflective process in which fans draw upon story worlds for self-expression." Fandom, and in particularly the purchasing and display of merchandise, provides fans the opportunities to explore and create their own identities out of the fictional universes provided by pop culture (59). In moving between worlds, fans, like players and social media posters, actively show their agency in a world shaped by consumer culture and mass media companies.

Agency and Choice in the American Multiverse?

As we explored each of the preceding themes, our conversations turned again and again to questions of agency, choice, and the possibilities of change. This is not surprising because nearly every element of the American Multiverse has been marketed as a form of consumer empowerment. That doesn't mean the multiverse can't be empowering, but it does mean that to truly understand its nature, we must examine the interplay between producers and consumers, structures and experiences. The multiverse isn't just a product foisted upon consumers by companies; its actively created as consumers navigate their own powers and identities amid the complexities of twenty-first century capitalism.

We ended the course with a reflection on two recurring conversations about agency that indicate the mixture of pessimism and optimism that pervaded our conversations throughout the semester. First, the pessimistic and nihilistic. Throughout the semester, we kept returning to the way that popular culture has often translated the multiverse into the idea that every choice creates a new universe, a narrative more prominently found in *Rick and Morty* (Egan 2023). What could be more indicative of the neoliberal ideology of choice than suggesting that every choice, regardless of any contingencies or power relations that might limit agency, creates its own separate reality. Of course, an alternative way of looking at such a metaphor—and one embedded within *Rick and Morty*—is that if every choice is critical enough to form its own universe, then no choice actually matters and nothing has consequences (Rick and Morty 2014, 17:10-18:10) Yes, such thinking suggested in our darker moments, we live in a multiverse of choice, but its not clear which choices, if any, actually mean anything.

And now, the optimistic. An interviewee in Joe Hunting's (2022) documentary *We Met in Virtual Reality* perhaps put the utopian possibilities of the multiverse the best:

> The ability to be lots of different avatars allows us to see different aspects of ourselves in a lot of different ways. In my case I am non-binary and being in VRChat means that I can run around as a space dog, as a … deer…as a Nargacuga, as…any number of different avatars that I collect across different worlds or that I make myself. And it doesn't mean that misgendering doesn't happen or that assumptions aren't made about gender or sexuality because they do. They certainly happen a lot, but there's a little bit of a sense of having just—just—just a little more control over how you might represent yourself and a little more fluidity in representation that is harder to achieve in real life because of the expectations of society and how people see you.
>
> *(41:00-43:00)*

It's a powerful, moving testament to the power of virtual worlds. But it also suggests a profound weariness at the failures of contemporary life to match the dreams of its rhetoric. It doesn't suggest that the speaker has given up on the physical world for they are explicitly using the documentary to argue for a more inclusive society. But it also isn't a solution, especially given the high pricing of VR systems. And therein lay the final questions we had to confront: is the multiverse an actual tool for social justice or simply a way to keep us consuming? Does the multiverse and what we do in it actually matter? I'm sure it's just a coincidence that that's also the central question of *Everything Everywhere All at Once* and one that each of us who navigate the American multiverse must answer. Hopefully, its one we don't try to answer on our own.

And that determination to not confront the multiverse on our own is ultimately what drives the course. As my students consistently expressed, they feel they are living in a terrifying moment in which American society lurches from crisis to crisis without resolving much of anything. They feel perpetually connected but always detached, powerless even though they are constantly told they are powerful. They might not worry about colliding universes, but they do worry about climate change, job prospects in an age of AI, and the future of their families in a culture that seems increasingly drifting toward violence. Learning about the American Multiverse doesn't solve those issues, but it does help them realize that they don't have to navigate this world alone.

References

Abnet, Dustin. 2020. *The American Robot: A Cultural History.* Chicago: University of Chicago Press.

Alexander, Connor. 2021. *Coyote and Crow: Core Rulebook v. 1.1.* USA: Coyote & Crow.

Betancourt, David"Marvel hired Gabby Rivera, a queer Latina writer, for its queer Latina superhero. That matters." *Washington Post*, March 8, 2017, https://www.washingtonpost.com/news/comic-riffs/wp/2017/03/08/marvel-hired-gabby-rivera-a-queer-latina-writer-for-its-queer-latina-superhero-that-matters/.

boyd, danah. 2014. *It's Complicated: The Social Lives of Networked Teens*, (New Haven: Yale University Press.

Chalmers, David J. 2022. *Reality+ Virtual Worlds and the Problems of Philosophy.* New York: W.W. Norton & Company.

Consalvo, Mia. 2009. "There is No Magic Circle" *Games and Culture* 4 no. 4, 408–417. https://doi.org/10.1177/1555412009343575

Egan, Toussaint. "Rick and Morty opened a portal to the multiverse and the rest of pop culture jumped through." *Polygon*, December 2, 2023. https://www.polygon.com/23983297/rick-morty-10th-anniversary-mcu-multiverse.

Eubanks, Eubanks. 2018. *Automating Inequality: How High-Tech Tools Profile, Police, and Punish the Poor.* New York: St. Martin's Press.

Fawaz, Ramzi. 2016. *The New Mutants: Superheroes and the Radical Imagination of American Comics.* New York: New York University Press.

Foucault, Michel. Alan Sheridan translator, 1995 second edition *Discipline and Punish: The Birth of the Prison*, New York: Vintage Books.

Godwin, Victoria L. 2018. "Hogwarts House Merchandise, Liminal Play, and Fan Identities." *Film Criticism* 42 no 2, 59–70. https://doi.org/10.3998/fc.13761232.0042.206

Greene, Brian. 2011. *The Hidden Reality: Parallel Universes and the Deep Laws of the Cosmos.* New York: Alfred A. Knopf.

Grossman, Lev. 2006. "You – Yes, You – Are TIME's Person of the Year," *Time Magazine*, December 25, 2006.

Hofstadter, Richard. 1964. "The Paranoid Style in American Politics." *Harper's Magazine.* https://harpers.org/archive/1964/11/the-paranoid-style-in-american-politics/.

Hon, Adrian. 2022. *You've Been Played: How Corporations, Governments, and Schools Use Games to Control Us All.* New York: Basic Books.

Hunting, Joe, dir. 2022. *We Met in Virtual Reality*, January 21, 2022.

Jeffries, Stuart. 2021. *Everything, All the Time, Everywhere: How We Became Postmodern.* London, New York: Verso.

Jenkins, Henry. 2006. *Convergence Culture: Where Old and New Media Collide.* New York: New York University Press.

Jenkins, Henry. 2013. *Textual Poachers: Television Fans and Participatory Culture.* New York: Routledge.

Kazan, Elia, dir. 1957. "A Face in the Crowd," May 28, 1957.

Kohler-Hausmann, Julilly. 2015. "Guns and Butter: The Welfare States, the Carceral State, and the Politics of Exclusion in the Postwar United States." *Journal of American History,* 102 no. 1, 87–99. https://doi.org/10.1093/jahist/jav239.

Marwick, Alice Emily. 2013. *Status Update: Celebrity, Publicity, and Branding in the Social Media Age.* New Haven, CT: Yale University Press.

McCulloch, Gretchen. 2019. *Because Internet: Understanding the New Rules of Language.* New York: Riverhead Books.

Miller, Katrina. "For Ytasha Womack, the Afrofuture is Now," *New York Times,* March 16, 2024. https://www.nytimes.com/2024/03/16/science/ytasha-womack-afrofuturism.html, accessed March 6, 2024.

Rick and Morty, season 1, episode 8, "Rixty Minutes," directed by Bryan Newton, written by Tom Kauffman and Justin Roiland, aired March 17, 2014, Amazon Prime Video, https://www.amazon.com/gp/video/detail/B0B8TNR5LJ/ref=atv_dp_season_select_s1.

Nguyen, C. Thi. 2020. *Games: Agency as Art.*(New York: Oxford University Press.

Page, Thomas. "Why the multiverse is the movie fantasy for our times," *CNN,* October 7, 2022. https://www.cnn.com/style/article/multiverse-movies-newfound-popularity/index.html.

Planas, Melissa Castillo. 2021 "Superhero Latinidad: The Diasporic Identitis of America Chavez and La Borinquena," *Journal of Popular Culture* 54, no 5., 1012–1030. https://doi.org/10.1111/jpcu.13055

Rasmussen, Mikkel Bolt. 2022. *Late Capitalist Fascism.* Medford, MA: Polity Press.

Rodgers, Daniel T. 2011. *Age of Fracture.* Cambridge, MA: The Belknap Press of Harvard University Press.

Rubenstein, Mary-Jane. 2014. *Worlds Without End: The Many Lives of the Multiverse.* New York: Columbia University Press.

Siegfried, Tom. 2019. *The Number of the Heavens: A History of the Multiverse and the Quest to Understand the Cosmos.* Cambridge, MA: Harvard University Press.

Spicer, Matt, dir. 2017. Ingrid Goes West, January 20, 2017.

Thomas, Ebony Elizabeth. 2019. *The Dark Fantastic: Race and the Imagination from Harry Potter to the Hunger Games.* New York: New York University Press.

29

REAFFIRMING JAPANESE IDENTITY THROUGH THE MULTIVERSE

A Response to Post-3.11 Uncertainties from *Your Name*

Erica Ka-yan Poon

Recent scholarly writings about the multiverse in films focus on Hollywood cinema, particularly superhero movies such as the Marvel Cinematic Universe (Proctor 2017; Berghofer 2018). However, the idea of a multiverse is not limited to Western media. This chapter speaks to current scholarship of the multiverse using anime *Your Name* (Makoto Shinkai 2016) as a critical example. *Your Name* achieved the third largest gross for a domestic film in Japan, behind the all-time favorite *Spirited Away* (Hayao Miyazaki 2001), which ranks second (Kōgyō Tsushinsha 2024). As of 2020, *Demon Slayer the Movie: Mugen Train* (Haruo Sotozaki 2020) is the highest-grossing film (Haruo Sotozaki 2020). It is important to note that *Demon Slayer the Movie* has a fanbase because it is part of an animated series based on comics. In contrast, *Your Name* is an animated feature-length film based on an original story. More importantly, *Your Name*'s allusions to the Great East Japan Earthquake on March 11, 2011 (3.11) make it culturally and historically significant. A fictional town called Itomori in *Your Name* vanished overnight because of the flooding of a lake caused by a meteor strike. This recalls the tsunami triggered by the earthquake that ravaged the towns along the coastline of the Northeast region of Japan. In *Your Name*, Itomori was still devastated years later, and blocked by barricades set up by the Reconstruction Agency. The scenes remind audiences of the area polluted by radioactive contaminants due to a nuclear power plant accident following the earthquake and tsunami. Some of those towns remained prohibited from entry by the Japanese government.

DOI: 10.4324/9781003480846-34

After 3.11, Director Shinkai considered "what if" as a new formation of thoughts that emerged in Japanese society (Andō 2017). What if the earthquake had not happened? What if I were you in such a situation? What if there were another me? This structure of feeling has inspired Shinkai's production of a multiverse in the narrative. *Your Name* follows a Shine maiden named Mitsuha Miyamizu, who lives in Itomori in the countryside, and a Tokyo high school boy called Taki Tachibana. The story begins with them waking up in one another's bodies and leading their counterparts' daily lives for a day, and then they revert during their sleep. Their body swap happens intermittently and involuntarily. Mitsuha in Taki's body enjoys a bustling metropolitan life in Tokyo, while Taki in Mitsuha's body appreciates the nature of Itomori. The plot twists when Taki tries to meet Mitsuha face-to-face and finds out that her death occurred three years ago. It is now revealed to the audiences that Mitsuha died in 2013 when a meteor struck Itomori, while Taki lives in 2016—the contemporary time of the diegesis. Strictly speaking, Mitsuha and Taki live in different universes. Taki's universe in 2016 is without Mitsuha because she has already died. Recalling his daily life as Mitsuha, whose family's religious and cultural traditions are rooted in Shinto, Taki realizes non-linear time and attempts to rescue Mitsuha and the whole Itomori town by traveling to the universe in which Mitsuha still survives. With Taki's assistance, Mitsuha manages to escape from the disaster. Hence, another universe with the continued existence of Mitsuha since 2013 is created. *Your Name* ends with a rendezvous between Taki and Mitsuha in 2021, although they have lost their memories of the body swap.

Although the idea of "what if" in *Your Name* offers a plausible impossibility and alternative realities, it is crucial to note that the multiverse in films can be limiting as they do not guarantee a suspension of ideological determination. This chapter argues that *Your Name* confirms Japanese identity by reassuring a sense of lineage from the past despite the multiverse narrative. On the one hand, *Your Name* incorporates Shinto's ideas to create different universes and temporalities. On the other hand, these concepts enable connections beyond time and space, allowing origins to be traced. This imagined continuation of history reinforces Japan's national discourse of unity for resolving post-3.11 uncertainties in society. The chapter will elucidate how *Your Name* approaches the multiverse, using the concept of mind-game films to analyze the narrative's causality and crisscrossing timelines. It will examine two critical moments in the anime that enable and justify Taki's journey to another universe. The first significant moment is when Taki transgresses the "border" between universes by drinking *kuchikamizake* (mouth-chewed sake/alcohol). The second pivotal moment is the *tasokare* (twilight) scene, in which Taki and Mitsuha meet face-to-face in their own bodies. Both moments are triggered by *Musubi*, meaning connection or referring to the deities of creation in Shinto.

Musubi *as the Agency for Causality of the Multiverse*

This section elucidates how *Musubi* causes and justifies the multiverse in the narrative. *Musubi* is a divine spirit in Shinto, which literally means the way of *kami* (deities). Deities can be spirits of places, natural forces, or figures of myth. Centered around the worship of these sacred powers, Shinto is one of the religious beliefs in Japan, and its doctrines of respecting living beings have been ingrained in Japanese culture (Hardacre 2017; Nobutaka, Teeuwen, and Breen 2003). Taki's journey to Mitsuha's universe occurs because he drinks Mitsuha's mouth-chewed sake at a cave-like repository where a deity resides. This follows the classical Hollywood narrative that the characters' actions advance the plot. However, the story revolves around the supernatural force behind the sake, namely *Musubi*. To create causal links between spaces and times, mind-game films "posit an agency inaccessible to the human actors and yet crucial to their story world" (Elsaesser 2021, 63). Quantum physics plays this role of agency in many films that involve the multiverse concept to engage with different possibilities. Given that such knowledge of science is incomprehensible to most audiences, it can fill the explanatory void (Elsaesser 2021). *Your Name* reserves this position of agency for *Musubi*, which also is beyond human understanding. First, *Your Name* deliberately showcases mouth-chewed sake through a scene in which Mitsuha dresses as a shrine maiden to perform a Shinto dance. Mitsuha's friends Tessie and Sayaka are chatting while watching Mitsuha's performance. Tessie explains that mouth-chewed sake is the oldest way of making alcohol by chewing on rice, spitting it out, and letting it ferment, while the visual is Mitsuha conducting this process on a Shinto stage. As depicted in this scene, mouth-chewed sake is not only a method of producing alcohol but also a Shinto ritual of dedicating it to deities (Wada 2015). The association with deities implies that a supernatural force behind the sake made by Mitsuha enables Taki's journey.

    Your Name further defines this unknown power as *Musubi*. When Taki (in Mitsuha's body) and the younger sister Yotusha offer the sake at the repository, Mitsuha's grandmother gives her grandchildren, as well as the audience, a lecture about *Musubi*:

Musubi is the old way of calling the local guardian god. This word has a profound meaning. Tying thread is *Musubi*. Connecting people is *Musubi*. The flow of time is *Musubi*. These are all the god's power. So, the braided cords that we make are god's art and represent the flow of time itself. They converge and take shape. They twist and tangle. Sometimes unravel, break, then connect again. *Musubi* – knotting. That's time.

(Shinkai 2016, at 34:49–36:25)

According to this scene, *Musubi* means knotting or god(s). *Your Name* draws the meaning of *Musubi* from *Kojiki* (Records of Ancient Matters), the oldest extant book written in the eighth century about Japanese myths, legends, and the imperial lineage. Many Shinto's ideas and customs are based on the interpretation of *Kojiki*. In this context, *Musubi* is the divine spirit(s) that establishes the world because it gives birth to everything on earth and in heaven. In short, *Musubi* is the origin of all life (Kōnoshi 1999, 89–94; and Nobutaka, Teeuwen, and Breen 2003, 34).

Along this line, *Your Name* assigns *Musubi* as the invisible but essential agency that interconnects the universes of Taki and Mitsuha. First and foremost, *Musubi* causes the two characters to meet through a fantastic body swap. After finishing the Shinto ritual dance, Mitsuha screams out her desire to be a handsome Tokyo boy in her next life at a *torii*, a gateway at a shrine that marks the transition from the secular to the sacred (Breen and Teeuwen 2010, 3). As if *Musubi* grants her wish at the *torii*, Mitsuha wakes up in Taki's body in the following scene. For Taki, instead of being a character whose actions propel a plot, he is "chosen." He gets involved because he is a cool Tokyo guy of the same age as Mitsuha.

Crisscrossing Timelines of the Universes

The body swap not only intertwines the lives of Mitsuha and Taki but also "opens up a potential confluence and coexistence of different timelines" (Elsaesser 2018, 21). It weaves different temporalities in the narrative: the 2013 Mitsuha in Taki's body leads a life in 2016 in one universe, while the 2016 Taki in Mitsuha's body survives in 2013. In other words, they are in distinct universes in which Mitsuha's present is Taki's past, while Taki's present is Mitsuha's future. Moreover, the sequence of Taki's search for the 2013 Mitsuha anchors Taki's journey as crossing universes, not only time travel. When Taki (in Mitsuha's body) and Yotsuha are about to enter the repository, their grandmother says *Anoyo* (underworld) is beyond this point. The underworld is the world for the afterlife and coexists with *Konoyo* (the world of present life). As *Anoyo* and *Konoyo* form one world, the absence of either will become a half-world (Iwata 2002, 300–1). The diegesis of *Your Name* omits an explanation as to why the mouth-chewed sake symbolizes "half of Mitsuha." Only by understanding the concept of a half-world can we make the inference about a lost half that might exist in another world. Taki repeats "half of Mitsuha" as a magic word before his transgression to the underworld. *Your Name* twists the idea of the underworld by adding the element of time travel so that Taki does not enter the afterlife world of Mitsuha but the one before her death. According to the convention of sci-fi genre films with a quantum multiverse, timelines converge rather than diverge when a character travels backward in time (Grazier and Cass 2015, 253–4). Following this agreement between genre and audiences, *Your Name* justifies that Taki can go to a particular moment in the past, that is, the day before Mitsuha died. However, moving forward in time will be along a different quantum path because the changes will create a new timeline, resulting in multiple universes (Grazier and Cass 2015). Therefore, Taki cannot return to his original universe in 2016; he exists in another one, and his memory about Mitsuha is erased.

However, the distinction between Taki's and Mitsuha's universes is difficult for the audiences to notice because *Your Name* has been based on a mistaken perceptual premise. The universes are not marked off as different by conventional film techniques such as change of color tone, soft focus, and superimposition (Elsaesser 2021, 92–94). The narrative strand of Mitsuha's universe and that of Taki's indeed have their respective status of reality (Elsaesser 2018, 19). For example, in Mitsuha's universe, a comet with an orbital period of 1200 years will pass by in a month and be visible to the naked eye. The TV at Mitsuha's house shows a news broadcast about this rare and exciting celestial phenomenon as the comet gets closer. Taki's universe has no such news due to different times and universes. Only after knowing the twist and watching *Your Name* again can the spectators understand why Taki in Mitsuha's body shows a confusing facial expression when Yotsuha talks about seeing the comet soon. Such a subtle hint is used to leave traces but not to mark off different realities.

The cinematic representation of their body swap further plays with the mistaken perceptual premise. It interweaves the timelines by generating an illusion that the two universes with distinct temporalities are happening simultaneously. Cinema creates "the spectatorial experience of presence"—spectators always experience moving images in the present tense (Doane 2002, 103). For example, even though flashbacks are related to the past in narratives, the images are presented to spectators at the current moment when they are watching the film. Flashbacks are indicated to spectators as the past by conventional means, such as color tone, by which cinema signals a timeline transition. In *Your Name*, the portrayal of one another's daily lives brought about by the body sway is condensed into a three-minute music video-like sequence. The sequence of their body swap plays with the spectatorial experience of presence by cross-cutting, dialogues, and music to reinforce simultaneity to conceal the three-year separation between the two worlds. It begins with Taki and Mitsuha screaming out their realization of switching bodies through their dreams on a split screen. The split screen that joins two scenes in distant places and times within a frame enhances a sense of concurrence. Then, the sequence depicts their parallel experiences. The shot in which Taki wakes up in Mitsuha's body is cross-cut with that of Mitsuha waking in Taki's. Cross-cutting edits separate scenes to establish actions that co-occur in a narrative. Also, their voiceover accompanies the cross-cutting of these scenes. The voiceover is in a dialogue format to intensify the feeling that Taki and Mitsuha are having a conversation without delay of time. For instance, when Taki complains that Mitsuha uses up his savings on desserts at cafés, it seems that Mitsuha replies to him immediately to defend that it is his body eating. The non-diegetic theme song additionally gives an impression of the uninterrupted passage of linear time, while the timelines of Taki and Mistuha are crisscrossing.

Despite the illusion of simultaneity, *Your Name* provides cues about different temporalities by visualizing the crisscrossing timelines of two universes using the motif of *kumihimo* (braided cords). When the grandmother explains *Musubi*, a close-up of braided cords is inserted (Figure 29.1). The denotation of the Japanese word *musubi* means tying threads, and the close-up showing the threads' weaving patterns matches

FIGURE 29.1 The motif of braided cords visualizes the crisscrossing timelines of the two universes.

this meaning. Additionally, the connotation of *musubi* means connection, and *Your Name* refers explicitly to the connection in terms of time. As a result, the close-up of braided cords also functions as a visual sign representing the concept of time. On the one hand, the shot visualizes the linear flow of time as a slight pan of camera movement follows straps of braided cords converging at the corner of the frame. On the other hand, the close-up of braided cords makes the confluence of timelines visible, like the twisting and tangling pattern of threads. The patterns of braided cords are out of focus, in focus, and again out of focus. This represents time unraveling, connecting, breaking, and converging. Therefore, time is represented not only as a linear progression but also as coexistent and interconnected relations. Further situating *musubi* in the diegesis of *Your Name*, the force behind these interweaving timelines of human life is the *Musubi* God. The grandmother designates that the braided cords are god's art. In short, the braided cords visualize the complicated temporal relations caused by *Musubi*, the sacred power.

Musubi, *Where the Origins Are Traced*

The motif of braided cords not only visualizes time but also symbolizes the invisible *Musubi*. The signification occurs in the scene that shows the first intersection of the timelines between Mitsuha and Taki. The scene concerns their initial face-to-face encounter in 2013 when Mitsuha travels to Tokyo to look for him. However, this Taki in Mitsuha's universe has not met her and cannot recognize her. For Taki, his body swap happens in 2016, and he is back in 2013 in another universe. Although Mitsuha runs into Taki on a train, he treats her indifferently. When Mitsuha is about to get off the train, Taki asks Mitsuha her name. Mitsuha answers while throwing him a braided cord, which she always uses to tie her hair. Taki catches the braided cord and never takes it off, even though he later forgets how he has received it. The braided cord is an indexical sign of Mitsuha because she always wears one, which has become inseparable from and part of her. In this sense, Mitsuha's throwing her braided cord can be imagined as extending her arm to connect with Taki. His catch of the braided cord then forms the linkage with Mitsuha. Given that the braided cord signifies Mitsuha, the connection between Taki and Mitsuha is less about this moment on the train than her whole life, that is, their intertwined fates.

As a result, when Taki travels across the universe to find Mitsuha, he is shown certain impactful moments of Mitsuha's life during the liminal space between two universes. This scene further reinforces the meaning of *Musubi* as the origin of life in a Shinto context. Curvy lines such as comets, waterspouts, and umbilical cords are other forms of braided cords that symbolize connection. After drinking the mouth-chewed sake, Taki is drawn to transgress to Mitsuha's universe. In the sequence depicting Taki's transgression, a comet moves swiftly through space, and waterspouts engulf Taki. The braided cord around Taki's wrist becomes loose and transforms into a comet, symbolizing the comet as another form of a braided cord. As the comet's nucleus splits, it turns into body cells. Subsequently, Mitsuha's umbilical cord is cut, followed by a shot of Taki's braided cord appearing to be cut as well, causing him to fall and float in the waterspouts. By showing body cells, umbilical cords, and Mitsuha's crucial upbringing moments, the sequence explicitly refers to life's origin. Various forms of braided cords in the sequence symbolize *Musubi*, and the cross-cutting further visualizes the intertwining relations between Taki and Mitsuha from the birth of life.

Despite a successful trip to Mitsuha's universe in 2013, Taki cannot meet Mitsuha. He only appears in her body. This passing each other makes sense because the 2016 Taki and the 2013 Mitsuha are not supposed to exist in the same universe. *Your Name* justifies their face-to-face encounter in their own bodies, that is, the confluences of different timelines and universes by the *tasokare* (twilight) scene, in which Taki communicates with Mitsuha about the rescue plan. Again, *Your Name* draws from traditional Japanese culture to make their coexistence possible. "Twilight" is a term from *Manyoshu*, a collection of classical Japanese poetries. To make it comprehensible to audiences, *Your Name* adds a scene in which a teacher explains *tasokare* to Mitsuha and her classmates. "*Tasokare* means 'Who is that' and is the origin of the word '*tasokare-doki*.' Twilight: when it's neither day nor night. When the world blurs and one might encounter something not human" (Shinkai 2016, at 08:49–09:26). "Twilight" in the context of classical Japanese is not a transition from day to night but *neither day nor night*. In this liminal time and space, another person cannot be recognized, leading to the meaning of "Who is that." It is essential to point out that *ōmagatoki* (twilight, but literally, it means encountering evils) is written on the blackboard in the scene. Although the teacher has not explained this term, she has mentioned that this twilight moment enables an encounter with spirits. Therefore, Taki can meet her in person, even though the twilight scene portrays a background of the vanished Itomori to remind audiences that Mitsuha is supposed to have died by this moment of encounter.

Due to the possibility of coming across spirits during this liminal twilight time, *Your Name* hints that the supernatural force *Musubi* causes Taki's and Mitsuha's encounter. More importantly, this twilight scene and the rescue sequences reveal the origin of the town, Itomori. *Musubi*, as the origin of life, refers not only to the birth of Mitsuha but also to the beginning of Itomori. Mitsuha's grandmother explains that *Musubi* is Itomori's local guardian god, but documents about its relationship with Itomori are lost in a big fire. However, *Your Name* suggests by visual cues of the landscape portrayal that *Musubi* caused the emergence of Itomori. The twilight scene portrays two crater-like lakes (Figure 29.2). The comet devastated Itomori in 2013 and formed a crater-like lake resembling the one beside it. It can be inferred that another crater-like lake was struck by the previous comet 1,200 years ago because, in the story, the comet that hit Itomori in 2013 has an orbital period of 1,200 years. The first strike nourished an environment with water and enabled people to live and form a new community, Itomori, while the second strike ruined the town. *Musubi* resides in a repository located in a crater-like basin. If comet strikes create crater-like shape landscapes, a comet strike thousands of years ago has also caused this basin where *Musubi* inhabits. Simply put, *Musubi* came to live in this basin. Right before Taki's "ride" to another universe, he slips and sees the mural painting of comets inside the repository. Paintings and statues in Shrines to represent deities are not uncommon under the influence of Buddhism, although deities in Shinto are not anthropomorphic. They can be spirits of natural forces such as trees, rocks, and, in this case, a comet as painted in the repository. In short, *Musubi* in *Your Name* is the spirit of a comet that has created and nourished Itomori.

FIGURE 29.2 Visual cues of two crater-like lakes imply that *Musubi* caused the emergence of Itomori.

While having the power to destroy Itomori, *Musubi* protects the town by delivering messages to its chosen residents. When Taki appears in Mistuha's body after traveling across the universe, Mitsuha's grandmother suddenly realizes that the person is not Mitsuha. This triggers memories of her body swap from her teenage years. The scene reveals that the maternal side of Mitsuha's family had body swap experiences. A body swap is a way for *Musubi* to message every generation of Mitsuha's family to protect Itomori against the disaster in 2013. In this sense, the genealogy of Mitsuha's family, who runs the local Shrine that sustains the culture of mouth-chewed sake and braided cords, can be traced from *Musubi*. In sum, the twilight sequence is essential not only because it drives toward the narrative closure but also because it reveals that *Musubi* is upon which the genealogy of Mitsuha's family and the origin of Itomori are predicated, creating a sense of lineage from the past.

Conclusion: A Response to Post-3.11

This chapter ends with a brief discussion about *Your Name*'s ideological implications related to 3.11. Japan's national discourse of unity aims to resolve post-3.11 societal uncertainties. After 3.11, the Japanese government mobilized *kizuna* (bond) for disaster recovery and national resilience (Koikari 2005, 4). *Kizuna* initially referred to personal bonds but has been used to sentimentalize the public relationships in the nation since 3.11 (Mihic 2020, 13). *Kizuna* was also publicized as a national characteristic to be proud of (Mihic 2020, 14). To unite its citizens after the disaster, the government attempted to reaffirm its citizens' identity by ascertaining their connection, grounded in a nation descended from a legitimate past.

Your Name resonates with the national discourse about a continuation of history with an acknowledged origin. First, a linear storyline establishes a sense of continuation from the past. Despite different universes and temporal relations, it is critical to emphasize that *Your Name* does not disrupt a linear storyline. As analyzed, *Your Name* utilizes *Musubi* to make possible the connection beyond time and space. After understanding *Musubi*'s role of causality in the multiverse narrative, the story of *Your Name* can actually be described linearly: *Musubi* has created Itomori while knowing it has the power to unintentionally destroy it someday. Therefore, it tells Mitsuha's family generation after generation to avoid the disaster by letting them foresee the future through body swaps. When its message reaches Mitsuha, she and Taki finally save Itomori.

Second, *Your Name* echoes the Shinto myths about the ancestry of Japan. Shinto myths tell stories of how *Musubi* and other gods created

Japan. The sovereign has been passed on to the Japanese emperors, and the lineage has continued (Ō No 2014). Approaching the multiverse through Shinto's ideas, *Your Name*'s story, in which *Musubi* created Itomori and is where Mitsuha's family traditions originated, is paralleled with the national discourse about a justifiable genealogy. Given that myths offer an imaginary resolution to societal contradictions (Lévi-Strauss 1955, 443), the imagined past with an origin serves to strengthen Japan's call for its citizens to be united for resilience after the national trauma.

Finally, revealing that *Your Name* conceals a bleak universe in which Mitsuha has died is significant to demonstrate a multiverse narrative may not provide room for ideological negotiation. The finale of *Your Name* only portrays the universe with Taki and Mitsuha living in 2021. They have lost memories of encountering each other in other universes but feel the need to search for each other. Their rendezvous occurs at places symbolizing connection, such as bridges, trains, and staircases. They finally run into each other face-to-face on the staircases to a shrine and ask each other for their name. Ending credits then follow. However, another universe in which Mitsuha died in 2013 has not disappeared. In fact, the story is based on Taki's acknowledgment of Mitsuha's death to propel the plot. This disheartening universe is concealed in the narrative to provide a happy ending and coherent story. The coherency limits other possibilities in different universes, which are supposed to be offered by a multiverse narrative. Therefore, *Your Name* does not provide alternatives to negotiate with the dominant post-3.11 discourse about a unified future.

References

Andō, Kenji. 2017. *"Kiminonaha Shinkai Makoto kantoku ga kataru: 2011 nen izen to wa, minna ga motomeru mono ga kawatte kita* [*Your Name* Director Makoto Shinkai says, What people want has changed after 2011]." *Huff Post Japan Edition.* Last modified February 3, 2023. https://www.huffingtonpost.jp/2016/12/20/makoto-shinkai_n_13739354.html

Berghofer, Philipp. 2018. "Doctor Strange, the Multiverse, and the Measurement Problem." In *Doctor Strange and Philosophy*, edited by Mark D., 151–63. Oxford: Wiley Blackwell.

Breen, John, and Mark Teeuwen. 2010. *A New History of Shinto.* Malden, MA: Wiley-Blackwell.

Doane, Mary Ann. 2002. *The Emergence of Cinematic Time: Modernity, Contingency, the Archive.* Cambridge, Massachusetts, and London: Harvard University Press.

Elsaesser, Thomas. 2018. "Contingency, Causality, Complexity: Distributed Agency in the Mind-Game Film." *New Review of Film and Television Studies* 16 (1): 1–39. http://doi.org/10.1080/17400309.2017.1411870.

Elsaesser, Thomas. 2021. *The Mind-Game Film: Distributed Agency, Time Travel, and Productive Pathology.* Edited by Warren Buckland, Dana Polan, and Seunghoon Jeong. London and New York: Routledge.

Grazier, Kevin R., and Stephen Cass. 2015. *Hollyweird Science: From Quantum Quirks to the Multiverse.* New York and London: Springer.

Hardacre, Helen. 2017. *Shinto: A History.* Oxford: Oxford University Press.

Iwata, Keiji. 2002. *Kami to Kami: Animizumu Uchūnotabi* [Kami and God: Animistic Space Journey]. Tokyo: Kōdansha.

Koikari, Mire. 2005. "Gender, Power, and U.S. Imperialism: The Occupation of Japan, 1945–1952." In *Bodies in Contact: Rethinking Colonial Encounters in World History*, edited by Tony Ballantyne and Antoinette Burton, 342–62. Durham and London: Duke University Press.

Kōnoshi, Takamitsu. 1999. *Kojiki and Nihon Shoki: Ten'nō Shinwa no Rekishi* [Records of ancient matters and the chronicles of Japan: History of the emperor myth]. Tokyo: Kōdansha.

Lévi-Strauss, Claude. 1955. "The Structural Study of Myth." *The Journal of American Folklore* 68 (270): 428–44. https://doi.org/10.2307/536768.

Mihic, Tamaki. 2020. *Re-Imagining Japan after Fukushima.* Action: Australian National University Press.

Miyazaki, Hayao. 2020. dir. 2001. *Spirited Away.* Tokyo: The Walt Disney Company (Japan) Ltd.

Ō No, Yasumaro. 2014. *The Kojiki: An Account of Ancient Matters.* Translated by Gustav Heldt. New York: Columbia University Press.

Nobutaka, Inoue, Mark Teeuwen, and John Breen. 2003. *Shinto: A Short History.* London and New York: Routledge Curzon.

Proctor, William. 2017. "The Quantum Seriality of the Marvel Multiverse." In *Make Ours Marvel: Media Convergence and a Comics Universe*, edited by Matt Yockey, 319–45. The University of Texas Press. Texas.

Shinkai, Makoto, dir. 2016. *Your Name.* Tokyo: Toho, 2017. DVD.

Sotozaki, Haruo. 2020. dir. 2020. *Demon Slayer the Movie: Mugen Train.* Tokyo: Aniplex.

Kōgyō Tsushinsha. 2024. "*Rekidai rakingu* [All time ranking]." Accessed April 6, 2024. https://www.kogyotsushin.com/archives/alltime/

Wada, Miyoko. 2015. *Nihonshu no kagaku: Mizu, kome, kōji no dentō no waza* [Science of sake: Traditional techniques of water, rice, and yeast]. Tokyo: Kōdansha.

30

HOW TO TRAVEL THE MULTIVERSE

A Text World Theory Guide

Matthew Voice

In spring 2014, during my Master's degree, I took a module entitled "Cognitive Poetics", which introduced an approach to literary linguistic analysis informed by cognitive science Our reading one week consisted of a journal article which, to this day, holds a special place in my heart for having one my favourite titles for a serious piece of academic research: "What Does Batman Think of Spongebob?" (Skolnick and Bloom 2006a). In it, the authors report a study in which children were asked whether characters from different series were fictional, as well as their relationships with one another. The tasks show that as well as being able to tell the difference between fantasy and reality, even preschool-aged children are capable of distinguishing between different fantasy worlds. In other words, they know that while Batman and Spongebob Squarepants are both fictional, they are fictional in different ways to one another, and occupy distinct fictional universes.

That a clear fantasy/fantasy distinction is present at such a young age leads Skolnick and Bloom to describe this as "our natural default, when exposed to a novel story" (2006a, B10), and this sentiment has been echoed elsewhere by prominent scholars of narrative. Most notably, Ryan (2006) observes that "for most of us, the idea of parallel realities is not yet solidly established in our private encyclopedias, and the text must give strong cues for us to suspend momentarily our belief in a classical cosmology" (671). In this view, the boundary between fictional universes—though permeable—is presented with an expectation of stability. As the present volume of collected

DOI: 10.4324/9781003480846-35

essays and creative works shows, however, the popularity of works in which these ontological expectations are challenged is expanding exponentially.

For instance, around the same time as I was first engaging with this research, I was also reading the recently published *Deadpool: Killustrated* (Bunn 2013). In this series, the eponymous antihero fights and kills literary heroes from across the Western canon, beginning with Don Quixote and Moby Dick. Concerned by the spree of literary killings, Sherlock Holmes and Beowulf team up to investigate, and confront Deadpool in a grand chase across literary worlds. For readers of this series, an understanding of parallel realties as in some way permeable is a functional necessity for making sense of this kind of plot which, while tongue in cheek, is not unusual among comic book narratives in its collision of universes (cf. Kukkonen 2010; see also Davis in this volume).

It takes only a cursory look at the recent box office to see the boundaries between fictional universes are frequently crossed in contemporary fiction. Major comic-inspired films like *Spiderman: Across the Spiderverse* (2022) and *Doctor Strange in the Multiverse of Madness* (2022) take the intersection between parallel universes as integral to their plots, while *Everything Everywhere All at Once* (2022) presented an original multiverse narrative to critical acclaim, winning numerous Oscars including Best Screenplay. In other media, the videogame *Fortnite* has been home to endless crossovers incorporating characters from multiple fictional universes, and has even been used as the site of canonical narrative development for the most recent *Star Wars* trilogy (Good 2019). Given the cultural popularity and commercial success of these works and their worlds, it is perhaps time to suggest that our private encyclopedias are in the process of being updated: for many members of a contemporary audience, the multiverse is an established, familiar concept.

But what does this surge in the popularity of multiverse fiction mean for the fantasy/fantasy distinction? Skolnick and Bloom recognize these sites of crossover and interaction as potential challenges for the understanding of fictional ontologies. Following Walton (1990), they suggest that one potential explanation for the relationship between these fantasy/fantasy distinction violations "is that these creative innovations involve the creation of a novel world" (2006b, 81). In other words, if I tell a story in which Batman and Spongebob are friends, that story is not set in the fantasy world you would typically evoke for stories containing either Batman or Spongebob, but rather an entirely new world with shared characteristics from both.

World creation is a part of all storytelling, and Text World Theory (Werth 1999; Gavins 2007) sets out the ways participants in any given discourse conceptualize these worlds, their contents, and their relationships between one another. Applying this theory of discourse to multiverse fiction, this chapter outlines and explains the cognitive processes through which readers and viewers make sense of multiverse narratives, and how these qualities might contribute to their enjoyment as literary texts. In doing so, it also establishes a typology of multiverse world-switching, providing a means to understand and delineate between the various cognitive processes that are currently bundled together in the guise of a singular notion of multiverse fiction.

What is Text World Theory?

First proposed by Werth (1999), and expanded by Gavins (2007), Text World Theory is a cognitive linguistic framework that describes the ways in which readers conceptualize knowledge and context, and interact through language. All discourse involves the creation of text-worlds, which mark the boundaries of conceptual knowledge relevant for understanding the meaning of a particular utterance, from single sentences to a novel-length texts. Text-worlds themselves are composed through the combination of "world-building" and "function-advancing" elements. World-builders can be thought of as elements that flesh out the details of the setting: cues as to the world's time, place, and occupants. Function-advancers, on the other hand, describe the actions and events that occur within a given text-world which "propel a discourse forward" (Gavins, 2007: 56).

So while most stories are set in one universe, they contain many different text-worlds, and switching between text-worlds is a natural part of both everyday storytelling and the way audiences make sense of them. In the anecdotal narrative which opened this chapter, the initial text-world builds a world set in a postgraduate classroom 2014, before briefly enacting a temporal world-switch to the present ("to this day"), then switching again to a narrative summary of Skolnick and Bloom (2006a). The relationship between these text-worlds is mapped in Figure 30.1, which shows the various worlds, settings, enactors, and actions which comprise the story. As can be seen, different text-worlds relate to one another in different ways: for example, text-world 4 reports a hypothetical imagining within the events of text-world 3, which in turn is embedded in the reading that takes place in text-world 1.

Text World Theory developed through the study of written discourse, and the application of the framework to multimodal discourses like film and television "remains a developing area" (Gibbons and Whiteley, 2021: 113). In these contexts, world-building and function-advancing are achieved

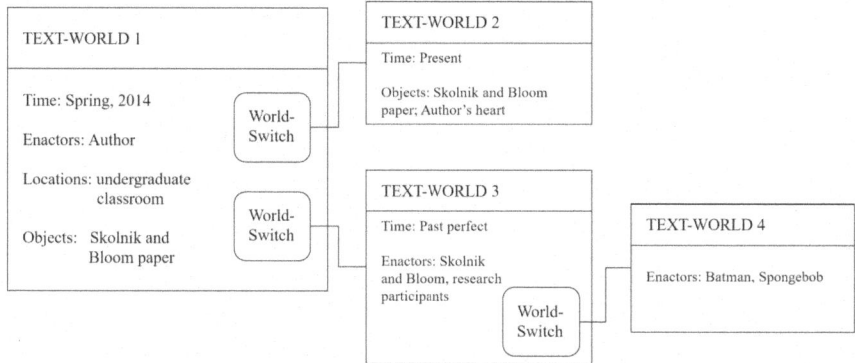

FIGURE 30.1 The Text-worlds of the Anecdote from this Chapter's First Paragraph.

not only through language, but through audio and visual cues, like cuts between camera shots (cf. Toolan, 2001; McIntyre, 2008). Though she does not reference Text World Theory directly, Kukkonen (2010) shows how in multiverse comic series like *Crisis on Infinite Earths* (Wolfman 1985), an iconography of visual differences in costume provides readers with different mental models (Johnson-Laird 1983) to cue the recognition of different versions of a character, and the universe to which they belong. In bringing Text World Theory to multiverses fiction, this chapter continues the development of a cognitive modeling of multiverse ontologies, and demonstrates the utility of cognitive approaches to literary linguistic scholarship for readers interested in mapping, describing, and making sense of the structure of as-yet unexplored multiverses for themselves.

Text Worlds in Multiverse Fiction

In fiction, there are many types of universes that can interact in many different ways. In fact, what constitutes "multiverse fiction" even differs between theorists. Boillat (2022) includes "mental" and "virtual" worlds (i.e., dream sequences and virtual realities, respectively) in his typology of styles of multiverses in fiction. These categories describe a certain type of movement between worlds, and such transitions can be accounted for through a process known as deictic shifting (Segal 1995). Think of the first time viewers of *The Matrix* (1999), following Neo's point of view, "pop" out of the simulated reality in which the protagonist has lived his life so far and into a new world occupied by the machines which produce the simulated Matrix. When characters jack back into the Matrix, the narrative "push" moves the viewer's attention back away from the framing "real world," and into the world of the simulation. In these instances, the narrative shifts between worlds which are ontologically distinct, with one existing inside of the other, and both occupying the same narrative universe.

In multiverse fiction as I define it here, the interactions between worlds do not involve a deictic "pop" or "push" between different ontological statuses. Rather, they exist at intersections: parallel worlds merging, combining, or existing side-by-side as "equally actualised realities" (Kukkonen 2010, 40). Even so, not all multiverse fiction involves the same kind of interaction between worlds. The task of the Text World Theory model below, then, is to provide a language through which to explain how readers conceptualize and make sense of these competing fictional realities.

Parallel Universes (or, Across the Multiverse)

Although the examples of multiverse fiction discussed above involve the collision and interaction between and across universes, not all multiverse fiction is quite so interactive. Consider, for instance, the film *Sliding Doors* (1998). Here, viewers are presented with multiple competing timelines of events, with no clear indication that any one represents a "true" version, or "textual actual world" (Ryan 1991, 113). While all multiverse fiction as defined for this chapter contains multiple universes which are ontologically equal, these "parallel universes" do not interact with one another in the way that characters and settings in other types of multiverse fiction are seen to cross between boundaries. Rather, the events and characters of each universe remain separate and unaware of each other, with only the viewer or reader aware of and moving between each world. Though perhaps less common as a style of multiverse fiction, these parallel universes provide a relatively straightforward starting point for a Text World Theory mapping of multiverses and their differing relationships.

The diagram in Figure 30.2 gives a general overview of the primary text-world structures accessed in a parallel universe narrative like *Sliding Doors*. On the far left, the model shows the discourse-world, representing the active participants in real-world discourse, for example, a speaker and listener, or author and reader. For *Sliding Doors*, or any other film, this is the interaction between actors, screenwriters, and directors as creators of the film as a text, and the viewer as its consumer. The initial world-switches into the primary text-worlds of the fiction take place in the discourse-world, as these changes are only accessible to participants in the discourse, and not to fictional enactors within the text-worlds themselves. While many world-switches involving changes of time, setting, and perspective will take place within text-worlds 1 and 2 over the course of the film, this

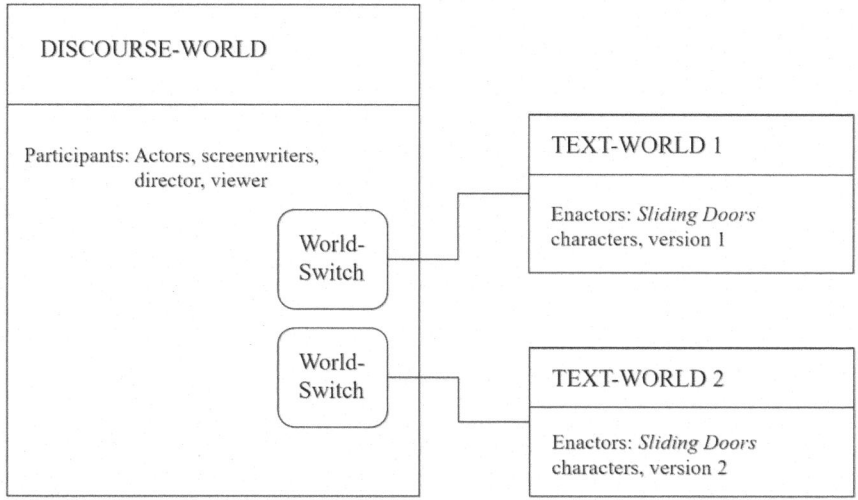

FIGURE 30.2 Parallel Text-worlds in *Sliding Doors.*

mapping shows how these text-worlds run in parallel to one another, available only to participants with access to the discourse-world, that is, the reader/viewer and the narrative's creator(s).

A further example of this kind of parallel multiverse fiction comes from Paul Auster's 2017 novel *4 3 2 1*. From the outset, four parallel versions of the protagonist Ferguson through parallel chapters recounting the same period of each of their lives (i.e., Chapters 1.1, 1.2., 1.3, 1.4 covering the early childhood of each). Although all four versions of Ferguson are presented equally at the outset, the reader later learns that the Ferguson in narrative 4 decided to "invent three other versions of himself and tell their stories along with his own" (862). With this revelation, the reader's conceptualization of the text-world ontology of the *4 3 2 1* multiverse necessarily shifts: rather than all four possibilities being equally real, following the structure presented in Figure 30.3, the world occupied by Ferguson 4 instead becomes the initial text-world which anchors the other three, with narratives 1, 2, and 3 now conceptualized as fictional *within* the text-world of narrative 4. The re-conceptualization of the reader's understanding of the relationship between these text-worlds is known as "world repair" (Gavins 2007, 141), in which new information requires the reassessment of prior knowledge about any element of text-world contents or relationships. As readers and viewers of multiverse fiction engage further with a given text, they may be required to revise their knowledge of the relationship between different text-worlds, and perhaps different universes.

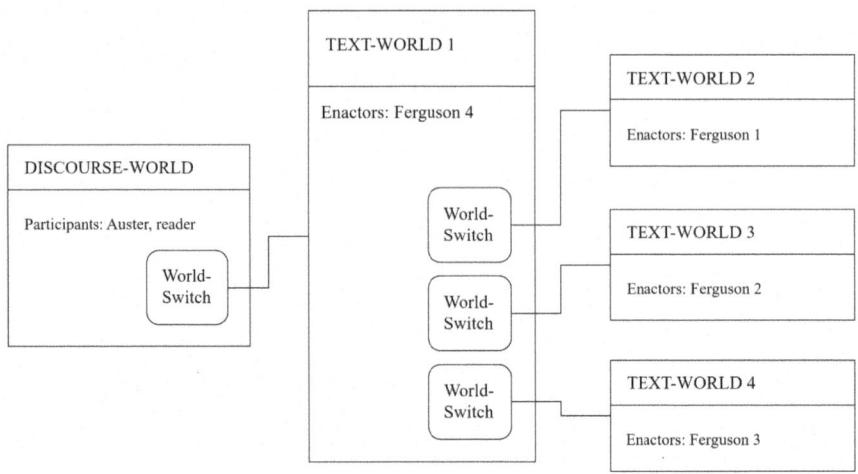

FIGURE 30.3 Revised Text-world Relationships in *4 3 2 1*, following World Repair.

World Hopping (or, Through the Multiverse)

Our ability to conceptualize different worlds is an essential part of understanding almost all stories. Many of these world-switches are subtle; a single adverb might cause a temporal shift, a pronoun might indicate a new point of view, while a conditional or negated clause might briefly require readers to imagine a range of hypothetical propositions even in the most conventional of narrative structures. How a reader or viewer might conceptualize the movement between universes in multiverse fiction, then, can be understood in Text World Theory as a particular kind of world-switch.

Fortunately for those of us interested in identifying and analyzing these switches, such transitions tend to be highly dramatic, and include clear visual or textual cues, unless the reveal of the transition is deliberately delayed for dramatic purposes. In a visual medium like films or comic books, such a transition might be marked visually, such as by Doctor Strange's portals, or audibly as the whirring of The Doctor's TARDIS often indicates. In *This is How You Lose the Time War* (El-Mohtar and Gladstone 2019) chapter breaks are used as a consistent marker of this transition. While the reader follows the same characters as enactors throughout the novel, they come to understand that other world-building elements around them (temporal continuity, in particular) change as the characters move between timelines. For stories that involve a particular emphasis on the exploration of a variety of potentially contradictory narrative universes, the process of following characters through a linear journey across worlds potentially "reduces the cognitive load" (Kukkonen 2010, 43) for readers by presenting these transitions in a clear, narrative sequence.

Consider, for instance, the first time Evelyn is transported between worlds in *Everything Everywhere All At Once (EEAAO)*. As she follows the instructions and presses the green button on her earpiece, the viewer watches from a fixed external perspective as the world around her distorts, with an accompanying shift in the pitch and pace of the music. Moments later, the viewer follows from the same perspective as she is launched backward, screaming as she is physically dragged away from the world she initially occupies. At this point, the camera "cracks" to show a split perspective: as the Evelyn we have followed to the janitor's closet looks around, the viewer also sees her sat in the initial position from which the shot began. Both Evelyns speak concurrently, and a dual audio of her voice further cues the reader to consider their similarity yet separation. As the camera pans, the viewer briefly joins the perspective of the Evelyn who remains at the tax inspector's desk, before physical intervention from Alpha-Waymond (the version of her husband who exists in a different universe) pulls at Evelyn. The camera turns, a graphic effect shimmers across the screen, and the scene completes its world-switch to the janitor's closet, with the viewer fully able to conceptualize Evelyn as present both at the inspector's desk and inside the closet simultaneously.

Dramatic transitions such as this have a dual purpose. First, the whiplash of the transition and its initial resistant return to the text-world of the initial shot simulate Evelyn's own disorientation, and resistance, to her first experience of "jumping." But it also serves to prime the viewer: a sustained and exaggerated focus on the experience of moving between text-worlds separated by ontological barriers beyond the conventions of time, space, and perspective makes future transitions of a similar nature more readily understood. While the concept of multiverse fiction might be increasingly familiar to a general audience, the continued presence of "strong cues" (Ryan 2006, 671) for this kind of world-switch foregrounds and prolongs the transition process, giving viewers time to adjust to a potentially unfamiliar conceptual transition. Indeed, by the time the film ends, montage sequences switch quickly between different Evelyns in different universes, with a sustained and consistent heterodiegetic soundtrack accompanying them. As the viewer becomes accustomed to the rules of *EEAAO*'s fiction, they require less cueing to interpret this kind of world-switch.

As well as temporal and spatial switches, Text World Theory also recognizes hypothetical worlds. "Modal-worlds" are a readily understood concept for most speakers: understanding the proposition "I would like to go for a walk" requires the conceptualization of an imagined situation, that is the modal-world in which the speaker goes for a walk, regardless of whether this actually occurs. Something potentially quite special about multiverse fiction, then, is the capacity for dramatic irony through text-world parallels. In one scene of *EEAAO*, in which Evelyn never emigrated with Waymond and instead became a movie star, Waymond says to her that "In another life, I would have really liked doing laundry and taxes with you." Situated at in a moment of particularly heightened tension in both the narrative and Evelyn and Wayond's relationship, this line and the surrounding scene is frame as a potential site of resonance: "a feeling of the affective power of an encounter with a piece of literature" (Stockwell 2009, 28). While resonance with literature is highly subjective, a Text World Theory model of the reader's relationship with the text-worlds and universes connected to this dialogue helps to account for the particular potential for resonance created here.

Figure 30.4 models the relationship between the modal-world enacted by Waymond's hypothetical statement and the previously accessed text-world which serves as *EEAAO*'s primary setting. The scene begins in the "original" universe in which the film opens, and which anchor's the viewer's sense of reality throughout. As Waymond pleads with Evelyn to control her powers, the camera cuts quickly between a five different shots of Waymond, typically from Evelyn's point of view. These visual cues evoke temporal world-switches, showing brief flashbacks to moments in the original universe's past, in which Evelyn and Waymond have shared happiness in their everyday lives. Another camera cut takes the scene briefly back to

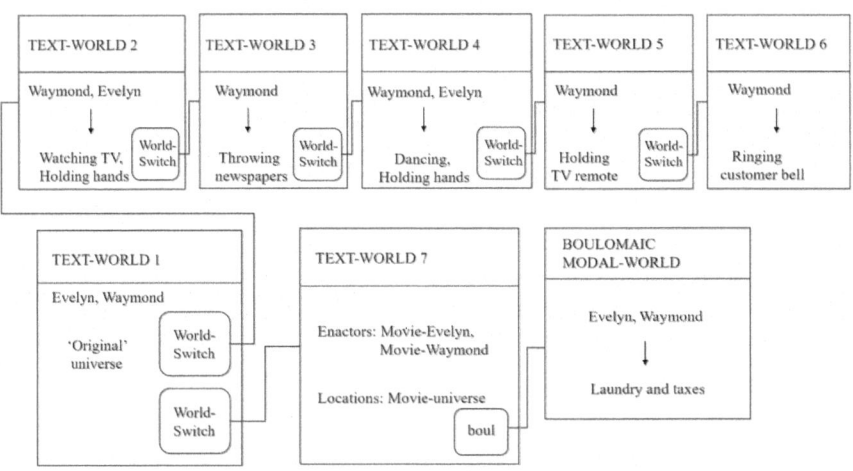

FIGURE 30.4   The Text-worlds of the "Laundry and Taxes" scene in *EEAAO*.

the present, before cutting again to the move-universe in which an alternate version of the two characters are in conversing.

Although movie-Waymond's modal-world is enacted only briefly, the few world-building elements and function-advancing processes (represented via arrows) incorporated into this line of dialogue closely parallel the central features to which the viewer's attention is drawn in the opening scenes of the film: Evelyn and Waymond preparing their taxes above their laundrette. Gavins (2007) and Gavins and Simpson (2015) note that the accessibility of a text-world is central to its perceived reliability. In this case, whereas access to a conventional modal-world relies on the enactor (movie-Waymond), and the viewer's evaluation of its truth is predicated on their trustworthiness, the multiverse narrative of *EEAAO* and its world-switching cuts between these universes foregrounds a comparable text-world which shares the values expressed in movie-Waymond's modal-world. Literary resonance in this case, then, arises from the viewer's conceptualization of this proposition not just as a hypothetical expression of belief, but as a world that has been independently accessed, and which the viewer knows to be true: in another life, Waymond really would be happy doing laundry and taxes with Evelyn.

World-Building Multiverses (or, Around the Multiverse)

The previous examples of multiverse fictions have considered world-building in contexts where readers and viewers are introduced to characters and settings for the first time within the parameters of the fiction with which they are engaging. But in some cases, crossovers occur with characters from pre-existing universes, breaching the fantasy/fantasy distinction boundary identified by Skolnick and Bloom (2006a). To make sense of *Deadpool: Killustrated*, for instance, readers are expected to recognize Sherlock Holmes and Beowulf as iconic heroes of the literary canon. However, doing so does not necessarily involve a the kind of switching between text-worlds explored above. Rather than the reader conceptualizing a perspectival movement across worlds within a multiverse, a sense of multiversal interconnectedness may be established via the sense that characters themselves have crossed not just between text-worlds, but across the boundaries of previously separated fictions.

Coincidentally, Stockwell (2020) discusses Sherlock Holmes as an example of a character who has particularly frequently appeared in a range of fictional contexts beyond his origins in Conan Doyle's detective stories, and could be described as having a "trans-world identity" (158). Where such characters are recognizable across universes, Stockwell argues that they "have achieved a degree of portability, in the sense that they can be removed entirely from their origin, or can even *jump across media*" (183, my emphasis). This portability arises from the "common ground" (160) of shared knowledge between participants in a given discourse-world interaction. If an author or filmmaker references a character from a pre-existing context, that character is only perceived to have a trans-world identity if they are recognizable to the reader or viewer.

For cognitive models of discourse like Text World Theory, shared knowledge between discourse participants is essential to understanding how meaning is shared and construed. Accordingly, the perception of fiction as multiversal occurs not only in direct switches between text-worlds situated in ontologically equal universes, but also in the world-building of individual text-worlds. Of course, these are not necessarily mutually exclusive. *Deadpool: Killustrated* (Bunn, 2013) involves near-continuous world-switching between literary universes, and both homodiegetic (costume, setting, styles of language) and heterodiegetic (text bubbles naming locations and dates) world-building devices which comprise each text-world signal intertextual points of reference to the reader, indicating the pre-existing contextual knowledge they are being asked to access in order to make sense of a scene.

As stories develop and worlds expand in ever-increasing detail, the common ground against which discourse-world participants measure their mutual understanding shifts: "new ideas are introduced and old concepts are disregarded as no longer relevant or fade away by no longer being mentioned" (Stockwell 2020, 160). One example of this from multiverse fiction would be the notion of the "canon event" from *Spider-Man: Across the Spider-Verse* (2023). As viewers are presented with a vast array of different iterations of Spiderman across the film, the framing of a supportive older relative's unpreventable death (Uncle Ben for Peter Parker, Uncle Aaron for Miles Morales) as a shared world-building element of each Spiderman's story adds depth to the connection between universes, as well as the characters with whom the audience is asked to relate.

The Non-Canonical Multiverse

In 1978, viewers of the *Star Wars Holiday Special* were introduced to the extended family of Chewbacca, the Wookie sidekick of Han Solo in the original *Star Wars* trilogy, as they celebrate Life Day on their home planet of Kashyyyk. Although critically derided, the *Holiday Special* was held by LucasArts to be a canonical entry in events of the *Star Wars* universe until Disney's acquisition of the franchise (Esterrich 2021). Ontologically, then, the *Holiday Special* provides an unusual example of a set of characters and events whose status within a fictional universe's ontology has shifted over time. Though the events of the narrative still occur within the *Special* itself, these are no longer considered world-building elements that have "happened" within the broader context of the official *Star Wars* universe, which thus comprises a multiverse of sometimes contradictory canonical and non-canonical narratives. In cases like these, audience members may bring their knowledge of these non-canonical worlds to inform their interpretation of texts and text-worlds, and vice versa.

Many other narratives occupy a similar position of non-canonical alternate universe fiction, such as Marvel's *What If…?* (Bradley, 2022) series. Presented from the outset as ontologically distinct from the "canonical" text world through the hypothetical framing of its title, the series nonetheless draws on discourse-world participants' expected knowledge of central, canonical events from which its stories deviate. As Barnette (2022) explains of the Marvel Cinematic Universe more generally: "Not only do the films pick up on storylines from one another, but they exist within a universe – a multiverse – that includes decades of comic-book iterations of these characters and their backstories (and retcons)" (41). Though familiarity with these pre-existing fictional worlds is not a requirement for making sense of each film or show, viewers aware of these stories are rewarded with narrative parallels and contrasts between different iterations of characters—and even retellings of events—across worlds and timelines.

As Kirby (2013) puts it, "the idea of the multiverse opens up the potential for massive intertextuality" (116). Whereas other interactions between worlds considered in this chapter have been *function-advancing*, intertextual references to equivalent universes might instead be considered *world-building* (cf. Mason 2019, 49–50). Rather than enacting a switch between two text-worlds, these references expand the common ground that readers might recognize across fictions. In fact, recognition of the ontological status of other fictional worlds within a discourse via world-building may be an important process for establishing boundaries of a given fantasy/fantasy distinction, similar to the way understanding everyday discourse requires a distinction between factual reporting and hypothetical abstraction.

Conclusions

The multiverse is a sprawling and complex concept, which challenges many of our expectations about the structure and nature of fictional worlds. In defining postmodernist fiction, McHale (1987) describes a range of ontological questions which the genre often invites:

What is a world?; What kinds of worlds are there, how are they constituted, and how do they differ?; What happens when different kinds of
worlds are placed in confrontation, or when boundaries between worlds
are violated?

(10)

All of these questions are at play when an audience is presented with a
multiverse of characters, stories, and ideas. In exploring this kind of fiction
through the lens of Text World Theory, this chapter has sought to recognize the places in which conceptual boundaries between worlds are tested
by different kinds of multiverse interaction. In doing so, I hope to have
shown that Text World Theory is a valuable model both for mapping these
relationships between worlds, and for connecting these structures to literary analyses, such as understanding the resonance of Waymond's modal-
world in *EEAAO*. Text World Theory is part of a 'tool-kit' of stylistic
resources (Wales 2014, 32), which can help scholars of multiverse fiction
to unpack the dynamics of a vast array of worlds and universes. For those
interested in further reading, Gavins (2007) remains the definitive textbook on Text World Theory, written in an accessible style suitable for a
comprehensive introduction.

While this chapter has been concerned with navigating the cognitive
complexities of multiverse fiction, I also hope that the examples discussed
here have highlighted the accessibility of the genre for audiences of all ages.
Films and comic books accessible to young audiences have been, and
remain, at the center of multiverse fiction. Wydrzynska (2021) warns of
the sidelining of children's literature in academic research, which "often
readily displays enough literary, linguistic, and narratological complexities
to rival even the most sophisticated literature for adults" (299). Multiverse
fiction is, at its heart, a playful experience exploring the permeability of
barriers between fictional worlds in a wide variety of literary contexts, and
one which seems to appear increasingly frequently across genres and
media. So, if you ever find yourself accessing a text-world in which Batman
and Spongebob *do* know each other (which, by reading this sentence, you
briefly have!), I hope Text World Theory can serve as a helpful guide
through making sense of this multiverse, and any others you find yourself
lost in.

References

Auster, Paul. 2017. *4 3 2 1*. Faber & Faber.
Barnette, Jane. 2022. "What is Wanda but Witches Persevering? Palimpsests of
American Witches in *WandaVision*." *Theatre Journal* 74(1): 41–57.

Binder, Steve, dir. 1978. *Star Wars Holiday Special*. Aired November 17 1978, on CBS.

Boillat, Alain. 2022. *Cinema as a Worldbuilding Machine in the Digital Era: Essays on Multiverse Films and TV Series*. Indiana University Press.

Bradley, A.C. 2022. *What If...?* Marvel Animation Studios.

Bunn, Cullen. 2013. *Deadpool: Killustrated*. Marvel.

Daniels, dirs.. 2022. *Everything Everywhere All at Once*. A24.

Dos Santos, Joaquim, Powers, Kemp, and Thompson, Justin, dirs. 2022. *Spider-Man: Across the Spider-Verse*. Sony Pictures.

El-Mohtar, Amal and Gladstone, Max. 2019. *This is How You Lose the Time War*. Simon & Schuster.

Esterrich, Carmelo. 2021. *Star Wars Multiverse*. Rutgers University Press.

Gavins, Joanna. 2007. *Text World Theory: An Introduction*. Edinburgh: Edinburgh University Press.

Gavins, Joanna and Simpson, Paul. 2015. "Regina v. John Terry: The Discursive Construction of an Alleged Racist Event", *Discourse and Society* 26(6): 712–732.

Gibbons, Alison, Vermeulen, Timotheus. and van den Akker, Robin. 2019. "Reality Beckons: Metamodernist Depthiness Beyond Panfictionality." *European Journal of English Studies* 23(2): 172–189.

Gibbons, Alison and Whiteley, Sara. 2021. "Do World Have (Fourth) Walls? A Text World Theory Approach to Direct Address in *Fleabag*." *Language and Literature* 30(2): 105–126.

Good, Owen. 2019. "*The Rise of Skywalker*'s Opening Crawl References an Event You Could Only Hear in Fortnite." *Polygon*, December 20. https://www.polygon.com/fortnite/2019/12/20/21031513/star-wars-the-rise-of-skywalker-fortnite-opening-crawl-palpatines-message

Howitt, Peter. 1998. *Sliding Doors*. Miramax Films.

Johnson-Laird, Philip. 1983. *Mental Models: Towards a Cognitive Science of Language, Inference and Consciousness*. Cambridge University Press.

Kirby, D. 2013. *Fantasy and Belief: Alternative Religions, Popular Narratives and Digital Cultures*. Equinox.

Kukkonen, Karin. 2010. "Navigating Infinite Earths: Readers, Mental Models, and the Multiverse of Superhero Comics." *Storyworlds: A Journal of Narrative Studies* 2(1): 39–58.

Mason, Jessica. 2019. *Intertextuality in Practice*. Amsterdam and Philadelphia: John Benjamins Publishing Company.

McHale, Brian. 1987. *Postmodernist Fiction*. London: Methuen.

McIntyre, Dan. 2008. "Integrating Multimodal Analysis and the Stylistics of Drama: A Multimodal Perspective on Ian McKellen's *Richard III*." *Language and Literature* 17(4): 309–334.

Raimi, Sam dir. 2022. *Doctor Strange in the Multiverse of Madness*. Walt Disney Studios Motion Pictures.

Ryan, Marie-Laure. 1991. *Possible Worlds, Artificial Intelligence, and Narrative Theory*. Indiana University Press.

Ryan, Marie-Laure. 2006. "From Parallel Universes to Possible Worlds: Ontological Pluralism in Physics, Narratology, and Narrative." *Poetics Today* 27(4): 633–674.

Segal, Erwin M. 1995. "Narrative Comprehension and the Role of Deictic Shift Theory." *Deixis in Narrative: A Cognitive Science Perspective*, edited by Judith Duchan, Gail Bruder, and Lynne Hewitt, 3–17. London and New York: Routledge.

Skolnick, Deena. and Bloom, Paul. 2006a. "What Does Batman Think about SpongeBob? Children's Understanding of the Fantasy/Fantasy Distinction." *Cognition* 101: B9–B18.

Skolnick, Deena. and Bloom, Paul. 2006b. "The Intuitive Cosmology of Fictional Worlds." *The Architecture of Imagination: New Essays on Pretence, Possibility, and Fiction*, edited by Shaun Nichols, 73–86. Oxford: Oxford University Press.

Stockwell, Peter. 2009. "The Cognitive Poetics of Literary Resonance." *Language and Cognition* 1(1): 25–44.

Stockwell, Peter. 2020. *Cognitive Poetics: An Introduction*. Second Edition. London: Routledge.

Toolan, Michael. 2001. *Narrative: A Critical Linguistic Introduction*. Routledge.

Wachowski, Lana and Wachowski, Lilly dirs. 1999. *The Matrix*. Warner Bros.

Wales, Katie. 2014. "The Stylistic Tool-Kit: Methods and Sub-Disciplines." *The Cambridge Handbook of Stylistics*, edited by Peter Stockwell and Sara Whiteley, 32–45. Cambridge University Press.

Walton, Kendall L. 1990. *Mimesis as Make-Believe*. Cambridge, MA: Harvard University Press.

Werth, Paul. 1999. *Text Worlds: Representing Conceptual Space in Discourse*. London: Longman.

Wolfman, Marv. 1985. *Crisis on Infinite Earths*. DC Comics.

Wydrzynska, Ella. 2021. "'I shouldn't even be telling you that I shouldn't be telling you the story': Pseudonymous Bosch and the Postmodern Narrator in Children's Literature." *Language and Literature* 30(3): 229–248.

31

THE MULTIVERSE OF FANFICTION

Julia Neugarten

In this universe, I am a fangirl. I am casting spells with Hermione at Hogwarts. I am hunting demons with the Winchester brothers in small-town America. I am making a cup of tea for Sherlock Holmes at his Baker Street apartment. I am Draco Malfoy, hopelessly in love with Harry Potter. I am Bella Swan, hopelessly in love with Edward Cullen. I am the Doctor's new companion, soaring across the universe together in the Tardis. I am a fly on the wall as Fox Mulder declares his love to Scully in their basement office at the FBI headquarters. I am a teenager holding an iPod touch close to my face in the dark of my childhood bedroom, hoping the battery won't die before I get to the happy ending. I am all of these things, all of these places, all of these loves. All at the same time.

As I become these different creatures, perspectives, and storyworlds in turn, I expand. I see myself become a person capable of imagining more, doing more, being more, and feeling more than I previously deemed possible. In the process, the stories I love also become capable of more. In the loving and interrogative custody of their fans, these stories can be used to question all sorts of things the mainstream media takes for granted; the ways our culture envisions gender roles, disability, love, friendship, happiness, sex, pleasure, leisure, and the role and nature of stories themselves.

In fanfiction (which we call *fic*), space emerges to explore emotion, not as it has been packaged and sanitized by Hallmark but as I feel it bursting

DOI: 10.4324/9781003480846-36

from my chest at unpredictable and strange moments, threatening to over-take me. I discover stories where, instead of *conquering* the limitations of their bodies and minds, characters live happily ever after by accepting themselves. As Francesca Coppa puts it in *The Fanfiction Reader*, I take from the buffet of stories "one Pringle, one Dorito, and one Oreo" (2017, vii). In the process, I learn to live differently in my own body. I pick and choose: this story helps me, makes me feel happy and good and compelled, and that one doesn't.

In another universe, I am an academic. I observe that multiverse fiction—narratives exploring contradictory or divergent realities within the con-fines of a single story—has become a staple of contemporary media, even though multiverses have, in some ways, been part of fanfiction communi-ties for a long time. Already in 2014, media scholar Anne Kustritz charac-terized fanfiction as a "transmedia serial multiverse" (par. 42). This essay explores this idea, that the enormous corpus of fanfiction that exists in the digital age can best be understood as one big multiverse. Of course, every work of fanfiction can be viewed as standing alone, as telling its own story. But this individuality is not where fanfiction gets its awesome power to move readers; that power comes from the endless reservoir of fannish ref-erences, knowledge, and feelings that each individual work takes from and adds to. Each fic gets its resonance from its position within and relation to the rest of the body of works that makes up all fanfiction.

The fabric of *fic* is best understood as a kind of multiverse. This multi-verse of fanfiction lends (sub)textual meaning to every new interpretation of a familiar trope, character, or universe. It allows diverse interpretations and truth claims about stories to exist side by side, to interrogate and complement each other. This essay explores the endless possibilities offered by the multiverse of fanfiction and the ways fanfiction transforms the meanings of different building blocks of story, such as plot, the spatiotem-poral and natural laws of storyworld, and character. Traveling back and forth between universes in which I am a fan and a scholar, I also reflect on the effects of fanfiction's multiversal qualities on my own reading experi-ences and make some suggestions about what a fanfictional mode of read-ing may have to offer you.

Exploring the Faniverse

If you enjoy a lot of popular media—as I did for the first twenty media-literate years of my life—you may share my experience of, one day, hitting a kind of wall. When browsing the offerings of the six different streaming services I currently have access to, nothing particularly stands out.

Nothing really appeals to me. Nothing seems refreshing, or capable of engendering surprise. Let me give you a tip, in case this feeling sounds familiar: try fanfiction.

I can always, always, always find fanfiction that caters to my exact narrative desires. This is in part because there is so much of it—*Archive of Our Own* (AO3), one of the most popular websites for reading and writing fanfiction in English, currently boasts over 12 million works—and in part because the interface of AO3 enables a lot of very specific selection. Because writers can choose tags to attach to their stories, readers can then filter for stories that meet their needs when it comes to fandoms, characters, relationships, storyworlds, genres, tropes, and word count. Essentially, it is this immense network of tags that makes fanfiction into one big multiverse. These tags allow me, the reader, to travel from one universe to another and back again.

This network of tags affords connection and ways of mapping and traveling the fanfiction universes—which I call the faniverse. Although the *Archive of Our Own* employs a small army of volunteer tag wranglers to keep the tagging system both freeform and usable, tags can present themselves to the initiated reader as imposing a kind of order. The reader who is in the know—both about the archive's interface and offerings and about their own tastes and preferences—can use the tagging system to curate their reading experience with attention to the finest detail. If fanfiction is a buffet, this tagging system is the tablet they give you at all-you-can-eat sushi restaurants that allows you to select different dishes, or to order the same dish hundreds of times over the course of your meal.

That's not an exaggeration. Once you have figured out that you crave stories where Sherlock Holmes and John Watson are forced to share a bed because there is only one room left at the inn, over 600 stories are at your disposal. If you are looking for fic where Harry Potter and Draco Malfoy are tied together by a magical soul bond, more than 200 stories are waiting for you. And if you desperately want to read about fictional characters who work in coffee shops, there are over 37,500 stories to choose from. These stories all point to some of the unifying principles of the fanfiction multiverse: its specific brand of serendipity, its tropes of interpersonal dynamics, and its shared locations.

Serendipity, of course, is the bread and butter of the romantic storyline. Consequently, the world of fanfiction is the world where every conceivable hotel only has a single bed available, where every barista is always incredibly attractive and flirting with you, where everyone's coffee order reveals something about their inner life, where those who are hurt can always count on being comforted, where your anonymous pen pal always turns out to be your long-lost love, where enemies and roommates are just lovers waiting to happen, and where you meet your favorite pop star in real life and they love you back.

In many corners of the faniverse, it is possible that your soulmate is bound to you the way we are bound to gravity on earth, through a mark or quotation on their skin, or through a mysterious curse that makes them feel ill when they are away from you. It is possible that monsters are real, that vampires exist, or that ordinary people suddenly start sprouting wings. There are worlds in this multiverse where gender is complicated by a set of secondary genders—alpha, beta, and omega or A/B/O,—or where romantic relationships are organized according to *sedoretu*, the polyamorous marriage system originally devised by Ursula Le Guin. You may find yourself part of a species that experiences *pon farr*, the madness-inducing mating cycle of the Vulcan people in *Star Trek*. All of these concepts—soul bonds, wing fics, alpha/beta/omega dynamics, sedoretu, pon farr, and many more—have their roots in particular fandom traditions but have since leaped into the fannish multiverse. The user-created wiki *TV Tropes* ("Fandom-Specific Plot," 2024) notes how "some plots are so embedded in the collective consciousness of netizens everywhere that no fandom is complete without them." Such tropes pop up in all kinds of narrative contexts, so that their underlying principles and the possibilities for interaction they offer can be explored, tested, and taken in new directions. This way, the fannish multiverse is ever-expanding.

And then there are the specific laws of space and time. In most parts of the fanfiction multiverse, the American high school is still a romantic and idyllic location where it doesn't smell of moldy lunchboxes. Demons may cross your path, but only when you have been training for ages to take them on. Narnia is in your wardrobe. Hogwarts is only ever a train ride away. And whenever something fun or cute happens, you can count on it to happen five times in a row, only to be changed or subverted the sixth time—we call this a 5+1 fic. These are just some of the natural laws that govern some of the universes in the fannish multiverse, but many of these are fanfiction staples that occur again and again, creating a united narrative framework or multiverse. None of these elements are present in every universe, all of them are present in some, and they can be combined, contrasted and subverted to your heart's content.

Nothing makes the multiversal nature of fanfiction clearer than the crossover fic, a subgenre where multiple storyworlds (and elements from them that often include plot, characters and tropes) are mixed, with most or all elements in the mix presumed familiar to the reader.[1] You need to be aware of the storyworld context of multiple fandoms to understand—and ideally, enjoy—a crossover story. To create a common ground between different fandoms, writers often take elements from the most popular and well-known fictional worlds, such as *Harry Potter* or the Marvel Cinematic Universe (MCU) and mix in a handful of the tropes that structure much of the faniverse. In crossovers, the shared cultural material of alternate universes and familiar character dynamics takes center stage, underlining the interconnectedness of the different storyworlds and elements that make up fanfiction.

Back to the Scholarverse

It is this multiversal quality that enables fanfiction to do its transformative work. By resting on a firm basis of shared knowledge and love, fanfiction can take liberties with some of the other elements of a story to create subversion, surprise, or outright outlandishness. This mix of familiarity and strangeness has been insightfully theorized by people in fandom as well. For example, in a 2023 Tumblr-post, fans @kayanem and @Ladylark (a.k.a. Naomi Jacobs, herself an academic studying fans) propose the theory of *concretes*. Concretes, they argue, are "certain aspects at the heart of what makes a character 'them', that have to be retained to make them recognisable across fic." Authors can pick and choose from the set of concretes that make up a character, and beliefs about which concretes are fundamental to a character can differ between individual writers and readers, but also between fanfiction (sub)communities. They conclude that concretes are a "useful way to refer to the character's essential defining components, and which of these elements it is possible or desirable to retain, change or remove across adaptation and transformation." Note how this theory, which offers up a theoretical concept for narrative analysis but is published on Tumblr, bridges the gap between the scholarly and fannish universes.

The theory of concretes directs attention to the final essential component of the fannish multiverse: characters. As *concretes* indicate, characters need to retain at least a subset of recognizable characteristics to be read and interpreted as themselves in a particular instance, and deviating too much from these concretes puts fanfiction writers at risk of committing the

---

1 In a survey conducted for the *Fansplaining* podcast, 29% of respondents (n = 6.744) said that they sometimes read fanfiction without being familiar with the source material (Minkel and Klink 2021).

worst fandom sin: an out-of-character portrayal. For many fanfiction readers, attachments to particular characters or the relationships between them (which we call *ships*) are the primary locus of fannish love and a strong motivator to read fanfiction. For example, in her analysis of a Russian-speaking *Harry Potter* fanfiction community, Natalia Samutina notes that "an emotional attachment to one of the canon characters" was a common motivator for story selection and enjoyment (2017, 263). This means that to meet the needs of fanfiction readers, stories must fulfill at least some of the expectations raised by the presence of a particular character or ship: who wants to read about a Sherlock Holmes who *isn't* dazzlingly clever, and *isn't* hopelessly, awkwardly in love with Dr. Watson? Not me. So, because fanfiction reading practices tend to privilege character, characters remain relatively stable across the multiverse of fanfiction.

Connecting My Universes

My fangirl self wants to end it here. I have shared with you some of the riches the faniverse has to offer, and that's the end of it. But, straddling the boundary of my fannish and academic worlds, I can hear you thinking: okay, that sounds like a pretty hilarious hobby you have got going on over there with the *fic* and the *ships* and the *sedoretu* or whatever, but so what? And my academic self agrees: what is the value of showing you that these things are being written and read online, of showcasing all the amazing things that fiction can do when stories are at the disposal of the audiences who love them? In short: why should you care?

I have shown you that everything is a story, and that every story can be rewritten, remade, and reinterpreted in many different ways that sprout separate but interconnected universes, together making up one big multiverse. By reading a lot of fanfiction and coming to terms with the malleability, narratability, and subjectivity of any event sequence, I have come to see that these adaptable qualities characterize the story of my life as well as the stories we commonly call fiction. Whether I view myself as limited or limitless, happy or troubled, fortunate or ill-fated depends on the perspectives I take, how I see my own positionality in relation to the events of my life, how I tell and read my story. It depends on where I locate my self within my own multiverse.

@kayanem and @Ladylark's theory of concretes offers a useful way of thinking through your life as a multiversal narrative, since you yourself are made up of concretes. Every day, you are selecting which concretes to retain, which ones to show the world, and which ones to bury or reject. What makes you, you? Is it that you are a loyal friend or a good baker? Is it that your hair is dyed blue, or that you always tell the same lame types of jokes? You can pick and choose from these concretes to adapt yourself to different situations, and you've probably done this many times over the course of your lifetime. In this process of self-creation, you have chosen a different branch of your own personal multiverse each time.

In academia, one theory that usefully describes this process of narrating the self is the idea of Storyworld Possible Selves or SPS, put forth by María-Ángeles Martínez (2018). She explains: "SPS is the term I use to refer to the hybrid mental construct (...) with which we inhabit the storyworlds projected by narratives" (section 1.1.1). SPS are hybrid because they combine elements of the real-world self with elements of fictional characters, and they are mental constructs because our minds create them in the process of reading. In this process, the self and the fictional other form a blend, which can point us to new insights and feelings. Narrative, within this framework, is an accessible, safe, easy, low-budget technology to try out possible selves, which can—if they prove useful or pleasurable in the playground of fiction—later be tried on in real-life situations. Storyworld Possible Selves thus account for the expansion of self I experience when I am reading fanfiction. Through fanfiction, I can be different versions of myself in different stories, and yet these are all interconnected in the network of the faniverse by unifying factor of me.

Fortunately, the multiverse of fanfiction is not the only place from which I can blend my possible selves. The universe I am writing from is one where I can be both fan and academic—an aca-fan. Fannish engagement and critical scholarship support each other in my writing—both academic and essayistic—and in my hobbies. From this dual vantage point, I can also see that despite the ever-changing media landscape and the move toward more multiversal modes of storytelling in the mainstream, fanfiction can and will continue to be special, to do what it does best: open our eyes to the endless experiences, possibilities, loves and pleasures available in our own universe and beyond.

References

Coppa, Francesca, ed. 2017. *The Fanfiction Reader: Folk Tales for the Digital Age*. Ann Arbor: University of Michigan Press.

"Fandom-Specific Plot." 2024. TV Tropes. https://tvtropes.org/pmwiki.pmwiki. php/Main/FandomSpecificPlot.

@kayanem, and @Ladylark (a.k.a Naomi Jacobs). 2023. "Concretes." Tumblr. *Tumblr* (blog). June 6, 2023. https://kayanem.tumblr.com/post/719409497577340928/ a-couple-of-years-ago-we-that-is-ladylark-and

Kustritz, Anne. 2014. "Seriality and Transmediality in the Fan Multiverse: Flexible and Multiple Narrative Structures in Fan Fiction, Art, and Vids." *TV/Series*, no. 6 (December). https://doi.org/10.4000/tvseries.331

Martínez, María-Ángeles. 2018. *Storyworld Possible Selves. Applications of Cognitive Linguistics*, Volume 37. Berlin Boston: De Gruyter Mouton.

Minkel, Elizabeth, and Flourish Klink. 2021. "The Fic and the Source Material." *Fansplaining.* 2021. https://www.fansplaining.com/articles/the-fic-and-the-source-material

Samutina, Natalia. 2017. "Emotional Landscapes of Reading: Fan Fiction in the Context of Contemporary Reading Practices." *International Journal of Cultural Studies* 20 (3): 253–69. https://doi.org/10.1177/1367877916628238

32

"LONG LIVE THE EMPIRE!"

Star Trek's Mirror Universe as a Form of Fanfiction

Lincoln Geraghty

In the original *Star Trek* (1966–69) episode "Mirror, Mirror" (1967) Captain Kirk and three of his crew are accidentally transported onto a parallel universe version of the *USS Enterprise* where the normally peace-loving Federation is replaced by the war-mongering Terran Empire. On board, familiar crew members are transformed into violent sadists willing to go to any length to assume power. Recognizing Vulcan logic remains intact, Kirk tries to persuade the "mirror" version of Spock to change in order for that universe to be more like his own. This brings about a short-term improvement for the alternate *Enterprise* but in the long-term Kirk's message of peace proves to be the downfall of the Empire. The first in what's become a popular series of *Star Trek* episodes that used the plot device of a parallel universe, "Mirror, Mirror" played with the concept of the multiverse and established an alternate canon for the franchise. This Mirror Universe (MU) has been used to pose ethical and moral dilemmas that stand in direct contrast with what audiences know and expect from the long-running science fiction series. Over the years, the MU has become a means through which writers can question established character motivations while at the same time using alternate canon to speculate on the nature of humanity, the repercussions of individual action, and the politics of difference.

Arguably the first science fiction series to regularly use the concept of a multiverse, *Star Trek*'s MU is also an important reminder that multiversal storytelling is not just confined to blockbuster franchises such as the

DOI: 10.4324/9781003480846-37

MCU. Recent movies such as *Spider-Man: No Way Home* (2021) and *Doctor Strange in the Multiverse of Madness* (2022), plus streamed content like *Loki* (2021–2023), would seem to suggest that Marvel alone invented the multiverse as a form of transmedia entertainment. However, Dan Falk's (2023) article in *Discover* magazine points out that the parallel universe story not only dates back to science fiction television like "Mirror, Mirror" but has also been a controversial topic within the science community. He argues that while science fiction films are "not particularly heavy on the physics [they] are definitely latching onto something" and that "something" allows "characters to explore a multitude of worlds with varying degrees of similarity to our own, as well as altered versions of themselves" (34). Indeed, we see the results of Kirk's visit to the MU in *Star Trek: Deep Space Nine* (1993–1999) crossover episodes where humans have become galactic slaves to the Klingon and Cardassian alliance. Subsequent prequels *Star Trek: Enterprise* (2001–2005) and *Star Trek Discovery* (2017–2024) have also used the MU to set up stories that merge timelines and cross over characters; often playing with and revising *Star Trek* canon. Steffen Hantke (2014) argues that the J.J. Abrams' "Kelvin Universe" films are part of *Star Trek*'s "alternative world" as they cleverly retell familiar stories within the wider universe (562). However, MU episodes play out distinct plotlines in which established characters are different to the originals and provide alternative dystopic versions of the *Star Trek* universe in which the moralistic characters are switched to tell often more dark and violent stories.

This chapter argues that *Star Trek*'s repeated return to and expansion of the MU resembles how fans have written and added to the universe in fanfiction. Just as fans continue to rewrite the text using different modes of storytelling that imitate the Vulcan edict of "Infinite Diversity in Infinite Combinations" (IDIC), the MU continually adds to the multiversal story world on screen. Indeed, while fans have written and read fanfiction for decades as a means to understand and get closer to their favorite characters—seeing them change, interact and encounter new scenarios not usually depicted on screen—the MU follows a strategy for building new worlds and expanding new storylines using very familiar, yet knowingly different, *Star Trek* visual and narrative tropes. Sheenagh Pugh (2005) describes a popular form of fanfiction called "Alternative Universe" (AU), and through this genre fans can explore the "what if" of their favorite fictional story (61). Yet, with the MU—which is already an alternative to the canon pitched in opposition to the original—it is more like a "what if" to the "what if," as *Star Trek* is already a fictional story of our potential future. Thus, I start this chapter with a discussion of *Star Trek* fanfiction's relationship to the MU. For Janet H. Murray (1997), mediated narratives at the end of the twentieth century (including fanfiction) have complicated more traditional linear modes of storytelling, on the one hand making them more playful and knowing but on the other demonstrating our continued fascination for "parallel possibilities" (37). As such, this chapter will also use critical work on parallel worlds, digital storytelling and networked narratives to consider *Star Trek*'s MU as emblematic of Murray's concept of the multiform story.

FanFiction and the Mirror Universe

Fanfiction is a means of reimaging the complex and changing relationship with a canonical text. Murray (1997) summarizes that "fans create their own stories by taking characters and situations from the series and developing them in ways closer to their own concerns," and identifies *Star Trek* fans in particular as having created "a vast literature of alternate adventures" (41). Indeed, modern fans studies were spawned from early studies of *Star Trek* fanfiction, including the works of Henry Jenkins (1992), Camille Bacon-Smith (1992), and Constance Penley (1997). These studies prioritized the notion that fanfiction was proactive, not just about repeating the pleasures of the original text but concerned with the transformation of the text by filling in what was lacking. However, histories of the practice are rather scarcer. Joan Marie Verba's *Boldy Writing* does offer us insight into the world of *Star Trek* fanfiction and reveals that the MU played an important creative role in expanding the canon while the series was no longer on our television screens. In 1974 editors of the fanzine *Alternative 4* explained their title: "We count our familiar *Star Trek* Universe – television's *Star Trek* – as number one; "Mirror, Mirror" Universe as number two; the Kraith universe as number 3; and Light Fleet's Universe as number 4" (quoted in Verba 1996, 18). The fanzine was known for publishing original stories in the parallel Alternative Universe 4 that focused on Captain Kirk working for Light Fleet, rather than Mr Spock (so often the major character for fanfiction writers). The pre-eminence of Spock in fanfiction is highlighted by the third universe—the Kraith universe. Created by fan writer Jacqueline Lichtenberg, Kraith was her own Alternate Universe canon, based on expanding what was known about Vulcan culture and telling the story of Spock, his family and the planet beyond the confines of Starfleet (see Bacon-Smith 1992, 58–61). Lichtenberg established an AU for Spock and Vulcan in which many other fans enjoyed writing. What is interesting in the contexts of this chapter is that the MU is identified as not only canon and an AU, but also a universe in which fans could set their own stories: A second universe, in fact, set behind the prime universe but nonetheless on a list containing other fan-created universes: thus, a form of fanfiction.

Often assumed to be of a sexual nature, fanfiction has a reputation for reversing power relationship and reveling in homoerotic depictions of heterosexual characters. The MU was not immune to such fan storytelling. Verba (1996, 35) mentions that in the 1976 *R&R* fanzine (sold in plain brown paper to hide the erotic stories and images contained inside) published a mirror story based on Sarek and Amanda. Also, in 1976, the adult-orientated *Grup V* fanzine printed a mirror Spock centerfold in an 89-page issue. Later, in 1980, a whole 251-page MU fanfiction novel was published by Poison Pen Press. Titled *One Way Mirror* and written by Barbara Wenk the story focused on an actual fan transported into the MU and becoming a sex slave to a Vulcan first officer serving aboard a starship. Verba recounts the popularity of this story and states it won a Fan Q Award (created to specifically reward the best fanfiction writers) in 1981 (52). These examples show us that not only was the MU an attractive universe in which fans could write and rewrite established characters but it was also a bonified subgenre of fanfiction, a form of AU that used the multiversal storyline to play with and expand what was first seen in "Mirror, Mirror."

However, in the case of the MU, the concept of the multiverse goes beyond differences between fiction and nonfiction, or, as above, canonical text and fan text; indeed, fictional worlds both on screen and in fanfiction can be defined by their similarities and differences with what is considered the real world. According to Marie-Laure Ryan (2017) in the theory of possible worlds, "a world is defined over a set of mutually compatible propositions [and] there can only be one actual or real world from a given point of view, one of these worlds will become actual, while the others will remain unrealized possibilities" (33). Yet, in the MU, as I will demonstrate, possibilities are not unrealized and variations of both the prime and mirror universe are made manifest in order to expand new ways of storytelling. The existence of the MU does not destabilize the original, or threaten to destroy it even—rather it serves to make the whole world more coherent, tangible, and immersive.

Inversions and Proliferations in the Original *Star Trek* and *DS9*

Important episodes such as "Mirror, Mirror" exemplify *Star Trek's* tendency to analyze human duality through the personification of the "evil twin" or "doppelganger" (see Hertenstein 1998, 7–16). Similarly, the earlier episode "The Enemy Within" (1966) has Kirk split into two after a transporter accident, one Kirk "good" and the other "evil." Louis Woods and Gray Harmon (1994) attempt to explain this by identifying Jung's concerns with opposites and the shadow in early episodes of the series, describing how "they interact in the formation of the nature, structure and functioning of the human psyche" (169). Considering that the series was only in its first season it demonstrates the powerful potential in using this narrative device of alternative or parallel versions to both highlight the lauded qualities of characters while at the same time telling new stories with familiar yet different "inversions." For Aiden Byrne and Mark Jones (2018), "inversions" used in *Star Trek* allow writers to play with social norms and invite audiences to question the order of the established universe. In the case of "Mirror, Mirror" and the MU as it was created in 1967, the notion of the multiverse demonstrates the narrative function of othering: "It is the narratorial act of inversion – the specific elements that are inverted and how this is achieved – that allow them to function as ideological critiques" (259). Inverting moral characters like Kirk and Spock, institutions like the Federation and Starfleet, and even objects like the *Enterprise* provides a blank slate for television but also for fans who can write in those new spaces and with those new characters. Something that fanfiction writers, as discussed earlier in this chapter, have been doing since *Star Trek*'s syndication.

While iterations of the MU flag the atrocities of authoritarian regimes and the lack of individual responsibility, thus holding a light to society in order to learn from their own mistakes, "Mirror, Mirror" also continued to offer problematic representations of race, gender and sexuality that scholars have long felt detracted from *Star Trek*'s message of equality and diversity. For example, Lt. Uhura's body in the MU is "eroticized" through close-up camera angles that focus on her legs, bare midriff, and breasts. Arguably, the decision to dress the female crew in revealing uniforms and have them as the subject of the male gaze underscores the MU's different attitudes toward women—humanity has regressed to see women only as sexual objects, prizes of conquest. However, as Daniel Bernardi (1998) points out, "It is as if her blackness is made safe and appealing when it is performing in fragmented and fetishized forms – when, in other words, it is exoticized as it is eroticized" (42). Problematical representations, although a challenge to the series' utopian credentials, offer further examples of the MU's narrative alterity and its potential to both expand and disrupt canon as in the case of fanfiction.

DS9 was the first series in the franchise to return to the MU. Episodes "Crossover" (1994), "Through the Looking Glass" (1995), "Shattered Mirror" (1996), "Resurrection" (1997), and "The Emperor's New Cloak" (1998) mapped out how changes to the Terran Empire instigated by Kirk and carried through by Mirror Spock altered the MU so humans were now slaves rather than galactic conquerors. These episodes focused on personal relationships and the moral ramifications of prime versions of characters meeting their mirror wives, partners, and friends (Barrett and Barrett 2001, 108). Chris Gregory's (2000) analysis of the series separates it out from others, being darker, more serious, and engaging with notions of family that previous incarnations of *Star Trek* had stopped short of doing. It played with the tropes of similarity and difference like the original series (17) but interweaved those tropes with stories that concentrated on emotion, trauma, and loss. MU episodes reinforced similarity and difference but the multiversal stories had real impact because characters were confronted with mirror versions that proved more familiar and attractive than their prime versions. For example, in "Shattered Mirror" Captain Sisko crosses to the MU to rescue his son Jake who believes that the MU version of Jennifer is his real mother. As had been established, Jennifer was killed in a Borg attack, leaving Sisko to raise his son as a single parent on board the space station. The episode's dramatic thrust is not really the notable differences between prime and mirror universes but rather the fact Jake and Sisko have to lose Jennifer all over again as she dies at the hands of the station Intendant (MU Major Kira). Gregory summarizes *DS9*'s narrative storytelling as presenting "its characters with challenging political and personal choices," emphasizing "the relative perspectives of different alien races," and grasping "the 'epic' form in a way unparalleled in TV history" (88).

While parts of Gregory's assessment seem hyperbolic considering how contemporary science fiction television has evolved on both streaming and network channels, it highlights the importance of multiple narratives and episodic media formats in the creation and development of fictional worlds and multiverses. Marie-Laure Ryan (2001) argues that "textual worlds function as imaginary counterparts or as models of the real world" (15) and as such we can read *Star Trek* as a future version of our own world that allows writers and audiences to imagine alternatives and better ways of living. However, as a mirror version of the imagined "textual world" the MU functions not just as a "counterpart" but also as another "model" of the world in which we live. The MU world is the result of war, conflict, and decisions taken to pursue violence rather than seek peace. Therefore, I would argue that as an example of multiversal storytelling the MU is part of what Ryan (2017) calls the "poetics of proliferation" where "the proliferation of texts around worlds, of worlds within texts, and of stories within worlds... has been explored, and elevated into an aesthetic" (43). The aesthetics of the MU allow for greater storytelling, expansion of the *Star Trek* canon, and the introduction of conflict which challenges familiar characters, and the emotional relationships audiences share with them. As with fanfiction, the *Star Trek* MU offers writers narrative space to build and create new without getting rid of the old.

Alternatives and World Management in *Enterprise* and *Discovery*

In the final season of *Enterprise*, the series returned to the MU in a two-part story called "In a Mirror, Darkly, Parts 1&2" (2005). In a departure from previous iterations the entire story is told in the mirror universe, all characters are their alternate selves. The MU Jonathan Archer is second in command to Captain Forrest on the *ISS Enterprise* and familiar crew members stand out as violent, scheming, and ultra-ambitious. The story involves the Terran Empire at war with Klingons and Romulans, having conquered planets such as Vulcan and Andor. Aliens are seen as inferior and Vulcans like T'Pol who still serve on the *Enterprise* are treated with contempt and suspicion. Rather than having prime universe characters crossing over the story involves the *USS Defiant*, last seen in the original series episode "The Tholian Web" (1968), falling back in time and between universes. After identifying that the ship is from the future, Archer mutinies and commandeers the *Defiant* so he can win the war and take control of the empire. Following several confrontations with both the MU Tholians who had captured the ship and members of his own crew, Archer is killed by MU Lt. Hoshi Sato who takes the *Defiant* and claims the throne as the new Empress. "In a Mirror, Darkly" combines *Star Trek* canon and the mirror universe to create depth to the multiverse. Filling in gaps to the story of what happened to the *USS Defiant* connects to "The Tholian Web," adds to the history of the Terran Empire, and informs a rereading of the original "Mirror, Mirror." Thus, more details lead to canon expansion and narrative proliferation.

Steffen Hantke (2014) describes the MU as "'alternative to the alternative'… an embedded diegetic space that simultaneously [sic] differs from and corresponds to the internally coherent world presented across all incarnations of *Star Trek*" (562). This is an apt description of the MU but in the case of *Enterprise* I would argue that it presents an "alternative to the alternative to the alternative" since the episode was only set in the MU and was aired with a new opening title sequence. When the series first aired in 2001 the opening titles were a departure for a *Star Trek* series; they were the first to use historical images and be accompanied by a song: Diane Warren's "Where My Heart Will Take Me" sung by Russell Watson. For Donna Minkowitz (2002), this change to *Star Trek* convention was regressive in that it set the titles "to a hymn that combines the first Christian references ever heard on *Star Trek* with some boasts about resisting alien domination" (36). In the same vein, I have argued that the images used depicting moments of human exploration, scientific innovation, and early spaceflight locate *Enterprise*, and therefore *Star Trek*, within a very specific tradition of American Exceptionalism (see Geraghty 2007, 18–19). However, going one step further, the opening titles for "In a Mirror, Darkly" present a history of predominantly white violence, colonialism, global conflict, and a future vision of humanity completely at odds with that represented by the Federation. The two episodes and specially made titles "have, by virtue of extended seriality, built a vast and narratively complex world that… encompasses several centuries, four quadrants of the galaxy, and alternate universes and timelines" (Pearson and Messenger Davies 2014, 147). It sets the scene for the rise of the Terran Empire, but, more importantly, adds depth and detail to its history and the MU as a whole.

The MU *Enterprise* titles add further complexity to the alternate world first seen in "Mirror. Mirror." As a fully formed vision of the MU depicted on television it retrovisions both established canon and alternative canon, creating an alternative to the alternative to the alternative as I described earlier. A "retrovision is a 'vision into or of the past' and implies an act of possessing the ability to read the past, in the way that one would possess a prophetic vision" (Neely 2001, 74). For Deborah Cartmell and I.Q. Hunter (2001) retrovisions are "makeovers of history" (7) and I apply the term to *Enterprise* here since it re-fashions *Star Trek*'s history and the history of the MU. Hantke (2014) argues that "alternative universe episodes" create "spaces of desublimation… spaces of the carnivalesque… [and] spaces in which the repressed can temporarily return… in short, spaces in which US militarism can encounter itself as other" (573). There is no denying that the MU is more militaristic in *Enterprise* and much of the two-part story is focused on humanity at war. However, as a means to expand the MU as part of a *Star Trek* multiverse, I would argue that it revels in using storytelling tropes to depict an alternate world just as convincing and detailed as the original. More than just a departure from the norm, "In a Mirror, Darkly" is the first example of the franchise moving toward a multiversal strategy for building new worlds and expanding new storylines that have long been the subject of fanfiction.

More recent series such as *Discovery* and the animated *Star Trek: Lower Decks* (2020–2025) display a more strategic effort to not just return to the MU once every two or three seasons, but develop a stable and consistent alternate universe that has its own established characters, stories and history. Near the end of *Discovery*'s first season, it is revealed in "Vaulting Ambition" (2018) that a major character, Captain Gabriel Lorca, is actually the mirror version of himself and that he was hiding in the prime universe after escaping from the Terran Emperor. The primacy of the MU is further underscored in "The Wolf Inside" (2018) by the fact the Emperor is revealed to be the mirror version of Captain Philippa Georgiou who was killed in the second episode, "Battle of the Binary

Stars" (2017). The second half of the season is largely spent in the MU with the *USS Discovery* and its crew having to traverse space and avoid conflict with the Terran Empire. In "Into the Forest I Go" (2017) and "Despite Yourself" (2018) the ship is transported to the mirror universe and assumes the identity *ISS Discovery*. MU Lorca is eventually killed trying to escape the Empire while the crew and Georgiou return to the prime universe in "What's Past is Prologue" (2018). Events in the mirror universe have a direct effect on the narrative direction of the first season and have long-lasting implications for character development. "The War Without, The War Within" (2018) sees Georgiou pretend to be her prime version and take over the *Discovery* and in the finale "Will You Take My Hand?" (2018) she is defeated but then recruited by Starfleet to work for their clandestine Section 31 (which is currently the title of a planned spin-off series that will expand MU Georgiou's story). More recently *Discovery* has returned to the MU in season three episodes "Terra Firma, Parts 1&2" (2020) and season five's "Mirrors" (2024); the latter including an earlier version of the *ISS Enterprise*.

The way writers on *Discovery* used the MU is symptomatic of the contemporary media industries that have adopted and adapted strategies for storytelling once confined to specific formats. Derek Johnson reminds us that the concept of the multiverse has been used to reboot and refresh characters and worlds present in comic books. Marvel's *Secret Wars* story was the first real attempt to rebuild a familiar world "as characters, publishers, and readers confronted and resolved the differences between the competing uses, interpretations, and iterations of Marvel's intellectual property over the past 50 years" (Johnson 2017, 129–30). Creating the concept of the multiverse meant different versions of heroes and villains could interact, kill off, and replace each other while reasserting Marvel's ownership and the industrial contexts in which they were originally conceived. Like with Marvel, *Star Trek*'s MU is an example of a multiverse where prime and alternative versions of familiar characters can mingle and merge within a managed world. Johnson (2017) affirms that "shared worlds are battleworlds to the extent that they require negotiation and management of their shared status by the multiple producers and industries that exploit them" (140). We not only see this in how several episodes of *Discovery* return to the MU, but also in the animated *Lower Decks* which has used the MU to connect the series more closely to the live action world and reward eagle-eyed fans keen to pick up on the intertextual references. This is perhaps unsurprising considering the self-conscious nature of the series that poked fun at, while revering, *Star Trek* canon in every episode. However, referencing the MU connects it to the alternative universe beyond a simple in-joke.

In "I, Excretus" (2021), the junior crew are put through a series of holopod tests with Ensign Beckett Mariner failing to infiltrate the MU *ISS Cerritos* because she did not know her Terran version was right handed. Albeit a small glimpse of the MU in *Lower Decks* so far, it highlights how recent *Star Trek* series are managed by producers keen to tap into established lore and canon—as if they are ticking off familiar world building elements so as to make connections with fans. On one level this corresponds to Matt Hills's (2014) discussion of transmedia storytelling strategies in the digital age, where narrative extensions not only cross media platforms but "have also moved across and between what can be understood as production discourse and fan discourse, with producers aiming to reward loyal fans via niche transmedia paratexts" (151). Yet, on another level, such examples drawn from *Discovery* and *Lower Decks* fit within what Veronica Innocenti and Guglielmo Pescatore (2017) call the "narrative ecosystem" defined as "a dynamic model that represents vast narratives, accounting for the interactions of agents, changes, and evolutions" (165–166). In this system "producers and viewers share the responsibility for the series' evolution" with producers creating content based on viewing figures and allocation of resources and fans acting as community with a "shared interest in the life, duration, and resilience of TV series" (170). The MU is therefore not only a multiversal story, it is now a multiplatform story that is managed so as to bring loyal audiences to new streaming platforms as well as across and between established narrative universes.

The Mirror Universe as Multiform Story

As the examples I have drawn from across multiple *Star Trek* series and episodes demonstrate, the MU offers countless potential for multiversal storytelling. Thus, as I eluded to earlier when discussing the relationship between fanfiction and the MU, it conforms to what Janet H. Murray (1997) calls the "multiform story" (30). She describes it as "a written or dramatic narrative that presents a single situation or plotline in multiple versions, versions that would be mutually exclusive in our ordinary experience." This seems to be a foundational tenet of the multiverse, a popular form of franchise storytelling now used in Hollywood and across multiple media industries. Whether managed as part of a wider "narrative ecosystem," inverted and proliferated, the MU is a dramatic narrative presented in multiple forms. Throughout the chapter I have argued that the MU is a shared parallel universe through which *Star Trek* told its stories; experimenting with characters and scenarios by depicting alternative views of the familiar turned upside down. The multiform story provides the means because it illustrates various alternative plots and outcomes originally and often played out in fanfiction. As Murray further states,

> [The multiform narrative's] alternate versions of reality are now part of the way we think, part of the way we experience the world. To be alive in the twentieth century is to be aware of the alternative possible selves, of alternative possible worlds, and of the limitless intersecting stories of the actual world.
>
> *(38)*

In *Star Trek*, the MU is an example of the multiform story because it plays with established narrative canon just like fanfiction; there is already a narrative framework in place where multiple plots can intersect and characters interact. For fans, *Star Trek*'s "alternate versions of reality" are part of the way they experience the franchise and, as a result, part of how they negotiate their relationship as fans with the *Star Trek* universe. This is most famously established and played out in "Mirror, Mirror" and through later incarnations including early fanfiction, *Enterprise*, *DS9*, *Discovery*, and even the animated series, *Lower Decks*.

References

Bacon-Smith, Camille. 1992. *Enterprising Women: Television Fandom and the Creation of Popular Myth*. Philadelphia, PA: University of Pennsylvania Press.

Barrett, Michèle, and Duncan Barrett. 2001. *Star Trek: The Human Frontier*. Cambridge: Polity Press.

Bernardi, Daniel L. 1998. *Star Trek and History: Race-ing Towards a White Future*. New Brunswick, NJ: Rutgers University Press.

Byrne, Aidan, and Mark Jones. 2018. "Worlds Turned Back to Front: The Politics of the Mirror Universe in *Doctor Who* and *Star Trek*." *Journal of Popular Television* 6, no. 2: 257–270.

Cartmell, Deborah, and I.Q. Hunter. 2001. "Introduction: Retrovisions: Historical Makeovers in Film and Literature." In *Retrovisions: Reinventing the Past in Film and Fiction*, edited by Deborah Cartmell, I.Q. Hunter, and Imelda Whelehan, 1–7. London: Pluto Press.

Falk, Dan. 2023. "Mapping the Multiverse." *Discover* 44, no. 2: 34–41.

Geraghty, Lincoln. 2007. "Eight Days That Changed American Television: Kirk's Opening Narration." In *The Influence of Star Tek on Television, Film and Culture*, edited by Lincoln Geraghty, 11–21. Jefferson, NC: McFarland Publishers.

Gregory, Chris. 2000. *Star Trek Parallel Narratives*. London: Macmillan Press.

Hantke, Steffen. 2014. "*Star Trek*'s Mirror Universe Episodes and US Military Culture through the Eyes of the Other." *Science Fiction Studies* 41, no. 3: 562–578.

Hertenstein, Mike. 1998. *The Double Vision of Star Trek: Half-Humans, Evil Twins and Science Fiction*. Chicago, IL: Cornerstone Press.

Hills, Matt. 2014. "Storytelling and Storykilling: Affirmational/Transformational Discourse of Television Narrative." In *Storytelling in the Media Convergence Age: Exploring Screen Narratives*, edited by Roberta Pearson and Anthony N. Smith, 151–173. London: Palgrave Macmillan.

Innocenti, Veronica, and Guglielmo Pescatore. 2017. "Narrative Ecosystems: A Multidisciplinary Approach to Media Worlds." In *World Building: Transmedia, Fans, Industries*, edited by Marta Boni, 164–183. Amsterdam: Amsterdam University Press.

Jenkins, Henry. 1992. *Textual Poachers: Television Fans and Participatory Culture*. New York: Routledge.

Johnson, Derek. 2017. "Battleworlds: The Management of Multiplicity in the Media Industries." In *World Building: Transmedia, Fans, Industries*, edited by Marta Boni, 129–142. Amsterdam: Amsterdam University Press.

Minkowitz, Donna. 2002. "Beam Us Back, Scotty!" *The Nation*, March 25: 36–37.

Murray, Janet H. 1997. *Hamlet on the Holodeck: The Future of Narrative in Cyberspace*. Cambridge, MA: MIT Press.

Neely, Sarah. 2001. "Cool Intentions: The Literary Classic, the Teenpic and the 'Chick Flick'." In *Retrovisions: Reinventing the Past in Film and Fiction*, edited by Deborah Cartmell, I.Q. Hunter, and Imelda Whelehan, 74–86. London: Pluto Press.

Pearson, Roberta, and Máire Messenger Davies. 2014. *Star Trek and American Television*. Berkeley, CA: University of California Press.

Penley, Constance. 1997. *NASA/TREK: Popular Science and Sex in America*. London: Verso.

Pugh, Sheenagh. 2005. *The Democratic Genre: Fan Fiction in a Literary Context*. Bridgend: Seren.

Ryan, Marie-Laure. 2001. *Narrative as Virtual Reality: Immersion and Interactivity in Literature and Electronic Media*. Baltimore, MD: The Johns Hopkins University Press.

Ryan, Marie-Laure. 2017. "The Aesthetics of Proliferation." In *World Building: Transmedia, Fans, Industries*, edited by Marta Boni, 31–46. Amsterdam: Amsterdam University Press.

Verba, Joan Marie. 1996. *Boldly Writing: A Trekker Fan and Zine History, 1967–1987*. Minnetonka, MN: FTL Publications.

Woods, Louis A., and Gary L. Harmon. 1994. "Jung and *Star Trek*: The Coincidentia Oppositorum and Images of the Shadow." *Journal of Popular Culture* 28, no. 2: 169–184.

INDEX